SHEAVES, GAMES, AND MODEL COMPLETIONS

TRENDS IN LOGIC
Studia Logica Library

VOLUME 14

SCOPE OF THE SERIES

Trends in Logic is a bookseries covering essentially the same area as the journal *Studia Logica* – that is, contemporary formal logic and its applications and relations to other disciplines. These include artificial intelligence, informatics, cognitive science, philosophy of science, and the philosophy of language. However, this list is not exhaustive, moreover, the range of applications, comparisons and sources of inspiration is open and evolves over time.

Volume Editor
Ryszard Wójcicki

The titles published in this series are listed at the end of this volume.

SILVIO GHILARDI
Università degli Studi di Milano, Milan, Italy

MAREK ZAWADOWSKI
Uniwersytet Warszawski, Warsaw, Poland

SHEAVES, GAMES, AND MODEL COMPLETIONS

A Categorial Approach to Nonclassical Propositional Logics

KLUWER ACADEMIC PUBLISHERS
DORDRECHT / BOSTON / LONDON

A C.I.P. Catalogue record for this book is available from the Library of Congress.

ISBN 1-4020-0660-8

Published by Kluwer Academic Publishers,
P.O. Box 17, 3300 AA Dordrecht, The Netherlands.

Sold and distributed in North, Central and South America
by Kluwer Academic Publishers,
101 Philip Drive, Norwell, MA 02061, U.S.A.

In all other countries, sold and distributed
by Kluwer Academic Publishers,
P.O. Box 322, 3300 AH Dordrecht, The Netherlands.

Printed on acid-free paper

Printed in the Netherlands.

To Jaś, Lavinia, Miś, and Virgilio.

Contents

Chapter 1

INTRODUCTION

This book is an example of fruitful interaction between (non-classical) propositional logics and (classical) model theory which was made possible due to categorical logic. Its main aim consists in investigating the existence of model-completions for equational theories arising from propositional logics (such as the theory of Heyting algebras and various kinds of theories related to propositional modal logic). The existence of model-completions turns out to be related to proof-theoretic facts concerning interpretability of second order propositional logic into ordinary propositional logic through the so-called 'Pitts' quantifiers' or 'bisimulation quantifiers'. On the other hand, the book develops a large number of topics concerning the categorical structure of finitely presented algebras, with related applications to propositional logics, both standard (like Beth's theorems) and new (like effectiveness of internal equivalence relations, projectivity and definability of dual connectives such as difference). A special emphasis is put on sheaf representation, showing that much of the nice categorical structure of finitely presented algebras is in fact only a restriction of natural structure in sheaves. Applications to the theory of classifying toposes are also covered, yielding new examples.

The book has to be considered mainly as a research book, reporting recent and often completely new results in the field; we believe it can also be fruitfully used as a complementary book for graduate courses in categorical and algebraic logic, universal algebra, model theory, and non-classical logics.

1. Motivating example

The origin of this work goes back to a surprising Theorem of A.M. Pitts, cf. [Pi2], stating that the second order intuitionistic propositional calculus IpC^2 can be interpreted into ordinary intuitionistic propositional calculus IpC. More precisely,

THEOREM 1.1 (A.M. PITTS) *For each propositional variable x and for each formula t of IpC, there exist formulas $\exists^x t$ and $\forall^x t$ of IpC (effectively computable from t) containing only variables not equal to x which occur in t, and such that for any formula u not involving x, we have*

$$\vdash_{IpC} \exists^x t \to u \qquad \textit{iff} \qquad \vdash_{IpC} t \to u$$

and

$$\vdash_{IpC} u \to \forall^x t \qquad \textit{iff} \qquad \vdash_{IpC} u \to t.$$

Although the above result looks like a purely proof-theoretical fact, it can be interpreted model-theoretically in a quite interesting way as a statement about the theory of Heyting algebras. We summarize the main point below. Using the identification of intuitionistic formulas with the terms in the first order theory of Heyting algebras we can characterize semantically the 'Pitts' quantifiers' $\exists^x$ and $\forall^x$, as follows. For a formula $t(\vec{y}, x)$ of IpC, and a tuple of elements $\vec{a}$ from a Heyting algebra H, we have that

$$H \models (\exists^x t)(\vec{a}) = 1 \quad \text{iff} \quad H[\mathbf{x}]/t(\vec{a}, \mathbf{x}) \text{ is an extension of } H$$

where $H[\mathbf{x}]/t(\vec{a}, \mathbf{x})$ is the Heyting algebra of polynomials $H[\mathbf{x}]$ divided by the congruence generated by the condition $t(\vec{a}, \mathbf{x}) = 1$. Moreover

$$H \models (\forall^x t)(\vec{a}) = 1 \quad \text{iff} \quad H[\mathbf{x}] \models t(\vec{a}, \mathbf{x}) = 1.$$

The proof of these characterizations easily follows from Pitts' Theorem using any presentation for H.

This explanation of Pitts' quantifiers in terms of Heyting algebras can be used in order to show that the first order theory of Heyting algebras admits a model completion. In fact, it turns out that the system of equations and inequations with parameters $\vec{a}$ from H

$$t(\vec{a}, x) = 1 \; \& \; u_1(\vec{a}, x) \neq 1 \; \& \; \ldots \; \& \; u_m(\vec{a}, x) \neq 1 \tag{1.1}$$

is solvable in an extension of H iff the quantifier-free formula

$$(\exists^x t)(\vec{a}) = 1 \; \& \; (\forall^x (t \to u_1))(\vec{a}) \neq 1 \; \& \; \ldots$$
$$\ldots \; \& \; (\forall^x (t \to u_m))(\vec{a}) \neq 1 \tag{1.2}$$

is true in H. If the formula (1.2) is true, we can take $H[\mathbf{x}]/t(\vec{a}, \mathbf{x})$ as an extension of H in which the system (1.1) has a solution. Conversely, if the system (1.1) is solvable in an extension H', then we have a factorization

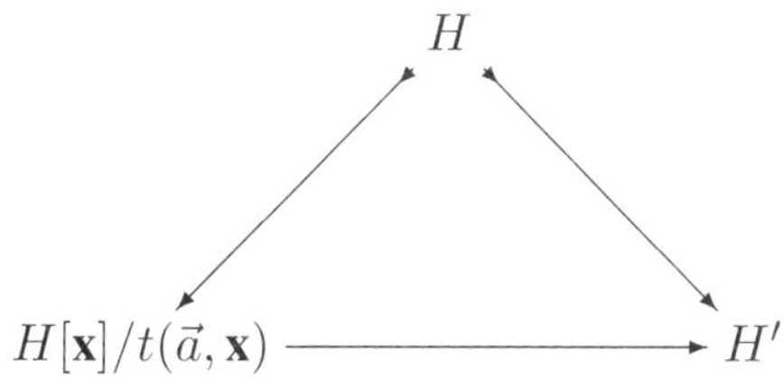

showing that $H[\mathbf{x}]/t(\vec{a},\mathbf{x})$ is an extension of H in which the system (1.1) has the solution $x = \mathbf{x}$. This, together with the above characterization of Pitts' quantifiers, shows that formula (1.2) is true in H. Thus the class of existentially closed Heyting algebras is an elementary class and, as the above quantifier-elimination procedure is effective, it can easily be shown that the related first order theory is decidable. In Section 4.7, we shall provide examples of this decision procedure, together with a list of some basic properties of existentially closed Heyting algebras.

In this way Pitts' Theorem implies that the first order theory of Heyting algebras admits a model completion. The interesting point is that the converse is also true, in a quite general setting. In order to explain what we mean by this, we need a category-theoretic formulation of Pitts' Theorem. In this equivalent formulation, Theorem 1.1 just says that the opposite of the category of finitely presented Heyting algebras is a *Heyting category*.

The notion of Heyting category ([MR1], [MR2] or logos in [Pi1]) is a quite standard notion in categorical logic: Heyting categories are just 'Lindenbaum categories' for many-sorted intuitionistic first-order theories. A Heyting category is a category with finite limits in which finite joins, images and dual images among subobjects exist and are pullback-stable. Such a structure is needed in order to interpret first-order intuitionistic logic: terms are interpreted as arrows, formulas as subobjects and images and dual images along projections correspond to quantifiers. With each first-order many sorted intuitionistic theory, a Heyting category, built up in a completely syntactic way, can be associated: objects are formulas, arrows are equivalence classes (with respect to provable equivalence) of formulas which are provably functional in the restricted domains given by the source and the target of the arrow they define. Conversely, with each Heyting category, a first-order many sorted intuitionistic theory can be associated: we have one sort for each object, one term for each arrow, no relation symbols, and, as axioms, all the formulas which are 'internally true' in the given Heyting category. The two inverse passages are bijective, modulo the standard notion of equivalence between categories and modulo some natural notion of equivalence between theories.

Thus, using this category-theoretic formulation of Pitts' Theorem, we can say that the fact that HA_{fp}^{op}, i.e. the opposite to the category of finitely presented Heyting algebras, has enough categorical structure to classify internally a first-order intuitionistic theory implies (and actually it is equivalent to, see below) the fact that the first-order theory of Heyting algebras admits a model completion.

This is a rather interesting kind of connection: it says that the existence of a *classical* theory (the model completion) is equivalent to the existence of a suitable *intuitionistic* theory. Notice that the connection is not completely trivial, in the sense that it can be shown that the first-order intuitionistic theory classified by HA_{fp}^{op} is a theory speaking about Heyting algebras, but it differs considerably from the model completion of the theory of Heyting algebras. The two theories are indeed almost contradictory, for instance the statement

$$\forall x \forall y\ (x \vee y = 1 \Rightarrow (x = 1 \text{ or } y = 1))$$

is false in any existentially closed (non degenerate) Heyting algebra, but it is true in the theory classified by the opposite to the category of finitely presented Heyting algebras.

2. An overview of the book

We describe here the main *strategy* of the book. In Chapter 3 there is the proof of a theorem which generalizes the above observations for Heyting algebras. We take into consideration an arbitrary equational theory T satisfying a certain assumption (see next Section) which is rather strong in general, but which is often satisfied in varieties of algebras arising from logic. Under this assumption, we prove (Theorem 3.11) that

T admits a model completion iff $\mathbf{T}$ is an r-Heyting category,

where $\mathbf{T}$ is the opposite of the category $Alg(T)_{fp}$ of finitely presented T-algebras. In other words T admits a model completion iff the category $\mathbf{T}$ derived from T has some nice categorical structure.

The notion of r-Heyting category is obtained from the notion of Heyting category by replacing 'subobject' by 'regular subobject' everywhere in the definition. This modification is due to the fact that we prefer not to assume that monos are all regular in $\mathbf{T}$ (i.e. that epis are quotients in $Alg(T)_{fp}$), an assumption which holds for Heyting algebras as a consequence of the Beth property (BP), cf. Theorem 2.14, but which may fail in other cases.

In the following three Chapters, we apply Theorem 3.11 to two kinds of varieties of algebras: Heyting algebras and modal algebras. In both cases we adopt a similar strategy. Theorem 3.11 says that, under suitable assumptions, the existence of a model completion for T is equivalent to the existence of a certain categorical structure in $\mathbf{T}$. Usually it is not easy to decide directly whether $\mathbf{T}$ is an r-Heyting category. But, as this is a purely categorical property, we can study it in any category equivalent to $\mathbf{T}$. The strategy we adopt for an equational theory T can be summarized in the following four steps:

1 *Embedding.* Find an r-Heyting category $\mathcal{E}$ and an embedding

$$\mathbf{\Phi_T} : Alg(T)^{op}_{fp} \longrightarrow \mathcal{E}$$

which is conservative, preserves finite limits and all the other r-Heyting category structure that exists in $Alg(T)^{op}_{fp}$.

Conservativity ensures that the operations that can be performed in $Alg(T)^{op}_{fp}$ and are preserved by $\mathbf{\Phi_T}$ satisfy automatically any exactness properties that these operations satisfy in $\mathcal{E}$. In particular the operations of left ($\exists_f$) and right ($\forall_f$) adjoint to the pullback functors f^* (operating on regular subobjects, see Section 2.3) in $Alg(T)^{op}_{fp}$, if they exist, they automatically satisfy the Beck-Chevalley condition.

In the applications the category $\mathcal{E}$ is (equivalent to) the category of sheaves on the opposite of the category of finite T-algebras with the canonical topology.

2 *Duality.* Identify the image of $\mathbf{\Phi_T}$ in $\mathcal{E}$, i.e. describe in a convenient way a subcategory $\mathbf{M_T}$ of $\mathcal{E}$ so that we have a factorization of $\mathbf{\Phi_T}$

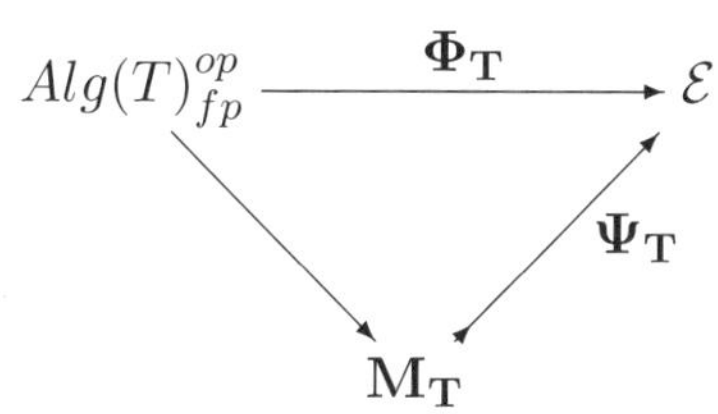

with the first component being an equivalence of categories and $\mathbf{\Psi_T}$ being an inclusion.

In the applications this point is slightly reversed. It is usually more natural to define a 'duality' functor in the opposite direction, i.e. $\mathbf{M_T} \longrightarrow Alg(T)^{op}_{fp}$.

3 *Combinatorial condition for existence of adjoints.* Now the existence of the adjoints is reduced to the verification whether the existing adjoints in $\mathcal{E}$ when applied to objects coming from $Alg(T)^{op}_{fp}$ give objects coming from $Alg(T)^{op}_{fp}$, as well.

In applications, with the help of an appropriate description of $\mathbf{M_T}$, this can be reduced to an equivalent condition of a combinatorial nature, expressed in terms of Ehrenfeucht-Fraissé games on finite Kripke models.

4 *Verification of combinatorial conditions.* Last, but not least, the combinatorial conditions should be verified to establish whether the adjoints do exist, if they do $Alg(T)^{op}_{fp}$ is an r-Heyting category.

We believe that this method is general and can be applied in other similar contexts.

3. How to read the book

The book can read in the obvious *sequential* way by those who are mostly interested in the main results concerning the existence of model completions in propositional logic. However the book contains much side information (for instance about the categorical structure of finitely presented algebras) as well as an introduction to research problems having some independent interest (see Chapter 8). Thus it can be used *as a reference book* as well. This is the reason why in this Section we shall provide some *extended abstracts* of the content of the various Chapters, so that readers may have direct access to what they might be interested in; for such direct access, just consult the present Section jointly with Chapter 2 and the Appendix (each of them only if needed). Certain Sections of Chapter 2 (e.g. Section 2.4) might also be needed for the explanation of some notations.

In **Chapter 3** the notion of r-Heyting category is introduced and it is related to the existence of model completions. Remember that if **C** is a category with finite limits (i.e. with binary products, terminal object and equalizers), then for every object A we can take all monomorphisms of codomain A which happen to be equalizers of some pair of arrows; if we consider two such monomorphisms equal in case they differ only by an A-isomorphism, we do get, in fact, a semilattice $Sub_r(A)$. Given an arrow $f : B \to A$, taking pullback induces a semilattice morphism $f^* : Sub_r(A) \longrightarrow Sub_r(B)$. We say that **C** is r-Heyting iff

(i) **C** has finite limits;

(ii) for every object A, $Sub_r(A)$ is a lattice with 0 and 1;

(iii) for every arrow $f : A \to B$ in **C**, the pullback functor f^* has a left and a right adjoint $\exists_f, \forall_f : Sub_r(A) \to Sub_r(B)$; this means that for every regular subobject $S \hookrightarrow A$, there are regular subobjects $\exists_f(S), \forall_f(S) \hookrightarrow B$ satisfying the conditions

$$f^*(T) \leq S \ \text{ iff } \ T \leq \forall_f(S)$$
$$\exists_f(S) \leq T \ \text{ iff } \ S \leq f^*(T)$$

for every regular subobject $T \hookrightarrow B$.

(iv) for every pullback square

$$\begin{array}{ccc} C & \xrightarrow{p_1} & B_1 \\ {\scriptstyle p_2}\downarrow & & \downarrow{\scriptstyle f_1} \\ B_2 & \xrightarrow[f_2]{} & A \end{array}$$

the Beck-Chevalley condition

$$\exists_{p_2}(p_1^*(S)) = f_2^*(\exists_{f_1}(S))$$

holds for every $S \in Sub_r(B_1)$.

Let us now consider an equational theory T and let $\mathbf{T}$ be the opposite of the category of finitely presented T-algebras; $\mathbf{T}$ clearly has finite limits and we suppose in addition that for every A in $\mathbf{T}$, $Sub_r(A)$ is not only a semilattice, but also a Heyting algebra and that the Heyting algebra structure is preserved by the semilattice morphisms f^*. Such assumption, which we call (EDO) (standing for 'equationally definable operations') corresponds to the conditions (EDPC) ('equationally definable principal congruences') and (EDPM) ('equationally definable principal meets') from universal algebra (this correspondence is fully explained in Section 2.4). As stated above, under this assumption, the main Theorem (Theorem 3.11) from Chapter 3 just says that T has a model completion iff $\mathbf{T}$ is an r-Heyting category.

Chapter 4 contains the main results concerning Heyting algebras. We first introduce some kinds of Ehrenfeucht-Fraïssé games over finite Kripke models in the following abstract way.

If a basic finite poset L (the poset of 'labels') is fixed, we call an *L-evaluation* or simply an *evaluation* a pair $\langle P, u\rangle$, where P is a rooted finite poset and $u : P \to L$ is an order-preserving map (a poset is rooted iff it has a greatest element).

We define for every $n \in \omega$ and for every pair of L-evaluations u and v, the notions of being *n-equivalent* (written $u \sim_n v$). We define also for two L-evaluations u, v the notions of being *infinitely equivalent* (written $u \sim_\infty v$).

Let $u : P \to L$ and $v : Q \to L$ be two L-evaluations. The *game*, we are interested in, has two players, player I and player II. Player I can choose either a point in P or a point in Q and player II must answer by choosing a point in the other poset and the only rule is that, if $\langle p \in P, q \in Q\rangle$ is the last move played so far, then in the successive move the two players can only choose points $\langle p', q'\rangle$ such that $p' \leq p$ and $q' \leq q$. If $\langle p_1, q_1\rangle, \ldots, \langle p_i, q_i\rangle, \ldots$ are the points chosen in the game, after an infinite number of moves, player II wins iff for every $i = 1, 2, \ldots$, we have that $u(p_i) = v(q_i)$. We say that

- $u \sim_\infty v$ iff *player II has a winning strategy*;

- $u \sim_n v$ (for $n > 0$) iff *player II has a winning strategy for the first n-moves*, i.e. he has a winning strategy provided we stipulate that the game terminates after n moves;

- $u \sim_0 v$ iff $u(\rho(P)) = v(\rho(Q))$ (here $\rho(P), \rho(Q)$ denote roots of P, Q).

In case $L = \mathcal{P}(x_1, \ldots, x_n)$, L-evaluations are just ordinary finite Kripke models over the language built up from the propositional variables $x_1, \ldots, x_n$; in such a case, two evaluations are $\sim_\infty$-equivalent (resp. $\sim_n$-equivalent) iff they force the same formulas (resp. the same formulas up to implicational degree n).

Remember that a map among posets is said to be *open* iff it is open in the topological sense (posets can be viewed as topological spaces whose open subsets are the downward closed subsets). Let $\mathbf{P_0}$ be the category of finite rooted posets and open maps between them, and let $\mathbf{J_0}$ be the canonical topology on $\mathbf{P_0}$ (i.e. $\{\alpha_i : P_i \longrightarrow P\}_{i \in I}$ is a $\mathbf{J_0}$-covering iff there is $i \in I$ such that α_i is a surjection). We have a functor

$$\mathbf{\Phi_H} : HA_{fp}^{op} \longrightarrow Sh(\mathbf{P_0}, \mathbf{J_0})$$

sending a finitely presented Heyting algebra H to a sheaf

$$\mathbf{\Phi_H}(H) \;=\; [\,P \mapsto HA(H, \mathcal{D}(P))\,]$$

(i.e. $\mathbf{\Phi_H}(H)$ associates to every finite rooted poset P the set of all Heyting morphisms from H to the Heyting algebra $\mathcal{D}(P)$ of downward closed subsets of P). Both $\mathbf{\Phi_H}(H)$ and $\mathbf{\Phi_H}$ act on morphisms in the obvious way, by composition. $\mathbf{\Phi_H}$ is conservative; in order to have a manageable description of its image we introduce a category $\mathbf{M_H}$ of some sheaves over $(\mathbf{P_0}, \mathbf{J_0})$ equipped with additional structure defined in terms of the above games. First we observe that for every finite poset L, h_L (which is the presheaf defined by associating with every finite poset P the set of all L-evaluations with domain P) is actually a sheaf.

Now we can give the definition of the category $\mathbf{M_H}$. An object of $\mathbf{M_H}$ is a pair $\langle L, C\rangle$ (often written simply as C^L or C) such that L is a finite poset and $C \subseteq h_L$ is a subsheaf with index (meaning that there exists some $n \in \omega$ such that if $u \in C$, $v \in h_L$ and $u \sim_n v$ then $v \in C$). A morphism $f : C \to C'$ in $\mathbf{M_H}$ is a natural transformation such that for some natural number k, if $v, v' \in C$ and $v \sim_k v'$ then $f(v) \sim_0 f(v')$. Any k satisfying the above condition is called an *index* of f.

In Section 4.3 it is formally proved that $\mathbf{M_H}$ is dual to the category of finitely presented Heyting algebras (Theorem 4.15), so that we get a sheaf representation of the latter category. The meaning of this duality can be easily explained in terms of Kripke models: any finitely presented Heyting algebra is isomorphic to one of form $F(x_1, \ldots, x_n)/t(x_1, \ldots, x_n)$ for some n and intuitionistic formula $t(x_1, \ldots, x_n)$, i.e. it is isomorphic to the free algebra on n generators $x_1, \ldots, x_n$ divided by the congruence generated by the condition $t(x_1, \ldots, x_n) = 1$. With such an algebra, our duality associates the collection of all finite Kripke models (seen *as a sheaf*) over $x_1, \ldots, x_n$ that force $t(x_1, \ldots, x_n)$. For morphisms among finitely presented algebras, a similar description can easily be obtained.

At the heart of the proof that $\mathbf{M_H}$ is a Heyting category is the Combinatorial Lemma 4.18. It allows us to conclude (Corollary 4.20) that $\mathbf{M_H}$ is closed in $Sh(\mathbf{P_0}, \mathbf{J_0})$ under the operation of taking both left ($\exists_f$) and right ($\forall_f$) adjoints to the pullback functor (f^*).

In Section 4.5 we prove (Theorem 4.24) that $\mathbf{M_H}$ and hence the opposite of the category of finitely presented Heyting algebras HA_{fp}^{op} is a Heyting category. Having such a good description $\mathbf{M_H}$ of HA_{fp}^{op}, we can easily establish other properties of HA_{fp}^{op} (some of them were already proved in [Pi2] proof-theoretically). Among other things we show (Theorem 4.25) that $\mathbf{M_H}$ is a bi-Heyting category satisfying the Lawvere condition. This is done again by showing that $\mathbf{M_H}$ is closed in $Sh(\mathbf{P_0}, \mathbf{J_0})$ under suitable operations and the Lawvere condition is inherited from $Sh(\mathbf{P_0}, \mathbf{J_0})$ via the conservative morphism $\mathbf{\Phi_H}$. This means that the structure on HA_{fp}^{op} discovered by A.M. Pitts is the restriction to HA_{fp}^{op} of a standard one existing on $Sh(\mathbf{P_0}, \mathbf{J_0})$. The good description of HA_{fp}^{op} allows us to prove negative results about this category, as well. In Section 4.8 we show for instance that HA_{fp}^{op} is not an exact category. Applications of these results to intuitionistic propositional calculus are explained in Sections 4.6 and 4.9. Finally Section 4.10 contains an analysis of finitely presented projective Heyting algebras.

In **Chapter 5**, we realize, for modal algebras, the first steps of the strategy outlined in the previous Section. Recall that $\mathbf{K4}$-*algebras* are Boolean algebras endowed with a unary operator $\Box$ preserving finite meets and satisfying the equation

$$\Box x \leq \Box\Box x.$$

There are many interesting equational classes of $\mathbf{K4}$-algebras. We shall mention just two of them; if also the equation

$$\Box x \leq x$$

is satisfied, we have the notion of *interior* or $\mathbf{S4}$-algebras; *diagonalizable* or *Magari* algebras are $\mathbf{K4}$-algebras satisfying the further equation

$$\Box(\Box x \to x) \leq \Box x.$$

We take into consideration finitely presented algebras for a subvariety $\mathbf{S}$ of $\mathbf{K4}$-algebras having the finite model property, which means that free $\mathbf{S}$-algebras embed into products of finite $\mathbf{S}$-algebras. Under this assumption, if the theory $\mathbf{S}$ admits a model completion, then the category of finite $\mathbf{S}$-algebras must enjoy the amalgamation property (AP) (see Section 2.6). Thus it is sufficient for us to consider only those theories $\mathbf{S}$ for which such an additional condition (the amalgamation on finite algebras) is satisfied.

A finite $\mathbf{S}$-*frame* (X, R) is a finite set X endowed with a transitive relation R whose associated power-set modal algebra $(\mathcal{P}(X), \Box_R)$ is an $\mathbf{S}$-algebra. A frame morphism (or *open* morphism) is just a set-theoretic map f such that inverse image along f is a modal algebras morphism. Thus we have, for every $\mathbf{S}$ as above, a category $\mathbf{F_S}$ of finite $\mathbf{S}$-frames. Extensive information of independent interest on such a category is given in Section 5.2. In particular, thanks to the amalgamation property for finite $\mathbf{S}$-algebras, we have that finite jointly surjective families of frame morphisms form a basis for the canonical topology $\mathbf{J_S}$ on $\mathbf{F_S}$. We have, similarly as before, a functor

$$\mathbf{\Phi_S} : Alg(\mathbf{S})_{fp}^{op} \longrightarrow Sh(\mathbf{F_S}, \mathbf{J_S}).$$

sending a finitely presented $\mathbf{S}$-algebra A to a sheaf

$$\mathbf{\Phi_S}(H) \;=\; [\,(X, R) \mapsto Alg(\mathbf{S})(A, (\mathcal{P}(X), \Box_R))\,]$$

We need to identify the essential image of this faithful functor; to this aim we define a category $\mathbf{M_S}$ which will turn out to be equivalent to $Alg(\mathbf{S})_{fp}^{op}$ (Theorem 5.37). Due to the fact that in modal algebras not all formulas correspond to congruences (only those of form $\Box t \wedge t$ actually do) the definition of the category $\mathbf{M_S}$ is slightly more involved. The objects of $\mathbf{M_S}$, as in case of $\mathbf{M_H}$, have a mixed geometric (sheaves) and combinatorial (games) structure, in the following sense.

An L-evaluation, where L is a finite set, is a pair given by a finite frame (X, R) in $\mathbf{F_S}$ and by a function $u : X \to L$. Games are introduced in a very similar way as for intuitionistic logic: for instance if (x, y) is the last move played, in the successive move only points (x', y') such that xRx' and yRy' can be chosen. These games allow us once again to define equivalence relations $\approx_n$ $(n \in \omega \cup \{\infty\})$ among L-evaluations. In addition, we also need equivalence relations $\sim_n$ $(n \in \omega \cup \{\infty\})$ among L-evaluations (u, x) and (v, y) with distinguished nodes: to define them, we just stipulate that the first move played must in any case be (x, y).

Let us now give a formal definition of $\mathbf{M_S}$. First, observe that, once again, the presheaf h_L of all evaluations is actually a $\mathbf{J_S}$-sheaf. An object of $\mathbf{M_S}$ is a pair $\langle L, C\rangle$ (denoted C^L for short), where L is a finite set and C is a closed subsheaf of h_L (meaning that there is $n \in \omega$ such that if C contains u and $u \approx_n v$, then C contains v as well). A morphism $f : C^L \to D^M$ in $\mathbf{M_S}$ is a natural transformation with an index, i.e. f is a natural transformation between sheaves C^L and D^M and there is $n \in \omega$ such that, for all L-evaluations v, u and all $x \in dom(v)$, $y \in dom(u)$

$$\text{if } (v, x) \sim_n (u, y) \text{ then } (f(v), x) \sim_0 (f(u), y).$$

In Section 5.6 it is shown that all monomorphisms in $\mathbf{M_S}$ are regular; in view of the duality Theorem (Theorem 5.37), this results actually proves Beth's Theorem for all subvarieties of $\mathbf{K4}$-algebras having the finite model property.

This is a new proof (having an exclusively combinatorial nature) of a result from [Ma5].

The category $\mathbf{M_S}$ can be easily mapped into the topos of sheaves $Sh(\mathbf{F_S}, \mathbf{J_S})$. In $\mathbf{M_S}$ there is an internal $\mathbf{S}$-algebra on the sheaf of 2-evaluations h_2, so that taking 'hom into it' we get a 'duality' functor

$$\mathbf{T_S} = \mathbf{M_S}(-, h_2) : \mathbf{M_S} \longrightarrow Alg(\mathbf{S})^{op}_{fp}$$

which is an equivalence of categories. We have a commuting triangle

$$\begin{array}{ccc} Alg(\mathbf{S})^{op}_{fp} & \xrightarrow{\ \mathbf{\Phi_S}\ } & Sh(\mathbf{F_S}, \mathbf{J_S}) \\ & {}_{\mathbf{T_S}}\nwarrow \quad \nearrow_{\mathbf{\Psi_S}} & \\ & \mathbf{M_S} & \end{array}$$

with $\mathbf{\Psi_S}$ being (essentially) an inclusion of a subcategory. The embedding $\mathbf{\Psi_S}$ preserves all the Heyting structure that exists in $\mathbf{M_S}$ and is conservative for monos. More specifically, $\mathbf{\Phi_S}$ preserves finite limits, finite joins and implications among subobjects and all existing left and right adjoints to the pullback functor operating on monomorphisms in $\mathbf{M_S}$, moreover a $\mathbf{\Phi_S}$ is conservative. In addition, $\mathbf{\Psi_S}$ preserves disjoint unions and subobject classifier (Theorem 5.50).

Finally, we find (Theorem 5.49) a combinatorial condition $(\forall\exists)$ which is equivalent to the existence of the Heyting category structure on $\mathbf{M_S}$. The statement is the following one (it concerns any arrow $f : C^L \longrightarrow D^M$ in $\mathbf{M_S}$):

$$\forall n\ \exists N\ \forall u \in C^L\ \forall v \in D^M$$

$$(u \approx_N f(v)\ \Rightarrow\ \exists u' \in C^L\ (u \approx_n u'\ \&\ f(u') \approx_\infty v)).$$

In other words, whenever u and $f(v)$ are 'similar enough' ($\approx_N$), we must be able to vary u up to a 'preassigned similarity' ($\approx_n$), so that, after applying f, it becomes 'almost the same as' ($\approx_\infty$) v.

In **Chapter 6** we obtain concrete results about the existence of model completions in modal logic. Most results or proofs in this Chapter are entirely new. We reach our goal in many cases: we show that diagonalizable algebras admit a model completion and we fully list (using Maksimova's results [Ma2], [Ma3] on amalgamability) the few systems extending **S4** that admit a model completion (Theorem 6.7). We also prove a negative result saying that the model completion of a system **S** extending **K4** and having the finite model property cannot exist if all finite linear reflexive frames and the three-element reflexive frame with a two-element internal cluster all belong to $\mathbf{F_S}$.

In **Chapter 7**, we take a closer look at the existentially and algebraically closed algebras that we studied earlier. This Chapter is independent from the rest of the book and needs only some background from Chapter 2 (and the categorical notions from the Appendix). We only briefly summarize it here. Since most of the results it contains are quite general, we isolated a class of equational theories to which they apply and called them FLEA's, standing for 'finitary, logical, equational and amalgamable theories'.

We present a set of axioms of the geometric theory $T^{\dagger}$ classified by the topos $Sh(Alg(T)^{op}_{fin}, \mathbf{J}^{\infty})$, for a FLEA T, where $\mathbf{J}^{\infty}$ is the topology consisting of jointly mono families of morphisms.

In case T is the theory of Heyting algebras T_{HA}, $T^{\dagger}_{HA}$ turns out to be the theory of algebraically closed, locally finite and prime Heyting algebras. We also prove that $T^{\dagger}_{HA}$ has only two countable models, differing from each other by the existence of a penultimate element (a condition which is elementary but not coherent).

In the last Section, we look at existentially closed models (where a classification is impossible) and prove the existence of a prime model.

Chapter 8 contains final comments as well as a discussion of relevant open problems to be addressed by future research.

At the end of each Chapter the reader will find **Exercises**. Some of them are meant to help the reader to understand the material in that Chapter and others show some problems that can be treated by the methods just developed. Finally, when the text of a Chapter refers to results from literature which cannot be fully reported, the exercises provide indications for proofs, at least under some simplifying but still relevant assumption (thus, for instance, Maksimova's results on amalgamation are used in Chapter 6 and in the exercises the reader can find out how to prove them if the finite model property is assumed).

4. Historical remarks

The dualities we develop in this book are not purely geometric, as for example the Stone duality for Boolean algebras. The objects of $\mathbf{M_H}$ have mixed combinatorial-geometric structure. They are almost 'language free' and, as we have said, they have already proved to be useful.

Duality theorems in logic, cf. [Pr] , [Jo], [MM1], establish equivalence between the algebraic counterpart of a logic on one side and a collection of models equipped with certain additional structure on the other (e.g. in the Stone duality we have Boolean algebras and Stone spaces). Our duality follows the general scheme. However, the structure on collections of models is of a new kind, having a combinatorial character. The only other duality we are aware of that is based on similar principles is a duality for free K-algebras, cf. [Gh3].

The introduction of Ehrenfeucht-Fraissé games in the context of modal logic is due to K. Fine (see [Fi1], [Fi2]).

The idea of proving exactness properties of logical doctrines using dualities is not new, cf. [MM1], [Za]. However not all dualities are equally good for this purpose. See [MM1] for some discussion of this matter. The reason why dualities may possibly help is the following. In reasonable categories finite limits are computed as in Set but colimits are usually more complicated. Passing to the dual category colimits become limits and vice versa. This conversion does not always guarantee the required result although it often reduces the problem to a simpler statement about geometric (in our case combinatorial and geometric) objects, which can be solved more easily. Note for example that from this point of view Priestley duality for distributive lattices is good and useful, cf. [Za], but its restriction to Heyting algebras is less promising since finite limits in the dual category are not computed in $Sets$ and can be complicated.

The results that we call 'Combinatorial Lemmas' (i.e. 4.18 and some of the technicalities in Chapter 6) already have a certain history, mainly connected to provability logic GL. Z.Gleit and W.Goldfarb proved in [GG] a Joint Satisfiability Theorem (JST) which immediately gives Craig interpolation for GL. V.Shavrukov in [Sh] proved an improved version of JST from which he deduced the existence of the least interpolant for GL. Our 'Combinatorial Lemmas' (first appearing in [GZ1] and, in a slightly different form, also in [Vi]) are successors of these results. However there are still some differences. For instance, the statements of our Lemmas are purely combinatorial, do not mention formulas and are extended to all (the duals of) morphisms among finitely presented algebras. Finally, our proofs use the notion of rank as in [Sh], however our arguments are simpler and more direct even in the case of the logic GL.

As we explained above, the book is self-contained. For readers who would like to have more information on the various areas touched upon in this book, we list some references: [CZ] for propositional logics, [CK], [MI] for model theory, [CWM] for general category theory, [MR1], [MR2] for categorical logic, and [CWL] for topos theory.

Acknowledgements

We would like to thank Bill Boshuck, Peter Johnstone, André Joyal, Jim Loveys, Mihaly Makkai and Richard Squire for the interesting and stimulating conversations we had while we were developing the theory contained in this book. The diagrams were prepared with the help of the *catmac* package by Michael Barr.

The work of the first author was carried out within the framework of MURST "Logica" project. The second author would like to express his thanks to KBN for financial support (grant 2 P03A 014 11).

Chapter 2

PRELIMINARY NOTIONS

1. Basic algebraic structures

In this section we recall the main algebraic structures which will be investigated within the book. They are structures that provides an algebraic semantics for propositional logics. They are usually obtained by enriching posets by some algebraic operations. We are mainly interested in Heyting and modal algebras i.e. those algebras that provide counterparts of superintuitionistic and modal logics.

A partially ordered set (*poset*, for short) is a set P equipped with a reflexive, transitive and antisymmetric binary relation $\leq$. For such a poset, the *infimum* (resp. *supremum*) of a family $\{a_i\}_{i\in I}$ of elements of P is an element (it may or may not exists, but if it exists it is unique) $\bigwedge_i a_i \in P$ (resp. $\bigvee_i a_i \in P$) such that for all $b \in P$, we have

$$(\forall i \in I\ b \leq a_i) \qquad \text{iff} \qquad b \leq \bigwedge_i a_i$$

(or

$$(\forall i \in I\ a_i \leq b) \qquad \text{iff} \qquad \bigvee_i a_i \leq b)$$

respectively). In case the index I is empty, the above conditions say that the infimum of the empty set is the maximum element of P and the supremum of the empty set is just the minimum.

We recall some facts about *adjoints among posets*, although they can be deduced from the general results about categories given in the Appendix, it is worth having a direct knowledge of what happens in this special case. The right adjoint f_* (resp. left adjoint f^*) to an order-preserving map $f : P \to Q$ among posets, is an order-preserving map in the opposite direction, satisfying

$$f(a) \leq b \qquad \text{iff} \qquad a \leq f_*(b)$$

(or

$$b \leq f(a) \qquad \text{iff} \qquad f^*(b) \leq a$$

respectively) for all $a \in P, b \in Q$. Such a right (left) adjoint may not exists, but if it exists it is unique. It is easily seen that left adjoints preserve existing suprema and right adjoints existing infima: the latter, for instance, is shown by an easy chain of equivalences as follows

$$\frac{\frac{\frac{\frac{a \leq f_*(\bigwedge_i b_i)}{f(a) \leq \bigwedge_i b_i}}{\forall i \; f(a) \leq b_i}}{\forall i \; a \leq f_*(b_i)}}{a \leq \bigwedge_i f_*(b_i)}$$

yieldying $f_*(\bigwedge_i b_i) = \bigwedge_i f_*(b_i)$ as a is arbitrary. If P is complete (i.e. iff all suprema -or equivalently all infima- exist), then any order-preserving map $f : P \to Q$ has a right adjoint iff it preserves suprema and has a left adjoint iff it preserves infima. Such adjoints are easily seen to be given by the following formulas:

$$f_*(b) = \bigvee_{f(a) \leq b} a \qquad f^*(b) = \bigwedge_{b \leq f(a)} a$$

for all $b \in Q$.

A (meet) *semilattice* is a commutative idempotent monoid, i.e. a structure $(M, \wedge, \top)$ satisfying the equations

$$a \wedge b = b \wedge a, \quad a \wedge \top = a, \quad a \wedge a = a, \quad a \wedge (b \wedge c) = (a \wedge b) \wedge c \quad (2.1)$$

for all $a, b, c \in M$. Putting

$$a \leq b \qquad \text{iff} \qquad a \wedge b = a$$

we can define a partial order in any semilattice; the operation $\wedge$ turns out to be the infimum (also called *meet*) of the pair $\{a, b\}$ and $\top$ turns out to be the maximum element. In fact, one can equivalently define a semilattice as a partially ordered set in which infima exist for all finite sets of elements (this includes the maximum element, seen as the infimum over the empty set).

Many important further operations can be characterized with respect to the partial order so introduced: in order to obtain the notion of a *lattice*[1] one simply has to require that also suprema (called *joins* as well) exist for all finite sets; equivalently, a lattice is a semilattice with another binary operation $\vee$

[1]Notice that we always require the presence of $\bot$ and $\top$ in a lattice (this is different from some common literature). Sometimes, we also use the notation **1** for $\top$ and **0** for $\bot$.

and another constant $\bot$ satisfying equations (2.1) (with $\wedge, \top$ replaced by $\vee, \bot$ respectively) and moreover the following absorption laws

$$a \wedge (a \vee b) = a, \quad a \vee (a \wedge b) = a.$$

A lattice is said to be *distributive* iff it satisfies one of the two (equivalent) equations

$$a \wedge (b \vee c) = (a \wedge b) \vee (a \wedge c), \quad a \vee (b \wedge c) = (a \vee b) \wedge (a \vee c).$$

In a given semilattice M it may happens that for $a, b \in M$ the supremum of the set $\{c \,|\, a \wedge c \leq b\}$ exists; such an element is called the *relative pseudocomplement* of a relative to b (or the *implication* of a and b, using logical terminology) and is written as $a \to b$. Otherwise said, $a \to b$, if it exists, it is the unique element satisfying the condition

$$a \wedge c \leq b \qquad \text{iff} \qquad c \leq a \to b$$

for all c. A *Brouwerian semilattice* is a semilattice in which all implications among pairs of elements exist and a *Heyting algebra* is a Brouwerian semilattice which is also a distributive lattice. Brouwerian semilattices (hence also Heyting algebras) may be equivalently introduced for instance through the equations

$$a \wedge (a \to b) = a \wedge b \qquad b \wedge (a \to b) = b$$
$$a \to (b \wedge c) = (a \to b) \wedge (a \to c) \qquad a \to a = \top.$$

This shows that Heyting algebras form an equational class, i.e. a *variety*. An important example of a Heyting algebra is given by the open sets of a topological space; here the partial order is inclusion, (finite) meets and joins are intersections and unions, whereas implication of the open subsets a and b is the interior of $a' \cup b$ (where a' is the complement of a). The most important example for us is given by the downward closed subsets $\mathcal{D}(P)$ of a poset P ($a \subseteq P$ is downward closed iff $p \in a$ and $q \leq p$ imply $q \in a$): here the partial order, joins and meets are again inclusion, intersections and unions, respectively, whereas the implication of a and b is

$$a \to b \;=\; \{p \in P \,|\, \forall q \leq p\, (q \in a \Rightarrow q \in b)\}.$$

A finite distributive lattice is always a Heyting algebra, because a finite distributive lattice is complete and, thanks to distributivity, for any element a, the order preserving map $a \wedge (-)$ preserves suprema, so that it has a right adjoint $a \to (-)$. For the same reason, a finite Brouwerian semilattice is always a Heyting algebra: in fact joins exists and are distributive as $a \wedge (-)$ preserves them (being a left adjoint).

In a Heyting algebra H, *negation* is introduced through

$$\neg a = a \to \bot;$$

such operation satisfies many usual laws, but not all the classical ones (for instance, only three of the four De Morgan identities hold). A *Boolean algebra* is a Heyting algebra in which we have $\neg\neg a = a$ (or, equivalently, $a \vee \neg a = \top$) for all a.

In a distributive lattice lattice D certain elements play a special role, they are the join-irreducible ones. We say that a is *join-irreducible* iff for all $n \geq 0, b_1, \dots, b_n \in D$, we have

$$\text{if } a \leq b_1 \vee \cdots \vee b_n \text{ then for some } 1 \leq i \leq n, \quad a \leq b_i$$

(notice that join-irreducible elements are non-zero, i.e. different from $\bot$, by taking $n = 0$ in the above definition). In distributive lattices of the kind $\mathcal{D}(P)$, where P is a finite poset, join-irreducible elements are those of the kind $\downarrow p$ for $p \in P$ (here $\downarrow p$ is $\{q \in P \,|\, q \leq p\}$). In a Boolean algebra B, join-irreducible elements are called *atoms* and turn out to be just the minimal non-zero elements. A Boolean algebra B may have no atoms (in this case we say that it is *atomless*), or, at the extreme opposite, it may happen that for any non-zero $b \in B$ there is an atom $a \leq b$ (in this case, we say that B is *atomic*).

Distributive lattices and Boolean algebras (also Brouwerian semilattices, but we shall not use this further result) are *locally finite varieties*, namely varieties in which finitely generated algebras are finite; this is easily seen, e.g. in the case of Boolean algebras, from the fact that if the set G generates the algebra B, then every element of B admits a representation of the kind $\bigwedge_i \bigvee_j x_{ij}$ where i, j range over finite sets of indices and where x_{ij} is either g or $\neg g$ for some $g \in G$. This is not true for Heyting algebras since the free Heyting algebra on one generator is infinite.

Modal algebras are just Boolean algebras endowed with a further unary 'necessity' operator $\Box$ satisfying the conditions:

$$\Box(a \wedge b) = \Box a \wedge \Box b \qquad \Box\top = \top;$$

the 'possibility' operator $\Diamond$ is introduced in any modal algebra through the definition $\Diamond a = \neg\Box\neg a$. We shall mainly deal with **K4**-*algebras*, i.e. modal algebras in which the operator $\Box$ satisfies the further axiom

$$\Box a \leq \Box\Box a.$$

We shall meet in the book many interesting varieties of $K4$-algebras; for the moment let us only mention **S4**-*algebras* (or interior algebras or topological Boolean algebras), which are characterized by the further axiom

$$\Box a \leq a.$$

The main examples of modal algebras we are interested in are obtained through frames. A *frame* is a pair (X, R), where X is a set endowed with a relation (the *accessibility* relation of the frame); a *transitive frame* is a frame in which R is assumed also to be transitive and a *preordered frame* is a transitive frame in which R is also reflexive. Given a frame (resp. a transitive frame, a preordered frame) (X, R), we can turn $\mathcal{P}(X)$ into a modal algebra (resp. into a $\mathbf{K4}$-algebra, into an $\mathbf{S4}$-algebra) by putting, for $a \subseteq X$

$$\Box_R a \;=\; \{p \in X \mid \forall q \in X(pRq \Rightarrow q \in a)\};$$

the corresponding definition of the possibility operator is

$$\Diamond_R a \;=\; \{p \in X \mid \exists q \in X(pRq \;\&\; q \in a)\}.$$

Let us mention how to describe quotients in Heyting and $K4$-algebras. The central notion to this respect is the notion of a *filter* F, which makes sense at the level of a semilattice R (although it becomes fully operative only when there are implications): this is a subset of R satisfying the following requirements

- $\top \in F$;
- if $a_1, a_2 \in F$, then $a_1 \wedge a_2 \in F$;
- if $a_1 \in F$ and $a_1 \leq a_2$, then $a_2 \in F$.

Given a subset $S \subseteq R$, there exists the minimum filter $[S]$ containing S, which is given by

$$[S] \;=\; \{b \in R \mid \exists n \geq 0, \exists a_1, \ldots, a_n \in S \text{ s.t. } a_1 \wedge \cdots \wedge a_n \leq b\}.$$

In particular, the minimum (or *principal*) filter containing an element a is just $[a] = \{b \mid a \leq b\}$. For the case of modal algebras, the relevant notion is the notion of *modal filter*, which is an ordinary filter satisfying the further condition

- if $a \in F$, then $\Box a \in F$.

A formula for the minimum modal filter $[S]_m$ containing a set S can be easily given; it simplifies considerably for the case of $K4$-algebras where we have

$$[S] \;=\; \{b \in R \mid \exists n \geq 0, \exists a_1, \ldots, a_n \in S \text{ s.t. } \Box^+ a_1 \wedge \cdots \wedge \Box^+ a_n \leq b\}$$

(here $\Box^+ a_i$ stands for $a_i \wedge \Box a_i$). Consequently, the principal modal filter corresponding to an element a is just $[a]_m = [\Box^+ a]$.

In Heyting algebras, the lattice of filters and the lattice of congruences are isomorphic; given a congruence $\simeq$, we can associate to it the filter $\{a \mid a \simeq \top\}$ and given a filter F we can associate to it the congruence $a \simeq b$ iff $a \leftrightarrow b \in F$

(where $a \leftrightarrow b$ is $(a \to b) \wedge (b \to a)$). A similar correspondence works for modal algebras too, with obviously filters replaced by modal filters.

One of the main tools of the book are the duality techniques; in this section we mention their 'finite' counterpart, which is more simple but nevertheless quite important. Recall (or see the Appendix) that a category $\mathbf{C}$ is dual to a category $\mathbf{D}$ iff $\mathbf{C}^{op}$ is equivalent to $\mathbf{D}$. An order-preserving map $f : P \to Q$ among posets is called *open* iff for all $p \in P, q \in Q$

$$q \leq f(p) \qquad \Rightarrow \qquad \exists p' \leq p \, (f(p') = q).$$

The name 'open' is due to topological reasons: a poset can be seen as a topological space with downward closed subsets as open subsets. In this case, order-preserving maps correspond to continuous maps and open maps correspond to continuous open maps. Surjective open maps are called *p-morphisms* (or bounded morphisms).

THEOREM 2.1 *The category of finite distributive lattices is dual to the category of finite posets and order-preserving maps; the category of finite Heyting algebras is dual to the category of finite posets and open maps and the category of finite Boolean algebras is dual to the category of finite sets.*[2]

Proof. Let $\mathbf{D}_{fin}$ and $\mathbf{P}_{fin}$ be the categories of finite distributive lattices and of finite posets, respectively. We define a functor

$$\mathcal{D} : \mathbf{P}_{fin} \longrightarrow \mathbf{D}^{op}_{fin}$$

and check that it is full, faithful and essentially surjective (we do this way, because it is quicker, we leave the reader to find in the exercises the appropriate functor in the other direction establishing the equivalence). For a finite poset P, $\mathcal{D}(P)$ is just the set of downward closed subsets of P; for an order-preserving map $f : P \to P'$, $\mathcal{D}(f)$ is the inverse image function. Faithfulness is immediate: if we have $\mathcal{D}(f') = \mathcal{D}(f)$ (for $f, f' : P \to P'$), then for all $p' \in P'$, $\mathcal{D}(f')(\downarrow p') = \mathcal{D}(f)(\downarrow p')$, which means

$$\forall p \in P \, (f(p) \leq p' \Leftrightarrow f'(p) \leq p'),$$

establishing $f = f'$, as p, p' are arbitrary. As to fullness, consider any distributive lattice morphism $h : \mathcal{D}(P') \longrightarrow \mathcal{D}(P)$. First notice that the left adjoint h^* (it exists as finite lattices are complete and as h preserves meets) maps join-irreducible elements to join-irreducible ones: in fact, if a is join-irreducible, then $h^*(a) \leq \bigvee_i b_i$ implies $a \leq \bigvee h(b_i)$ and hence that there is b_i such that

[2] For a finite duality for Brouwerian semilattices, see an exercise at the end of this Chapter and more in [Kö].

$h^*(a) \leq b_i$. So h^* maps elements of the kind $\downarrow p$ into elements of the same kind; this suggest defining $f : P \to P'$ so that we have

$$h^*(\downarrow p) = \downarrow f(p)$$

for all $p \in P$. We now have the following chain of equivalences for $a \in \mathcal{D}(P')$ and $p \in P$

$$\frac{\frac{\frac{\frac{\frac{p \in \mathcal{D}(f)(a)}{f(p) \in a}}{\downarrow f(p) \subseteq a}}{h^*(\downarrow p) \subseteq a}}{\downarrow p \subseteq h(a)}}{p \in h(a)}$$

establishing that $h = \mathcal{D}(f)$. It remains to show essential surjectivity. Let D be any finite distributive lattice and let $J(D)$ be the poset of its join-irreducible elements. We prove that D is isomorphic to $\mathcal{D}(J(D))$. First, the map $D \to \mathcal{D}(J(D))$ associating with $a \in D$ the set $\{b \in J(D) \,|\, b \leq a\}$ is easily seen to be a distributive lattice morphism. We need to find its inverse: take any $s \in \mathcal{D}(J(D))$ and associate with it its supremum. We have for $a \in D$, $s \in \mathcal{D}(J(D))$

$$\bigvee_{b \in J(D), b \leq a} b = a \qquad s = \{b \in J(D) \,|\, b \leq \bigvee s\}$$

where the latter holds by the definition of join-irreducible element and the former because either $a \in J(D)$ or $a = a_1 \vee a_2$ with $a_1, a_2 < a$, hence we can apply well-founded induction to a_1, a_2 (recall that D is finite).

In order to extend finite duality to Heyting algebras, we only have to show that for $f : P \to P'$, $\mathcal{D}(f)$ preserves implications iff f is open. For $p \in P'$, let $p^c = \{q \in P' \,|\, p \not\leq q\}$ (notice that $p^c \in \mathcal{D}(P')$). Any element a of $\mathcal{D}(P')$ can be represented in two ways, namely as the union of all $\downarrow p$ for $p \in a$ and as intersection of all p^c for $p \notin a$. Taking into consideration the fact that implication preserves meets in the second variable and change joins into meets in the first variable (in other words, the equations $a \to \bigwedge b_i = \bigwedge_i (a \to b_i)$ and $\bigvee_i a_i \to b = \bigwedge_i (a_i \to b)$ hold in any Heyting algebra whenever the mentioned suprema and infima exist), we have that $\mathcal{D}(f)$ preserves implications iff

$$\mathcal{D}(f)(\downarrow p) \to \mathcal{D}(f)(q^c) \subseteq \mathcal{D}(f)(\downarrow p \to q^c) \tag{2.2}$$

holds for every $p, q \in P'$ (the other side $\supseteq$ being always true). However equation (2.2) says that[3]

$$\forall p, q \in P' \; \forall r \in P$$

[3] Just unravel the definitions from 'r does not belong to the second member implies r does not belong to the first member'.

$$((\exists s \leq f(r)\ (q \leq s \leq p)) \Rightarrow \exists r' \leq r\ (q \leq f(r') \leq p));$$

moving the quantifier 'there is an s' to the front, we equivalently get that

$$\forall p, q, s \in P'\ \forall r \in P$$

$$((s \leq f(r)\ \&\ q \leq s \leq p) \Rightarrow \exists r' \leq r\ (q \leq f(r') \leq p)).$$

But this is just the definition of open map (just consider that p, q, s may coincide).

For the Boolean case, it is easy to check that for a finite poset P, $\mathcal{D}(P)$ is a Boolean algebra iff P is discrete. □

A map $f : (X, R) \to (X', R')$ between frames is said to be:

- *relation-preserving* iff $p_1 R p_2$ implies $f(p_1) R' f(p_2)$, for all $p_1, p_2 \in X$;
- *open* iff it is relation-preserving and moreover $f(p) R' q$ implies that there is $p' \in X$ with pRp' and $f(p') = q$, for all $p \in X$ and $q \in X'$.
- a *p-morphism* iff it is surjective and open.

THEOREM 2.2 *The category of finite modal algebras is dual to the category of finite frames and open maps; the category of finite* **K4**-*algebras is dual to the category of finite transitive frames and open maps and the category of finite* **S4**-*algebras is dual to the category of finite preordered frames and open maps.*

Proof. By the above proof, it is clear that the category of finite modal algebras is dual to the category having: i) as objects, finite sets X endowed with a join-preserving operator $\Diamond : \mathcal{P}(X) \to \mathcal{P}(X)$ (considering possibility operators here -instead of necessity operators- makes calculations easier to follow); ii) as arrows $X \to Y$ the set-theoretic functions whose inverse image preserves $\Diamond$. Now the correspondence $R \mapsto \Diamond_R$ is easily seen to be a bijection from the relations on a set X and the join-preserving operators $\mathcal{P}(X) \to \mathcal{P}(X)$ (the inverse correspondence associates with $\Diamond$ the relation $pR_\Diamond q$ iff $p \in \Diamond\{q\}$).[4]

Given that any subset is the set-theoretical union of its singletons and given that inverse image preserves unions, the fact that the inverse image $\mathcal{D}(f)$ along $f : (X, R) \to (X', R')$ preserves the possibility operators induced by the two relations is equivalent to the condition

$$\forall p \in P'\ \ \mathcal{D}(f)(\Diamond_{R'}\{p\}) = \Diamond_R \mathcal{D}(f)(\{p\}).$$

[4] Another more conceptual way of seen the correspondence is the following: $\mathcal{P}(X)$ is the free $\bigvee$-semilattice over the set X, so $\bigvee$-preserving maps $\mathcal{P}(X) \to \mathcal{P}(X)$ corresponds to functions $X \to \mathcal{P}(X)$, which are nothing but relations on X, by the elementary topos axioms.

Unravelling the definitions, one simply recover that this is the condition of being an open map between frames.

For the case of finite **K4** and **S4**-algebras, one has to prove that, given a frame (X, R), the equations

$$\Diamond_R \Diamond_R a \subseteq \Diamond_R a, \qquad a \subseteq \Diamond_R a \tag{2.3}$$

hold (for every $a \subseteq X$) iff R is transitive and, respectively, reflexive. To this aim, one first observes that equations (2.3) hold for arbitrary a iff they hold for a equal to a singleton and then unravels the definitions. □

An important tool in the book is the *finite model property*; we show here how to prove it for Heyting, **K4** and **S4**-algebras; for the other varieties we shall meet in the book, we refer the reader to the specialized textbooks ([CZ], [Kr] for instance contain all that is needed).

THEOREM 2.3 *Every free Heyting algebra embeds into a product of finite Heyting algebras.*

Proof. Let $\mathcal{F}(G)$ be the free Heyting algebra over the generators G and let $\top \neq a \in \mathcal{F}(G)$. According to the general construction of free algebras, there is a term $t(\vec{x})$ and there are $\vec{g} \in G$ such that $t(\vec{g}) = a$ (see for instance section 2.4 below). Let D be the distributive sublattice of $\mathcal{F}(G)$ generated by the elements $t'(\vec{g})$, where t' is a subterm of t. D is finite, hence a Heyting algebra, as remarked above and contains $\vec{g}$. If we succeed in proving that $t'(\vec{x})$ interpreted over $\vec{g}$ both in $\mathcal{F}(G)$ and in D gives the same element, we are done: for, the inclusion $G \hookrightarrow D$ extends uniquely to a morphism $\mathcal{F}(G) \to D$ not mapping a to $\top$, hence we simply have to repeat this construction for all possible a's and take the product. We can easily show inductively that for every $t'(\vec{x})$ subterm of $t(\vec{x})$, $t'(\vec{g})$ (in D) is equal to $t'(\vec{g})$ (in $\mathcal{F}(G)$), provided we are able to establish the following fact:

$(*)$ for $a, b \in D$, if the implication of $a \to b$, taken in $\mathcal{F}(G)$, is equal to an element $c \in D$, then c is also the implication $a \to b$, taken in D.

But this fact is trivial: 'c is the implication of a and b' is equivalent to 'for all d, $a \wedge d \leq b$ holds iff $d \leq c$ holds' and if the latter is true with the quantifier 'for all d' ranging over $\mathcal{F}(G)$, then it is certainly true with the quantifier 'for all d' ranging only over $D \subseteq \mathcal{F}(G)$. □

THEOREM 2.4 *Every free* **K4** *or* **S4**-*algebra embeds into a product of finite algebras.*

Proof. We use the above schema once again; let $\mathcal{F}(G)$ be the free **K4** or **S4**-algebra over the generators G and let $\top \neq t(\vec{g}) \in \mathcal{F}(G)$. Let B be the finite

Boolean algebra generated by the elements $t'(\vec{g})$, where t' is a subterm of t. For the above argument to work, we need to turn B into a **K4** or **S4**-algebra $(B, \Box')$ satisfying the following condition (let i be the inclusion of B into $\mathcal{F}(G)$):

$(*)$ for $b \in B$, if there is $c \in \mathcal{F}(X)$ such that $\Box i(b) = i(c)$, then $\Box' b = c$.

In order to define $\Box'$, consider the following composition of semilattice morphisms (recall that the right adjoint i_* exists because B is finite, hence complete, and because i preserves joins):

$$\otimes : B \xrightarrow{i} \mathcal{F}(G) \xrightarrow{\Box} \mathcal{F}(G) \xrightarrow{i_*} B$$

and put for $b \in B$:

$$\Box'(b) = \bigwedge_{k \geq 1} \otimes^k(b)$$

(for the **S4** case, iteration of $\otimes$ should begin at stage 0, instead of at stage 1). By construction $(B, \Box')$ is a **K4**-algebra. It remains to show $(*)$. Assume $\Box i(b) = i(c)$ and let us first show that $\Box' b \leq c$. This will follow from $i_*(\Box i(b)) \leq c$ (notice that only first iteration is used): as we have $i_*(\Box i(b)) = i_*(i(c))$, we need only $i_*(i(c)) \leq c$, which follows from $i(i_*(i(c))) = i(c)$ (a general fact directly coming from adjointness) and from the fact that i is mono. Let us show the converse, namely that $c \leq \Box' b$, i.e. that

$$c \leq \otimes^k(b). \tag{2.4}$$

for all k. For $k = 1$,[5] we have $c \leq i_*(\Box i(b))$ iff $i(c) \leq \Box i(b)$ which is true, by assumption on c. Now assume that (2.4) holds for some $k > 0$. Since both $\Box$, i are monotone and $i \vdash i_*$, we have the following sequence of inequalities, each one implying the next

$$\frac{\dfrac{\dfrac{\dfrac{\dfrac{\dfrac{\dfrac{c \leq \otimes^k(b)}{i(c) \leq i \otimes^k (b)}}{\Box\, i(b) \leq i \otimes^k (b)}}{\Box\, \Box\, i(b) \leq \Box\, i \otimes^k (b)}}{\Box\, i(b) \leq \Box\, i \otimes^k (b)}}{i(c) \leq \Box\, i \otimes^k (b)}}{c \leq i_* \Box\, i \otimes^k (b)}}{c \leq \otimes^{k+1}(b)}$$

The first inequality holds by assumption and the last shows that (2.4) holds for $k + 1$, as well. $\Box$

[5]For the **S4** case, we have also to prove that $c \leq b$, which follows from $i(c) = \Box i(b) \leq i(b)$ and the fact that i is mono.

2. Model theoretic background

In this section we give the necessary background about model completions. We shall sketch the relevant notions, leaving the interested reader to consult some more complete introduction from the literature (like e.g. [MI] , [Wh], [Sa]; [Ro] gives excellent examples and algebraic applications of model completeness).

We make the following conventions for first-order languages: we indicate the boolean connectives as $\sim$ (negation),& (conjunction), v (disjunction), $\Rightarrow$ (implication); we use the symbols $\top$ and $\bot$ for empty conjunction (or 'true') and empty disjunction (or 'false'), respectively. For existential and universal quantifiers we keep the symbols $\exists, \forall$. Terms will be indicated with the letters $t, u, \ldots$ and formulas will be indicated with the letters $\varphi, \psi, ..$; $t(x_1, \ldots, x_n)$ and $\varphi(x_1, \ldots, x_n)$ or, for short, $t(\vec{x})$, $\varphi(\vec{x})$, mean that t and φ contain free variables from the list $\vec{x} = x_1, \ldots, x_n$. Notations like $t(\vec{u}/\vec{x}), \varphi(\vec{u}/\vec{x})$ (or simply $t(\vec{u}), \varphi(\vec{u})$) are used for simultaneous substitutions.

We say that a first order theory T admits *quantifier elimination* (EQ) iff for every formula $\varphi(\vec{x})$ there is a quantifier-free formula $\psi(\vec{x})$ such that

$$T \models \varphi(\vec{x}) \Leftrightarrow \psi(\vec{x}).$$

One can easily show by induction that we can equivalently restrict the definition to *primitive* formulas, i.e. to formulas $\varphi(\vec{x})$ which are of the kind $\exists x(\beta_1(x, \vec{x})\& \cdots \& \beta_n(x, \vec{x}))$ where the β_i are *basic* (that is, atomic or negated atomic).

On the semantic side, we say that T is *submodel-complete* iff given two models $\mathcal{M}_1, \mathcal{M}_2$ of T and given a common substructure A[6] of them, we have that $\mathcal{M}_1$ is elementarily equivalent to $\mathcal{M}_2$ in the language augmented with names for elements of the support of A.

We recall that, by definition, the *diagram* Δ_A of a structure A is the set of basic sentences in the language augmented by names for elements of the support of A which are true in A; the *positive* (resp. *elementary*) diagram is the same, but with 'atomic sentences' (resp. 'all first-order sentences') instead of 'basic sentences'. Taking in mind diagram Theorem (saying that a structure B is a model of the diagram of A iff there is an embedding from A into B), we may equivalently say that the theory T is submodel-complete iff $T \cup \Delta_A$ is complete whenever A is a substructure of a model of T. We have that:

PROPOSITION 2.5 *T admits (EQ) iff it is submodel-complete.*

---

[6]One should not exclude the case in which A is empty if the language of T does not contain any constant symbol.

Proof. One side is obvious; suppose, for the other side, that T is submodel-complete and let $\varphi(\vec{x})$ be an arbitrary formula. For new constants $\vec{a}$ consider the set of sentences

$$T \cup \{\varphi(\vec{a})\} \cup \{\neg\psi(\vec{a}) \mid \psi \text{ is quantifier free and } T \models \psi(\vec{a}) \Rightarrow \varphi(\vec{a})\}. \quad (2.5)$$

If this set is inconsistent, then we have $T \models \varphi(\vec{a}) \Rightarrow \psi_1(\vec{a}) \text{v} \cdots \text{v} \psi_n(\vec{a})$ for quantifier-free ψ_i implying φ, so that we have

$$T \models \varphi(\vec{x}) \Leftrightarrow \psi_1(\vec{x}) \text{v} \cdots \text{v} \psi_n(\vec{x})$$

yielding the fact that T admits (EQ). So it suffices to show that the set of sentences (2.5) cannot be consistent. Suppose it is and let $\mathcal{M}$ be a model of it. Let A be the substructure of it generated by the $\vec{a}$ (notice that A is finitely generated); we must have

$$T \cup \Delta \models \varphi(\vec{a})$$

where Δ is the diagram of A, because otherwise we would be able to build a model of T containing A as a substructure and falsifying $\varphi(\vec{a})$ (which cannot be because T is submodel-complete and $\varphi(\vec{a})$ is true in $\mathcal{M}$ according to (2.5)). This means that for some quantifier-free sentence $\psi(\vec{a})$ true in A we have that $T \models \psi(\vec{a}) \Rightarrow \varphi(\vec{a})$. According to (2.5), $\neg\psi(\vec{a})$ is true in $\mathcal{M}$ and also in A (because it is quantifier-free), contradiction. $\square$

From the proof it is clear that we may equivalently modify the definition of submodel-completeness by limiting ourselves to finitely generated common substructures of a couple of models of the theory. If moreover the theory does not have finite models (which is the rule in the relevant examples), we may furthermore limit ourselves to pairs of models of any fixed cardinality greater or equal to the cardinality of the language (by an easy application of Löwhenheim-Skolem theorems). These modifications to the definition of submodel-completeness will be used without explicit mention when needed.

The other ingredient for the notion of model completion is given by one of the two equivalent notions of the next Proposition:

PROPOSITION 2.6 *Let T_1, T_2 be first-order theories in the same language; the following conditions are equivalent:*

(i) *each model of T_1 embeds into a model of T_2;*

(ii) *for every quantifier-free formula $\psi(\vec{x})$ we have that*

$$T_2 \models \psi(\vec{x}) \quad \textit{implies} \quad T_1 \models \psi(\vec{x}).$$

Proof. Ad (i)$\Rightarrow$ (ii): if $T_1 \not\models \psi(\vec{x})$ (where ψ is quantifier-free), we have $\mathcal{M} \not\models \psi(\vec{a})$ for some model $\mathcal{M}$ of T_1 and for some $\vec{a}$ taken from the support of

$\mathcal{M}$. Take a model $\mathcal{M}'$ of T_2 such that $\mathcal{M} \subseteq \mathcal{M}'$. As ψ is quantifier-free, $\psi(\vec{a})$ is false in $\mathcal{M}'$ too, showing that $T_2 \not\models \psi(\vec{x})$.

Ad (ii)$\Rightarrow$ (i): let $\mathcal{M}$ be a model of T_1 and consider $T_2 \cup \Delta$, where Δ is the diagram of $\mathcal{M}$. It is sufficient to show that this set of sentences is consistent. Suppose not; then we have that $T_2 \models \psi(\vec{a})$, where ψ is quantifier-free and false in $\mathcal{M}$. It follows that $T_1 \models \psi(\vec{a})$, which is clearly impossible. $\square$

We are ready for the main definition. Let T be a universal theory;[7] we say that a theory $T^* \supseteq T$ (in the same language) is a *model completion of* T iff T^* admits quantifier-elimination and T^* proves the same quantifier-free formulas as T does. We first show that a model completion is unique provided it exists:

PROPOSITION 2.7 *Let T_1^*, T_2^* be two model completions of the same first-order theory T; then, for every formula φ, we have that $T_1^* \models \varphi$ iff $T_2^* \models \varphi$.*

Proof. We show that any model $\mathcal{M}_0$ of, say T_2^*, is also a model of T_1^*. As $\mathcal{M}_0$ is a model of $T \subseteq T_2^*$, we can embed it into a model $\mathcal{M}_1$ of T_1^*; for the same reason $\mathcal{M}_1$ embeds into a model $\mathcal{M}_2$ of T_2, etc. Let $\mathcal{M}$ be the union of the chain so produced. In fact, we have two chains: the odd indices chain (entirely formed by models of T_1^*) and the even indices chain (entirely formed by models of T_2^*). Both chains are elementary, because T_1^*, T_2^* are submodel complete,[8] hence by the elementary chain theorem $\mathcal{M}$ is elementarily equivalent to both $\mathcal{M}_0$ and $\mathcal{M}_1$; it follows that $\mathcal{M}_0 \models T_1^*$. $\square$

An important feature (which we will use within the book) of model-completable theories is amalgamation:

PROPOSITION 2.8 *If a theory T admits a model completion then for every pair of monomorphisms $f_1 : \mathcal{M}_0 \longrightarrow \mathcal{M}_1$, $f_2 : \mathcal{M}_0 \longrightarrow \mathcal{M}_2$ among models of T, there exists a model $\mathcal{M}$ of T and monomorphisms $g_1 : \mathcal{M}_1 \longrightarrow \mathcal{M}$, $f_2 : \mathcal{M}_2 \longrightarrow \mathcal{M}$ such that the following square commutes.*

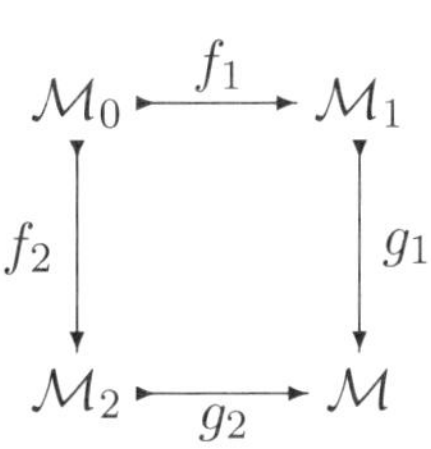

[7]The definition can be extended also to theories which are not universal, but we do not insist on that since such extension is not needed for the purposes of this book.

[8]As a special case in the definition of submodel completeness, any monomorphism among models of a submodel-complete theory is elementary (this is the special case in which the common substructure specified in the definition coincides with one of the two models).

Proof. Let us first embed $\mathcal{M}_1, \mathcal{M}_2$ into models $\mathcal{M}'_1, \mathcal{M}'_2$ of the model completion T^*; for simplicity, suppose that $\mathcal{M}_0$ is a a common substructure of them. If Δ is the diagram of $\mathcal{M}_0$, it happens that $\mathcal{M}'_1, \mathcal{M}'_2$ are both models of the theory $T^* \cup \Delta$ which is complete (by the submodel-completeness of T^*). But any two models of a complete theory jointly elementarily embed into a third model $\mathcal{M}$ of the same theory (because the union of their elementary diagrams is consistent); such third model fits our $\mathcal{M}$. □

A model $\mathcal{M}$ is said to be *existentially closed* for a theory T iff for every embedding $\mathcal{M} \subseteq \mathcal{M}'$ into a model of T, we have that every existential sentence (possibly with parameters from the domain of $\mathcal{M}$) which holds in $\mathcal{M}'$ holds in $\mathcal{M}$ too.

PROPOSITION 2.9 *If T has a model completion T^*, then the class of models of T^* is just the class of models of T which are existentially closed for T.*

Proof. As T^* is submodel-complete, any monomorphism among models of T^* is elementary, hence, by the elementary chain theorem, T^* is preserved under union on chains. By a well-known preservation theorem (see e.g. [CK]), this implies that T^* has a set of $\forall\exists$ axioms. Given that, let $\mathcal{M}$ be an existentially closed model of T and let $\forall\vec{x}\exists\vec{y}\psi(\vec{x},\vec{y})$ be an axiom for T^*, where ψ is quantifier-free. Let us embed $\mathcal{M}$ into a model $\mathcal{N}$ of T^*; we need to show that $\exists\vec{y}\psi(\vec{a},\vec{y})$ is true in $\mathcal{M}$ for every $\vec{a}$ taken from the support of $\mathcal{M}$ itself. Such a sentence is clearly true in $\mathcal{N}$ (because $\mathcal{N} \models T^*$) and hence also in $\mathcal{M}$, as $\mathcal{M}$ is existentially closed for $T \subseteq T^*$.

Conversely, suppose that $\mathcal{M}$ is a model of T^* and let us show that every existential sentence φ with parameters from $\mathcal{M}$ holding in an extension $\mathcal{N} \supseteq \mathcal{M}$ such that $\mathcal{N} \models T$, holds in $\mathcal{M}$ as well. First embed $\mathcal{N}$ into $\mathcal{M}' \models T^*$; then observe that $\mathcal{M}' \models \varphi$ (as φ is existential) and finally recall that the embedding $\mathcal{M} \subseteq \mathcal{M}'$ is elementary. □

There are many interesting mathematical examples of model-completable theories: the classical example is the theory of integral domains (the model completion is the theory of algebraically closed fields): in fact on one hand every integral domain embeds into the algebraic closure of its field of fractions and on the other hand recall that there is only one (up to K-isomorphism) algebraically closed extension in a sufficiently large (but fixed) cardinality m of a given field K (it is the algebraically closed extension having transcendence degree m over K). Other examples are provided by torsion-free abelian groups (the model completion is the theory of non-zero divisible torsion-free abelian groups), ordered integral domains (the model completion is the theory of ordered real closed fields). linear orders (the model completion is the theory of dense linear order without endpoints). In the sequel we analyze the case of Boolean algebras (the model completion is the theory of atomless Boolean algebras), which is a paradigm for much more complicated examples we shall meet afterwards.

First, let us show how to embed every non degenerate[9] Boolean algebra B_0 into an atomless Boolean algebra. Let T be the theory of Boolean algebras and let Δ be the diagram of B; let c_b be new constants (we add a new constant c_b for each $b \in B$, $b \neq \perp$); consider the set of sentences

$$T \cup \Delta \cup \{\perp < c_b < b\}. \tag{2.6}$$

If this set is consistent, we may use it in order to produce a Boolean algebra $B_1 \supseteq B_0$; repeating, we produce a chain whose union is an atomless Boolean algebra extending B_0. It remains to show that (2.6) is consistent; it is sufficient to show that any set like (2.6) is consistent, where now Δ is the diagram of a finite non degenerate Boolean algebra (this is by the compactness theorem and because finitely generated Boolean algebras are finite). A finite non degenerate Boolean algebra is of the kind $\mathcal{P}(X)$ for some non-empty set X, by finite duality; now $\mathcal{P}(X)$ embeds into $\mathcal{P}(X \times \{0, 1\})$ through inverse image along the first projection. This makes clear how to get a model of (2.6): simply interpret c_b as $\{(x, 0) \mid x \in b\}$.

Now let us show that the theory of atomless Boolean algebras is submodel-complete; we prove that, given a finite Boolean algebra B_0 and given two countable atomless Boolean algebras $C \supseteq B_0$, $D \supseteq B_0$, there is an isomorphism $f : C \to D$ fixing B_0. Let $c_1, c_2, \ldots$ and $d_1, d_2, \ldots$ be enumerations of the elements of $C \backslash B_0$ and $D \backslash B_0$. Let C_i be the Boolean algebra generated by B_0 and $c_1, \ldots, c_i$ and let D_i be the Boolean algebra generated by B_0 and $d_1, \ldots, d_i$. We shall establish below the following fact:

$(*)$ given an atomless Boolean algebra A, given finite Boolean algebras A_1, A_2 and given monomorphisms $m : A_1 \to A$, $n : A_1 \to A_2$, there exists a monomorphism $m' : A_2 \to A$ such that $m' \circ n = m$.

Using $(*)$ and a back-and-forth construction it is easy to exhibit the required isomorphism: at stage i (for odd i), define an isomorphism $f_i : C'_i \to D'_i$ fixing B_0, where C'_i is a subalgebra of C including C_i and D'_i is a subalgebra of D including B_0. At stage i (for even i) just reverse the role of C and D; everything should be done, of course, in such a way that the sequence $\{f_i\}_i$ is increasing (we may start with f_0 equal to $1_{B_0} : B_0 \to B_0$).

It remains to check fact $(*)$. Once again, we use finite duality. The reader is asked to show by himself that for any nonzero element y of an atomless Boolean algebra and for any $k \geq 2$ there exist nonzero elements $y_1, \ldots, y_k$ such that

$$y_1 \vee \cdots \vee y_k = y \qquad \text{and } y_i \wedge y_j = 0 \text{ (for } i \neq j)$$

---

[9] For the one-element Boolean algebra there is nothing to prove: such algebra is atomless (it has no atoms, because $\perp$ is not an atom).

(such $y_1, \ldots, y_k$ are called a k-partition of y). Let A, A_1, A_2, m, n be as in the statement of $(*)$; by finite duality, $n : A_1 \to A_2$ is the dual of a surjective map $h : X_2 \to X_1$ among finite sets. Let k be big enough so that each element of X_1 has less than k pre-images along h and choose $A' \subseteq A$ big enough so that every element of the kind $m(a)$ has a k-partition in A'. It is clearly sufficient to show $(*)$ with A replaced by such A'. Again by finite duality, $m : A_1 \to A'$ is the dual of a surjective map $l : X \to X_1$ among finite sets; here, by the choice of A', each element of X_1 has more than k pre-images along l, that is it has more pre-images along l than it has along h. This is sufficient to find a surjective map $l' : X \to X_2$ such that $h \circ l' = l$. The dual of l' is the required m'.

3. Subobjects and regular subobjects

We summarize here some basic material concerning subobject and regular subobject semilattices in a category $\mathbf{C}$ with finite limits. All the proofs are quite simple, so that we omit or merely sketch them. On the other hand, the content of this Section will be used without explicit mention throughout the whole book.

Given an object A in $\mathbf{C}$, we first consider the class of monomorphisms

$$S \stackrel{s}{\hookrightarrow} A$$

having target A (according to common practice, we often indicate such monos by the name of their domain, i.e. simply by S in this case); two such monos $S_1 \stackrel{s_1}{\hookrightarrow} A$ and $S_1 \stackrel{s_2}{\hookrightarrow} A$ are considered equivalent iff there is an iso $k : S_1 \longrightarrow S_2$ such that the triangle

$$\begin{array}{ccc} S_1 & \xrightarrow{\quad k \quad} & S_2 \\ & {}_{s_1}\searrow \quad \swarrow_{s_2} & \\ & A & \end{array}$$

commutes; we indicate by $Sub(A)$ the related quotient set, which is called the set of subobjects of A.[10] $Sub(A)$ is actually a poset if we define $S_1 \leq S_2$ iff there is $k : S_1 \longrightarrow S_2$ commuting the related triangle over A (notice that we directly indicate equivalence classes of monos by their representatives, notice also that such a k must be mono, although not necessarily iso). An easy but important fact is that monos are pullback stable, that is for any $f : B \longrightarrow A$ in the pullback

[10] In relevant examples, $Sub(A)$ is a set, not a proper class.

$$\begin{array}{ccc} f^*(S) & \longrightarrow & S \\ {\scriptstyle s'}\downarrow & & \downarrow{\scriptstyle s} \\ B & \xrightarrow[f]{} & A \end{array}$$

if s is mono so is s'. As a consequence, we have that $Sub(A)$ is actually a semilattice: top element is the identity, meet is obtained by pullback, that is $S_1 \wedge S_2$ is the subobject of A obtained by the diagonal of the pullback square:

$$\begin{array}{ccc} S_1 \wedge S_2 & \longrightarrow & S_1 \\ \downarrow & & \downarrow \\ S_2 & \longrightarrow & A \end{array}$$

Moreover, again by taking pullback, we have a semilattice morphism

$$Sub(f) : Sub(A) \longrightarrow Sub(B)$$

for any $f : B \longrightarrow A$ in $\mathbf{C}$ (we usually indicate $Sub(f)(S)$ as $f^*(S)$). To summarize:

PROPOSITION 2.10 *For any category* $\mathbf{C}$ *with finite limits, we have that* $Sub : \mathbf{C}^{op} \longrightarrow \mathbf{SemiL}$ *is a contravariant functor taking values into the category of semilattices and related morphisms.* □

We now consider regular subobjects, that is equivalence classes of monos $S \hookrightarrow A$ which happens to be equalizers of some parallel pair of arrows of domain A. We shall meet such regular subobject in the universal algebra context of next Section, here we analyze them in an appropriate general categorical framework. The reason why we need regular monos and not just monos is clear: we are mainly concerned with opposite categories of kind of algebras, so regular monos must stand for quotients (this may not happen with plain monos: for instance, the embedding of an integral domain into its field of fractions is both epi and mono but it is evidently not a quotient). We denote by $Sub_r(A)$ the subposet of $Sub(A)$ formed by regular subobjects. The important point is that the previous Proposition still holds:

PROPOSITION 2.11 *For any category* $\mathbf{C}$ *with finite limits, we have that* $Sub_r : \mathbf{C}^{op} \longrightarrow \mathbf{SemiL}$ *is a contravariant functor taking values into the category of semilattices (it is actually a subfunctor of* Sub*).*

Proof. We need to show that if $S, T \hookrightarrow A$ are regular subobjects of A and if $f : B \longrightarrow A$ is any arrow, then $S \wedge T$ and $f^*(S)$ are regular subobjects too. In fact if $S \hookrightarrow A$ is an equalizer of $h_1, h_2 : A \longrightarrow C$ and if $T \hookrightarrow A$ is an equalizer of $g_1, g_2 : A \longrightarrow D$, then $S \wedge T \hookrightarrow A$ is an equalizer of $\langle h_1, g_1\rangle, \langle h_2, g_2\rangle : A \longrightarrow C \times D$ and $f^*(S) \hookrightarrow B$ is an equalizer of $h_1 \circ f, h_2 \circ f : B \longrightarrow C$. □

4. Finitely presented algebras

We fix in this section an *equational* first order theory T. We use the letters $E, F, \dots$ for formulas which are (possibly empty) conjunctions of equations (briefly, *e-formulas*). $Alg(T)$ is the category of algebras which are models of T (briefly, T-algebras) and $Alg(T)_{fin}$ is the full subcategory determined by finite algebras. T-algebras will be indicated by $A, B, \dots$ and morphisms among T-algebras by $f, g, \dots$. We shall make a systematic confusion between an algebra and its carrier set. Elements of an algebra are indicated with letters $a_1, a_2, \dots, b_1, b_2, \dots$; the letters $\vec{a}, \vec{b}, \dots$ are used for strings of elements. When we expand the language of T with parameters arising from a given algebra, we do not distinguish between elements and their names in the expanded language. A pair given by a set of new constants and a set of closed e-formulas in the related expanded language is called a *presentation*. Given such a presentation $\langle X, \Delta\rangle$, there are an algebra (unique up to isomorphism) denoted by $\mathcal{F}(X)/\Delta$ and an interpretation for the new constants $i : X \to \mathcal{F}(X)/\Delta$ such that $\mathcal{F}(X)/\Delta \models \Delta$ and such that for every other pair $\langle A, i'\rangle$ with the same property, there is a unique morphism $f : \mathcal{F}(X)/\Delta \to A$ such that $f \circ i = i'$. We recall how $\mathcal{F}(X)/\Delta$ is built; we take all closed terms $t(\vec{a})$ with parameters from X and we introduce the equivalence relation:

$$t_1 \sim t_2 \quad \text{iff} \quad E_1, \dots, E_n \vdash_T t_1 = t_2$$

for some $E_1, \dots, E_n \in \Delta$ (provability is always intended in T, unless remarked by a different subscript on $\vdash$). Equivalence classes can be turned into an algebra $\mathcal{F}(X)/\Delta$ which is a model of T by interpreting all function symbols on representative elements. The interpretation $i : X \to \mathcal{F}(X)/\Delta$ is obtained by associating every $a \in X$ with its own equivalence class. The above mentioned universal property can be easily checked.

We say that $\langle X, \Delta\rangle$ is a presentation for (any algebra isomorphic to) $\mathcal{F}(X)/\Delta$. As a first application of the existence of algebras matching each presentation in a universal way, we have the following:

PROPOSITION 2.12 *If* $E_1(\vec{x})$ & $\cdots$ & $E_n(\vec{x}) \Rightarrow F_1(\vec{x})v\cdots$v$F_m(\vec{x})$ *is provable in* T*, then there is an* $i = 1, \dots, m$ *such that* $\vdash_T E_1$ & $\cdots$ & $E_n \Rightarrow F_i$.

Proof. Take the universal algebra matching the presentation $\langle \vec{x}, \{E_1, \dots, E_n\}\rangle$; as F_1v$\cdots$vF_m must be true in it, we have that F_i is true in it for some i. By the universal property, $F_i(\vec{a})$ is true in any model of T provided $E_1(\vec{a}), \dots, E_n(\vec{a})$ are true in such a model. □

Each algebra admits a presentation, the canonical one is given by the carrier set and the positive diagram. If Δ is empty, $\mathcal{F}(X)$ is the free algebra over X; if both X and Δ are finite (in this case Δ can be taken to be a singleton because e-formulas are closed under finite conjunctions) the presentation is said to be finite. An algebra is *finitely presented* iff it admits a finite presentation: we can represent a finitely presented algebra as $\mathcal{F}(\vec{x})/E(\vec{x})$, meaning that we see its elements as terms in the variables $\vec{x}$ quotiented by provable equality under assumption $E(\vec{x})$. The notion of finitely presented algebra can be introduced also in a purely categorical way, cf. [GU], [MaPi]: A is finitely presented iff the set-valued representable functor represented by A preserves filtered colimits. We denote by $Alg(T)_{fp}$ the full subcategory of $Alg(T)$ given by the finitely presented algebras. We use also the following notation: if $A \in Alg(T)$ and $E(\vec{a})$ is the e-formula with parameters from A given by $t_1(\vec{a}) = t'_1(\vec{a})\ \& \cdots \&\ t_n(\vec{a}) = t'_n(\vec{a})$, $A/E(\vec{a})$ is the quotient of A by the congruence generated by the pairs $\langle t_i(\vec{a}), t'_i(\vec{a})\rangle$ $(i = 1, \ldots, n)$. Notice that if A admits the presentation $\mathcal{F}(X)/\Delta$ and if $\vec{a} \subseteq X$, then $A/E(\vec{a})$ admits the presentation $\mathcal{F}(X)/(\Delta \cup \{E(\vec{a})\})$. Any (single) morphism $f : A \to B$ in $Alg(T)_{fp}$ can be presented as $\bar{f} : \mathcal{F}(\vec{x})/E(\vec{x}) \longrightarrow \mathcal{F}(\vec{x}, \vec{y})/F(\vec{x}, \vec{y})$ for some e-formulas $E(\vec{x})$ and $F(\vec{x}, \vec{y})$, where $\bar{f}$ sends generators $\vec{x}$ in $\mathcal{F}(\vec{x})/E(\vec{x})$ to the same generators in $\mathcal{F}(\vec{x}, \vec{y})/F(\vec{x}, \vec{y})$ so that there is a commuting square in $Alg(T)_{fp}$

$$\begin{array}{ccc} A & \xrightarrow{\ f\ } & B \\ {\scriptstyle\cong}\downarrow & & \downarrow{\scriptstyle\cong} \\ \mathcal{F}(\vec{x})/E(\vec{x}) & \xrightarrow[\bar{f}]{} & \mathcal{F}(\vec{x}, \vec{y})/F(\vec{x}, \vec{y}) \end{array}$$

with vertical morphisms, being isomorphisms. We often do not define and even name a morphism $\mathcal{F}(\vec{x})/E(\vec{x}) \longrightarrow \mathcal{F}(\vec{x}, \vec{y})/F(\vec{x}, \vec{y})$ if we mean that it sends $\vec{x}$ to $\vec{x}$.

It is well known that $Alg(T)_{fp}$ has finite colimits. We describe some of them in terms of their presentations. The initial object is the free algebra on empty set of generators $\mathcal{F}(\emptyset)$. For any algebras $\mathcal{F}(\vec{x})/E(\vec{x})$ and $\mathcal{F}(\vec{y})/F(\vec{y})$

$$\mathcal{F}(\vec{x})/E(\vec{x}) \longrightarrow \mathcal{F}(\vec{x}, \vec{y})/E(\vec{x})\&F(\vec{y}) \longleftarrow \mathcal{F}(\vec{y})/F(\vec{y})$$

is a coproduct diagram. For algebras $\mathcal{F}(\vec{x})/E(\vec{x})$, $\mathcal{F}(\vec{x}, \vec{y})/F(\vec{x}, \vec{y})$ and $\mathcal{F}(\vec{x}, \vec{z})/G(\vec{x}, \vec{z})$ the diagram

$$\begin{array}{ccc} \mathcal{F}(\vec{x})/E(\vec{x}) & \longrightarrow & \mathcal{F}(\vec{x},\vec{y})/F(\vec{x},\vec{y}) \\ \downarrow & & \downarrow \\ \mathcal{F}(\vec{x},\vec{z})/G(\vec{x},\vec{z}) & \longrightarrow & \mathcal{F}(\vec{x},\vec{y},\vec{z})/F(\vec{x},\vec{y})\&G(\vec{x},\vec{z}) \end{array}$$

is a pushout. Finally, for morphisms $f, g : \mathcal{F}(\vec{y})/F(\vec{y}) \to \mathcal{F}(\vec{x})/E(\vec{x})$ the diagram

$$\mathcal{F}(\vec{y})/F(\vec{y}) \underset{g}{\overset{f}{\rightrightarrows}} \mathcal{F}(\vec{x})/E(\vec{x}) \longrightarrow \mathcal{F}(\vec{x})/E(\vec{x})\&E'(\vec{x})$$

is a coequalizer, where, if $\vec{y} = y_1, \ldots, y_n$, $E'(\vec{x})$ is $f(y_1) = g(y_1)\& \cdots \&f(y_n) = g(y_n)$. By definition (see the Appendix), a regular epi is a coequalizer. Thus by the above description regular epis are surjective in $Alg(T)_{fp}$ (for general reasons this is also true in $Alg(T)$ and in $Alg(T)_{fin}$).

The converse is true, as well:

PROPOSITION 2.13 *If a morphism in $Alg(T)_{fp}$ is a surjection, then it is a regular epi.*

Proof. Suppose that a morphism

$$\mathcal{F}(\vec{x})/E(\vec{x}) \longrightarrow \mathcal{F}(\vec{x},\vec{y})/G(\vec{x},\vec{y}) \tag{2.7}$$

in $Alg(T)_{fp}$ is a surjection. Then for every y_i there is $t_i(\vec{x})$ such that

$$G(\vec{x},\vec{y}) \vdash y_i = t_i(\vec{x}),$$

moreover $G(\vec{x},\vec{y}) \vdash E(\vec{x})$. This makes evident that the morphism (2.7) can be presented as

$$\mathcal{F}(\vec{x})/E(\vec{x}) \longrightarrow \mathcal{F}(\vec{x})/F(\vec{x})\&E(\vec{x})$$

(where $F(\vec{x}) = G(\vec{x}, \vec{t}(\vec{x})/\vec{y})$), thus it is a coequalizer of the pair

$$\mathcal{F}(\vec{y}) \underset{g}{\overset{f}{\rightrightarrows}} \mathcal{F}(\vec{x})/E(\vec{x})$$

where, if $F(\vec{x})$ is $t_1(\vec{x}) = s_1(\vec{x})\& \cdots \&t_n(\vec{x}) = s_n(\vec{x})$, we take $\vec{y} = y_1, \ldots, y_n$ and $f(y_i) = t_i(\vec{x})$, $g(y_i) = s_i(\vec{x})$, for $1 \leq i \leq n$. $\square$

We say, that T has the *Beth property*, BP for short, iff for any e-formula $E(\vec{p}, x)$, if

$$E(\vec{p}, x), E(\vec{p}, y) \vdash x = y \tag{2.8}$$

then

$$\text{there is a term } t(\vec{p}) \text{ such that } E(\vec{p}, x) \vdash x = t(\vec{p}). \tag{2.9}$$

We have

PROPOSITION 2.14 *T has the Beth property iff epimorphisms are regular in* $Alg(T)_{fp}$.

Proof. Let $E(\vec{p}, \vec{x})$ be an e-formula. In the following pushout

$$\begin{array}{ccc}
\mathcal{F}(\vec{p}) & \xrightarrow{\quad f \quad} & \mathcal{F}(\vec{p}, \vec{x})/E(\vec{p}, \vec{x}) \\
\Big\downarrow f & & \Big\downarrow \kappa \\
\mathcal{F}(\vec{p}, \vec{x})/E(\vec{p}, \vec{x}) & \xrightarrow[\kappa']{} & \mathcal{F}(\vec{p}, \vec{x}, \vec{y})/E(\vec{p}, \vec{x})\&E(\vec{p}, \vec{y})
\end{array}$$

where $\kappa'(x_i) = y_i$, both κ and κ' are split mono with the same splitting

$$h : \mathcal{F}(\vec{p}, \vec{x}, \vec{y})/E(\vec{p}, \vec{x})\&E(\vec{p}, \vec{y}) \longrightarrow \mathcal{F}(\vec{p}, \vec{x})/E(\vec{p}, \vec{x})$$

where $h(y_i) = x_i$. It is a general fact, that f is epi iff both κ and κ' are iso. If

$$E(\vec{p}, \vec{x}), E(\vec{p}, \vec{y}) \vdash \vec{x} = \vec{y} \tag{2.10}$$

then both κ and κ' are epi and hence iso (epi and spilt mono is iso). Hence f is epi. On the other hand, if f is epi then both κ and κ' are iso, and since $\kappa' \circ h(x_i) = y_i$ we have (2.10). Thus we have shown that *f is epi iff (2.10) holds.*

Clearly, (2.9) holds iff the morphism $f : \mathcal{F}(\vec{p}) \to \mathcal{F}(\vec{p}, x)/E(\vec{p}, x)$ is a surjection, i.e. it is a regular epi.

Thus, if epimorphisms are regular in $Alg(T)_{fp}$ then the Beth property holds for T.

Now assume that the Beth property holds for T. We shall show by induction on number of variables in $\vec{x}$ that if $f_1 : \mathcal{F}(\vec{p})/F(\vec{p}) \to \mathcal{F}(\vec{p}, \vec{x})/E(\vec{p}, \vec{x})$ is epi then it is a surjection. Since $f_2 : \mathcal{F}(\vec{p}) \to \mathcal{F}(\vec{p})/F(\vec{p})$ is always a surjection, we can simply prove that $f = f_1 \circ f_2 : \mathcal{F}(\vec{p}) \longrightarrow \mathcal{F}(\vec{p})/E(\vec{p}, \vec{x})$ is a surjection.

If $n = 0$ then f is clearly a surjection. Suppose that we have proved the claim for $n - 1$ and let $\vec{x} = x_1, \ldots, x_n$. Since f is epi, (2.10) holds. By the Beth property, there is a formula $t(\vec{p}, x_1, \ldots, x_{n-1})$ such that

$$E(\vec{p}, \vec{x}) \vdash x_n = t(\vec{p}, x_1, \ldots, x_{n-1}) \tag{2.11}$$

Hence we have a commuting triangle

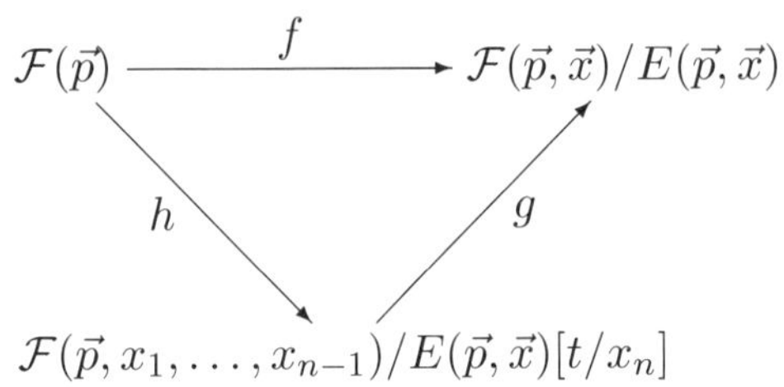

with g being a surjection by (2.11), and h being a surjection by inductive hypothesis, since, if we substitute $t(\vec{p},\vec{x}')$ for x_n and $t(\vec{p},\vec{y}')$ for y_n in (2.10) (where $\vec{x}' = x_1, \ldots, x_{n-1}$, $\vec{y}' = y_1, \ldots, y_{n-1}$), we obtain

$$E(\vec{p},\vec{x})[t(\vec{p},\vec{x}\,')/x_n], E(\vec{p},\vec{y})[t(\vec{p},\vec{y}\,')/y_n] \vdash x_i = y_i$$

for $i = 1, \ldots, n-1$ Thus f is a surjection as required. $\square$

We shall be mainly interested in the category $\mathbf{T}$, the opposite category of $Alg(T)_{fp}$. Since $Alg(T)_{fp}$ has finite colimits, $\mathbf{T}$ has finite limits.

For e-formulas $E(\vec{x})$, $E'(\vec{x})$ the morphism $\mathcal{F}(\vec{x})/E \longrightarrow \mathcal{F}(\vec{x})/E'$, sending $\vec{x}$ to $\vec{x}$ exists iff $\vdash_T E' \Rightarrow E$, and if this is the case it a regular epi. By our characterization of coequalizers (see above), all regular epis with domain $\mathcal{F}(\vec{x})/E$ arise in this way. Thus regular monos in $\mathbf{T}$ with codomain $\mathcal{F}(\vec{x})/E(\vec{x})$ corresponds to (equivalence classes of) e-formulas $E'(\vec{x})$ such that $\vdash_T E' \Rightarrow E$. We shall write $[\![E']\!]$ for the regular subobject represented by E' (a more precise, but cumbersome notation would be $[\![E']\!]_{\mathcal{F}(\vec{x})/E}$). Conversely, given a regular subobject S, we use the notation $[S]$ in order to indicate an e-formula representing it. Notice that $[\![E_1]\!] \leq [\![E_2]\!]$ iff $\vdash_T E_1 \Rightarrow E_2$; also the meets in the meet semilattice of regular subobjects $Sub_r(A)$ of a fp algebra A (see the Appendix) are represented by conjunction, that is $[\![E_1]\!] \wedge [\![E_1]\!] = [\![E_1 \& E_2]\!]$. However, if in $Sub_r(A)$ there are other lattice-theoretic operations like joins, implications, etc. they have *nothing to do* with the corresponding propositional logical connectives. Moreover, given an arbitrary morphism $\mathcal{F}(\vec{x})/E \longrightarrow \mathcal{F}(\vec{y})/F$ in $Alg(T)_{fp}$ which is identified, say, by the correspondence $\vec{x} \mapsto \vec{t}(\vec{y})$, the pullback functor in $\mathbf{T}$ (pushout in $Alg(T)_{fp}$) maps the regular subobject $[\![E'(\vec{x})]\!]$ into the regular subobject $[\![E'(\vec{t}(\vec{y}))\& F]\!]$, so it acts as substitution from a logical point of view.

5. Principal congruences

Let T and $\mathbf{T}$ be as in the previous section.

We show here that the fact that $\mathbf{T}$ has stable finite joins and implications among regular subobjects is equivalent to the fact that $Alg(T)$ satisfies some well-known conditions from universal algebra. These conditions say that there exist suitable terms describing membership to principal congruences and their intersections. The material presented here follows quite freely some arguments from [KP] and [BP] (notice however that we do not have reasons for investigating condition (EDPM) below separately, in this sense proofs may not be optimal).

We say that a variety has *equationally definable principal congruences*, (EDPC) for short,[11] iff there exists an e-formula $I(x_1, x_2, x_3, x_4)$ such that for every algebra A and for every 4-tuple of elements a_1, a_2, a_3, a_4 from it we have that

$$A/a_1 = a_2 \models a_3 = a_4 \text{ iff } A \models I(a_1, a_2, a_3, a_4).$$

We say that a variety has *equationally definable principal meets*, (EDPM) for short,[12] iff there exist e-formulas $J(x_1, x_2, x_3, x_4)$ and J_0 (J_0 is variable-free) such that:

(i) for every algebra A and for every 6-tuple of elements $a_1, a_2, a_3, a_4, b_1, b_2$ from it we have that

$$(A/a_1 = a_2 \models b_1 = b_2 \text{ and } A/a_3 = a_4 \models b_1 = b_2)$$
$$\text{iff} \quad A/J(a_1, a_2, a_3, a_4) \models b_1 = b_2;$$

(ii) for every algebra A and for every pair of elements b_1, b_2 from it we have that

$$A/J_0 \models b_1 = b_2.$$

Finally, we say that a variety has *equationally definable operations*, (EDO) for short, iff it has both (EDPC) and (EDPM).

We begin with a Lemma:

LEMMA 2.15 *Suppose that R is a meet-semilattice and that S is a generating subset of it such that all implications and finite joins among members of S exist in R. Then:*

(i) *R is a Heyting algebra;*

(ii) *for every other Heyting algebra H, we have that any semilattice morphism $f : R \to H$ preserving implications and finite joins among members of S is in fact a Heyting algebras morphism.*

Proof. Ad (i). Let $r = \bigwedge_{i=0}^{n} s_i$ and $r' = \bigwedge_{j=0}^{m} s'_j$ be two elements of R expressed as meets of elements coming from S. $r \to r'$ is introduced by induction on n as follows. For $n = 0$, it is equal to $\bigwedge_{j=0}^{m}(s_0 \to s'_j)$ and for $n > 0$, it is equal to $s_0 \to (\bigwedge_{i=1}^{n} s_i \to r')$.

[11]In [KP] the name (REDPC) is used for the same concept.

[12]The definition given in [BP] is slightly different because it requires only condition (i) below.

For binary joins (empty join exists by hypothesis), $r \vee r'$ is defined as $\bigwedge_{i=0}^{n} \bigwedge_{j=0}^{m}(s_i \vee s'_j)$. In order to prove that it is really the join of r and r' the relevant point is to show by induction on $n+m$ that for every $u \in R$

$$(r \leq u \text{ and } r' \leq u) \text{ imply } \bigwedge_{i=0}^{n} \bigwedge_{j=0}^{m}(s_i \vee s'_j) \leq u.$$

For $n+m=0$ the claim is ensured by the hypotheses; for $n+m>0$ we use the fact, just proved, that there are implications in R. In fact, we can apply induction to

$$\bigwedge_{i=1}^{n} s_i \leq s_0 \to u \text{ and to } \bigwedge_{j=0}^{m} s'_j \leq s_0 \to u$$

in order to get $\bigwedge_{i=1}^{n} \bigwedge_{j=0}^{m}(s_i \vee s'_j) \leq s_0 \to u$. By standard properties of implications and by the fact that $\bigwedge_{i=0}^{n} \bigwedge_{j=0}^{m}(s_i \vee s'_j) \leq \bigwedge_{i=1}^{n} \bigwedge_{j=0}^{m}(s_i \vee s'_j)$ we obtain $s_0 \leq \bigwedge_{i=0}^{n} \bigwedge_{j=0}^{m}(s_i \vee s'_j) \to u$. The same argument, repeated for s'_0, gives $s'_0 \leq \bigwedge_{i=0}^{n} \bigwedge_{j=0}^{m}(s_i \vee s'_j) \to u$, whence

$$s_0 \vee s'_0 \leq \bigwedge_{i=0}^{n} \bigwedge_{j=0}^{m}(s_i \vee s'_j) \to u$$

and finally $\bigwedge_{i=0}^{n} \bigwedge_{j=0}^{m}(s_i \vee s'_j) \leq u$.

Ad (ii): the statement trivially follows from the way we introduced joins and implications during the proof of (i). □

PROPOSITION 2.16 *The following are equivalent for an equational theory T:*

(i) *$Alg(T)$ has (EDO);*

(ii) *in $\mathbf{T}$ there are finite joins and implications among regular subobjects and moreover such finite joins and implications are stable under taking pullbacks.*

Proof. ($\Leftarrow$ side) Suppose that $\mathbf{T}$ has stable implications and joins among regular subobjects. Let $I(x_1, x_2, x_3, x_4)$ be an e-formula such that in $Sub_r(\mathcal{F}(x_1, x_2, x_3, x_4))$ we have

$$[\![I(x_1, x_2, x_3, x_4)]\!] = p^*_{x_1,x_2}[\![x_1 = x_2]\!] \to p^*_{x_3,x_4}[\![x_3 = x_4]\!]$$

(here e.g. p_{x_1,x_2} denotes the pair given by the first and the second projection in $\mathbf{T}$ from the domain $\mathcal{F}(x_1, x_2, x_3, x_4) \simeq \mathcal{F}(x_1) \times \mathcal{F}(x_2) \times \mathcal{F}(x_3) \times \mathcal{F}(x_4)$). We show that such I is an e-formula suitable for condition (EDPC). Suppose that for $A \in Alg(T)$ and $a_1, a_2, a_3, a_4 \in A$, we have that $A/a_1 = a_2 \models a_3 = a_4$. Let

$\mathcal{F}(X)/\Delta$ be a presentation for A (we can take distinct names for a_1, a_2, a_3, a_4 in this presentation, even if they are not distinct); now we have

$$\vdash_T \Delta_0(a_1, a_2, a_3, a_4, \vec{b}) \,\&\, a_1 = a_2 \;\Rightarrow\; a_3 = a_4$$

for some finite conjunction Δ_0 of members of Δ. Taking variables instead of constants and interpreting this relation in regular subobjects of $\mathcal{F}(x_1, x_2, x_3, x_4, \vec{y})$ in $\mathbf{T}$, we get

$$[\![\Delta_0]\!] \wedge p^*_{x_1,x_2,x_3,x_4} p^*_{x_1,x_2}([\![x_1 = x_2]\!]) \leq p^*_{x_1,x_2,x_3,x_4} p^*_{x_3,x_4}([\![x_3 = x_4]\!]).$$

By existence and stability of implications, we get

$$[\![\Delta_0]\!] \leq p^*_{x_1,x_2,x_3,x_4}([\![I]\!]),$$

from which we obtain that $A \models I(a_1, a_2, a_3, a_4)$. The same passages, read in the opposite sense, show that if $A \models I(a_1, a_2, a_3, a_4)$ then $A/a_1 = a_2 \models a_3 = a_4$.

For (EDPM) an analogous argument can be applied: we take as $J(x_1, x_2, x_3, x_4)$ any e-formula representing the regular subobject $p^*_{x_1,x_2}[\![x_1 = x_2]\!] \vee p^*_{x_3,x_4}[\![x_3 = x_4]\!]$ and as J_0 any e-formula representing the minimum regular subobject of $\mathcal{F}(\emptyset)$. Notice that, in the relative verifications, distributivity of the regular subobject lattices (a consequence of the existence of implications) is used.

($\Rightarrow$ side) Suppose that (EDO) holds; we show that there are stable joins and implications among regular subobjects in $\mathbf{T}$. Given an object $\mathcal{F}(\vec{x})/F$ of $\mathbf{T}$, we have that the semilattice of its regular subobjects is generated by the set S of regular subobjects of the form $[\![F \,\&\, u_1 = u_2]\!]$, so we can apply Lemma 2.15 to it. We take $[\![F \,\&\, I(u_1, u_2, u_3, u_4)]\!]$ as $[\![F \,\&\, u_1 = u_2]\!] \rightarrow [\![F \,\&\, u_3 = u_4]\!]$: we need to prove that for every e-formula $E(\vec{x})$ such that $\vdash_T E \Rightarrow F$, we have that

$$[\![E]\!] \wedge [\![F \,\&\, u_1 = u_2]\!] \leq [\![F \,\&\, u_3 = u_4]\!] \text{ iff}$$
$$\text{iff } [\![E]\!] \leq [\![F \,\&\, I(u_1, u_2, u_3, u_4)]\!].$$

Using the fact that $\vdash_T E \Rightarrow F$, we get that $[\![E]\!] \wedge [\![F \,\&\, u_1 = u_2]\!] \leq [\![F \,\&\, u_3 = u_4]\!]$ is equivalent to $\vdash_T E \,\&\, u_1 = u_2 \Rightarrow u_3 = u_4$. But this means that $\mathcal{F}(\vec{x})/(E \,\&\, u_1 = u_2) \models u_3 = u_4$ or again that $(\mathcal{F}(\vec{x})/E)/u_1 = u_2 \models u_3 = u_4$. By (EDPC), this is equivalent to $\mathcal{F}(\vec{x})/E \models I(u_1, u_2, u_3, u_4)$, which is the same as $[\![E]\!] \leq [\![F \,\&\, I(u_1, u_2, u_3, u_4)]\!]$, in view of the fact that $\vdash_T E \Rightarrow F$.

For finite joins the situation is similar: we take $[\![F \,\&\, J(u_1, u_2, u_3, u_4)]\!]$ as $[\![F \,\&\, u_1 = u_2]\!] \vee [\![F \,\&\, u_3 = u_4]\!]$ and $[\![F \,\&\, J_0]\!]$ as the minimum regular subobject.

Let us now check stability under pullbacks. Suppose we are given an arrow $f : \mathcal{F}(\vec{y})/G \longrightarrow \mathcal{F}(\vec{x})/F$ in $\mathbf{T}$. This arrow is induced by a substitution, in the sense that, as a morphism in the opposite sense among algebras, it maps the equivalence classes of $\vec{x}$ into the equivalence classes of some terms $\vec{t}$ in the variables $\vec{y}$, more generally it will map the equivalence class of $u(\vec{x})$ into the equivalence class of $u(\vec{t}/\vec{x})$.[13] As mentioned in the previous section, at the level of regular subobjects, we have that $f^*([\![F']\!]) = [\![G \;\&\; F'(\vec{t})]\!]$ for every $[\![F'(\vec{x})]\!] \in Sub_r(\mathcal{F}(\vec{x})/F)$. Let us now check stability of implications: by Lemma 2.15(ii), we must only check that

$$f^*([\![F \;\&\; u_1 = u_2]\!]) \to f^*([\![F \;\&\; u_3 = u_4]\!])$$

is equal to

$$f^*([\![F \;\&\; u_1 = u_2]\!] \to [\![F \;\&\; u_3 = u_4]\!]),$$

which is trivial as both members are equal to

$$[\![G \;\&\; I(u_1(\vec{t}), u_2(\vec{t}), u_3(\vec{t}), u_4(\vec{t}))]\!].$$

The same argument goes for stability of binary joins. Stability of minimum regular subobject is immediate: $f^*([\![F \;\&\; J_0]\!])$ is directly equal to $[\![G \;\&\; J_0]\!]$ (J_0 is variable-free). $\square$

Proposition 2.16 makes evident that Theorem 3.11 applies to varieties of Heyting and **K4**-algebras (our main topic in Chapters 4,5 and 6), because such varieties have (EDO). In fact, for **K4**-algebras we can take:

- $\square^+(x_1 \leftrightarrow x_2) \leq x_3 \leftrightarrow x_4$ as $I(x_1, x_2, x_3, x_4)$;
- $\square^+(x_1 \leftrightarrow x_2) \vee \square^+(x_3 \leftrightarrow x_4) = \top$ as $J(x_1, x_2, x_3, x_4)$;
- $\bot = \top$ as J_0.

For Heyting algebras, we can do the same (omitting $\square^+$ everywhere). Further examples of varieties satisfying (EDO) are given in the exercises.

6. The amalgamation property

The amalgamation property and the related notions are usually formulated for categories of algebras and for this reason they are formulated in terms which are not categorical (like injections, congruences, etc). Some of them have clear purely categorical meaning. The categorical notion corresponding to the universal algebra notions have broader scope of application but more importantly they have clear dual notions. This is essential to us since we shall study properties of some categories via their duals. So only categorical properties can be easily transferred back and forth between these categories.

[13]Notice that not every substitution $\vec{x} \mapsto \vec{t}(\vec{y})$ defines such a morphism, we must have $\mathcal{F}(\vec{y})/G \models F(\vec{t})$ or, which is the same, $\vdash_T G \Rightarrow F(\vec{t})$.

We recall some notions from universal algebra in categorical formulation so that they keep the usual meaning in categories of algebras.

Let **C** be a category. We say that **C** has the *amalgamation property* (AP) iff for any monomorphisms $u : A \to B$, $f : A \to C$ in **C** there are monomorphisms $v : C \to D$, $g : B \to D$ in **C** such that $g \circ u = v \circ f$, i.e. the square

$$\begin{array}{ccc} A & \xrightarrow{u} & B \\ {\scriptstyle f}\downarrow & & \downarrow{\scriptstyle g} \\ C & \xrightarrow[v]{} & D \end{array}$$

commutes. We say that **C** has the *strong amalgamation property* (AP_{st}) iff for any monomorphisms $u : A \to B$, $f : A \to C$ in **C** there are monomorphisms $v : C \to B$, $g : B \to D$ in **C** such that the diagram above is a pullback. We say that *injections are transferable* in **C** (IT) iff for any morphism $f : A \to C$ and any monomorphism $u : A \to B$ in **C** there are a morphism $g : B \to D$ and a monomorphism $v : C \to B$ in **C** such that $g \circ u = v \circ f$.

We say that **C** has the *congruence extension property* (CEP) iff for any regular epimorphism $f : A \to C$ and any monomorphism $u : A \to B$ in **C** there are a regular epimorphism $g : B \to D$ and a monomorphism $v : C \to B$ in **C** such that $g \circ u = v \circ f$.

The above notions have clear dual notions, e.g. the dual to AP is obtained by replacing in the definition of AP mono by epi and reversing the direction of the arrows. We denote the dual notions with a superscript $*$, i.e. AP^*, AP^*_{st}, IT^*, CEP^*. In view of the existence of pushouts, for any equational theory T, in the categories of algebras $Alg(T)$ and $Alg(T)_{fp}$, we have that (IT) is equivalent to the conjunction of (AP) and (CEP). As we shall see in the next Chapter, Lemma 3.9, if $Alg(T)^{op}_{fp}$ has stable implications among regular subobjects then (CEP) holds in $Alg(T)_{fp}$. Thus, in view of Proposition 2.16, if T has (EDO) then (AP) is equivalent to (IT) in $Alg(T)_{fp}$.

LEMMA 2.17 *If* **C** *satisfies* AP_{st} *then monomorphisms are regular in* **C**.

Proof. Let $m : A \to B$ be a monomorphism in **C**. Then, by AP_{st}, we can form a pullback in **C**

$$\begin{array}{ccc} A & \xrightarrow{m} & B \\ {\scriptstyle m}\downarrow & & \downarrow{\scriptstyle k_1} \\ B & \xrightarrow[k_0]{} & C \end{array}$$

It is easy to see that the diagram

$$A \xrightarrow{m} B \overset{k_0}{\underset{k_1}{\rightrightarrows}} C$$

is an equalizer in $\mathbf{C}$, i.e. m is a regular mono. □

The following technical lemma will be used later in this section and in the following Chapter.

LEMMA 2.18 *Assume* $\mathbf{T}$ *has stable implications among regular subobjects. Then, for every* T*-algebra* A *and for every e-formulas* $E_1(\vec{a}), E_2(\vec{a})$ *with parameters* $\vec{a}$ *from* A*, we have that*[14]

$$A/E_1(\vec{a}) \models E_2(\vec{a}) \text{ iff } A \models [[\![E_1]\!] \to [\![E_2]\!]](\vec{a}).$$

Proof. Recall that $\mathcal{F}(A)/\Delta$, where Δ is the positive diagram of A, is a presentation for A. Thus $A/E_1(\vec{a})$ has presentation $\mathcal{F}(A)/\Delta \cup \{E_1(\vec{a})\}$. We have that $A/E_1(\vec{a}) \models E_2(\vec{a})$ iff for some finite conjunction $\Delta_0(\vec{a}, \vec{b})$ of formulas from Δ, we have

$$\vdash_T \Delta_0(\vec{x}, \vec{y}) \ \& \ E_1(\vec{x}) \ \Rightarrow E_2(\vec{x}).$$

Taking internal meaning in $\mathbf{T}$, this is the same as

$$[\![\Delta_0]\!] \wedge p^*_{\vec{x}}([\![E_1]\!]) \leq p^*_{\vec{x}}([\![E_2]\!])$$

i.e., by the fact that $p^*_{\vec{x}}$ preserves implications (as implications are stable), this is the same as

$$[\![\Delta_0]\!] \leq p^*_{\vec{x}}([\![E_1]\!] \to [\![E_2]\!]).$$

So, $A/E_1(\vec{a}) \models E_2(\vec{a})$ iff for some $\Delta_0(\vec{a}, \vec{b})$ in Δ, we have $\vdash_T \Delta_0(\vec{a}, \vec{b}) \Rightarrow [[\![E_1]\!] \to [\![E_2]\!]](\vec{a})$ iff $A \models [[\![E_1]\!] \to [\![E_2]\!]](\vec{a})$. □

We end this section with a little remark about amalgamation in finite algebras which will be useful in the sequel, because it will absolve us from providing a duality Theorem for finitely presented modal algebras in cases in which sheaf representation might be problematic.

We say that our equational theory T has *the finite model property* iff finitely generated free algebras embed into products of finite algebras. This means that $Alg(T)$, as a variety, is generated by its finite members.

In case $\mathbf{T}$ has stable implications, we have an equivalent stronger characterization:

LEMMA 2.19 *If* T *has the finite model property and if* $\mathbf{T}$ *has stable implications among regular subobjects, then every finitely presented algebra embeds into a product of finite algebras.*

[14]Notations $[\![-]\!]$ and $[-]$ are explained in the last paragraph of Section 2.4.

Proof. Let $\mathcal{F}(\vec{x})/E_1(\vec{x})$ be a finitely presented algebra. It is sufficient to show that for every e-formula $E_2(\vec{x})$, if $\mathcal{F}(\vec{x})/E_1(\vec{x}) \not\models E_2(\vec{x})$, then there exist a finite algebra B and a morphism $f : \mathcal{F}(\vec{x})/E_1(\vec{x}) \longrightarrow B$ such that $B \not\models E_2(f(\vec{x}))$. But, by the above Lemma, $\mathcal{F}(\vec{x})/E_1(\vec{x}) \not\models E_2(\vec{x})$ implies that $\mathcal{F}(\vec{x}) \not\models [\llbracket E_1 \rrbracket \to \llbracket E_2 \rrbracket](\vec{x})$, so by the finite model property there are a finite algebra C and a morphism $g : \mathcal{F}(\vec{x}) \longrightarrow C$ such that $C \not\models [\llbracket E_1 \rrbracket \to \llbracket E_2 \rrbracket](g(\vec{x}))$. This means that $C/E_1(g(\vec{x})) \not\models E_2(g(\vec{x}))$. By the universal property of quotients, we can form a commutative square

$$\begin{array}{ccc} \mathcal{F}(\vec{x}) & \xrightarrow{\quad g \quad} & C \\ \downarrow & & \downarrow \\ \mathcal{F}(\vec{x})/E_1(\vec{x}) & \longrightarrow & C/E_1(g(\vec{x})) \end{array}$$

yielding the claim. □

PROPOSITION 2.20 *Suppose that the language of T is finite, that T has the finite model property and that* **T** *has stable implications among regular subobjects. If the amalgamation property holds in $Alg(T)$ then it holds in $Alg_{fin}(T)$, as well. In particular (by Proposition 2.8), if T has a model completion then the amalgamation property holds in $Alg_{fin}(T)$.*

Proof. As the amalgamation property holds in $Alg(T)$, given finite algebras A_0, A_1, A_2 and monomorphisms $f_1 : A_0 \to A_1$, $f_2 : A_0 \to A_2$ there are an algebra A and monomorphisms $g_1 : A_1 \to A$, $g_2 : A_2 \to A$ such that the square

$$\begin{array}{ccc} A_0 & \xrightarrow{\quad f_1 \quad} & A_1 \\ {\scriptstyle f_2}\downarrow & & \downarrow{\scriptstyle g_1} \\ A_2 & \xrightarrow[\quad g_2 \quad]{} & A \end{array}$$

commutes. By the universal property of pushouts and by the fact that the first component of a monomorphism is a monomorphism, we can freely suppose that A is just the pushout of A_1 and A_2 over A_0. As A_0, A_1, A_2 are finite, they are finitely presented because the language of T is finite[15], hence A, being the pushout, is finitely presented as well. By the finite model property and Lemma 2.19, there are finite algebras $\{B_i\}_{i\in I}$ and an embedding $< h_i >_{i\in I}$:

[15] This is a standard fact (see the Exercises).

$A \longrightarrow \prod_{i\in I} B_i$. For every pair a_1, b_1 of different elements of A_1, select $i_{a_1,b_1} \in I$ such that $h_{i_{a_1 b_1}} g_1(a_1) \neq h_{i_{a_1 b_1}} g_1(b_1)$. Do the same for every pair a_2, b_2 of different elements of A_2. Let $I_0 \subseteq I$ be a finite subset of I containing all $i_{a_1 b_1}, i_{a_2 b_2}$. It is clear that $< h_i >_{i\in I_0} \circ g_1 : A_1 \longrightarrow \prod_{i\in I_0} B_i$ and $< h_i >_{i\in I_0} \circ g_2 : A_2 \longrightarrow \prod_{i\in I_0} B_i$ are monomorphisms amalgamating A_1 and A_2 over A_0 into the finite algebra $\prod_{i\in I_0} B_i$. $\square$

7. Exercises

1 We gave two definitions of lattices, Brouwerian semilattices, Heyting algebras, etc. (one is in term of properties of posets, the other one is equational): show that they are equivalent.

2 Describe explicitly a pair of functors establishing a duality between the category of finite posets and of the category of finite distributive lattices.

3 (Köhler) Show that the category of finite Brouwerian semilattices is dual to the category of finite posets and strict open partial maps (a partial map $f : P \longrightarrow Q$ among posets is said to be strict open iff once restricted to its domain is open and satisfies the condition $p < p' \Rightarrow f(p) < f(p')$).

4 A theory T^* is said to be a *model companion* of a theory T iff i) they both prove the same quantifier-free formulas; ii) T^* is model-complete, i.e. any monomorphisms between models of T^* is elementary. Show that any theory has at most one model companion.

5 Show that the model companion T^* of a theory T is the theory of the class of models of T which are existentially closed for T. Show that T^* exists iff such a class is elementary.

6 Show that in any category a morphism which is either regular epi and mono or epi and regular mono is an iso. Find an equational theory T such that in $Alg(T)$ there is a morphism which is both epi and mono but not iso.

7 Prove Propositions 2.10 and 2.11 in all the details.

8 [KP] Recall that a variety of algebras is *congruence distributive* iff the lattice of congruences of every algebra in the variety is distributive. Show that if T has EDPC then $Alg(T)$ is a congruence distributive variety.

9 A Lawvere algebra is a Heyting algebra endowed with a unary operator l satisfying the axioms: i) $x \leq lx$; ii) $llx = lx$; iii) $l(x \wedge y) = lx \wedge ly$. Show that the Lawvere algebras enjoy (EDO).

10 A monadic Heyting algebra is a structure $\langle H, \forall, \exists\rangle$, where H is a Heyting algebra and the following equations hold

$$\forall x \leq x \qquad\qquad x \leq \exists x$$

$$\forall(x \wedge y) = \forall x \wedge \forall y \qquad \exists(x \vee y) = \exists x \vee \exists y$$
$$\forall 1 = 1 \qquad \exists 0 = 0$$
$$\forall \exists x = \exists x \qquad \exists \forall x = \forall x$$
$$\exists(\exists x \wedge y) = \exists x \wedge \exists y$$

Show that monadic Heyting algebras enjoy (EDO).

11 A bi-modal algebra with universal modality is a structure $\langle B, \Box, U\rangle$, where: i)$\langle B, \Box\rangle$ is a modal algebra; ii) $\langle B, U\rangle$ is an **S5**-algebra (namely an **S4**-algebra with the additional axiom $x \leq U\neg U\neg x$); iii) the condition $Ux \leq \Box x$ holds. Show that bi-modal algebras with universal modality enjoy (EDO).

12 Show that finitely generated congruences on any **K4**-algebra A correspond to the elements of A such that $x = x \wedge \Box x$. Show that the elements satisfying the above equation in A form a Heyting algebra.

13 Show that the e-formula $J(x_1, x_2, x_3, x_4)$ introduced in section 2.5 can be characterized up to provable equivalence as the only e-formula such that for any T-algebra A and for any four elements a_1, a_2, a_3, a_4 from A the canonical morphism

$$A \longrightarrow A/a_1 = a_2 \times_{A/a_1=a_2,a_3=a_4} A/a_3 = a_4$$

factorizes through the quotient morphism

$$A \longrightarrow A/J(a_1, a_2, a_3, a_4)$$

followed by a monomorphism.

14 Prove that a finite algebra A is finitely presented in case the language is finite (i.e. in case the signature contains only finitely many function and constant symbols).

15 Let T be an equational theory. Show that the embedding $Alg(T)_{fp} \rightarrow Alg(T)$ preserves all the finite colimits, in particular it preserves epimorphisms.

16 Let T be an equational theory in a finite language, with finite model property and EDPC. Show that the embedding $Alg(T)_{fin} \rightarrow Alg(T)$ of finite T-algebras into all T-algebras preserves existing finite colimits.

17 Let T be an equational theory. Show that the following conditions are equivalent

(a) all monos in $Alg(T)$ are regular;

(b) any morphism in $Alg(T)$ can be factored as a regular epi followed by a regular mono;

(c) the regular and the coregular factorizations of all morphisms in $Alg(T)$ coincide.

18 Let T be an equational theory. We say, that T has the *strong Beth property*, (SBP) for short, iff for any e-formula $E(\vec{p},\vec{q})$ and any term $u(\vec{p},\vec{q})$, if

$$E(\vec{p},\vec{q}), E(\vec{p},\vec{q}\,') \vdash u(\vec{p},\vec{q}) = u(\vec{p},\vec{q}\,')$$

then there is a term $t(\vec{p})$ such that $E(\vec{p},\vec{q}) \vdash u(\vec{p},\vec{q}) = t(\vec{p})$.

Show that T has SBP iff all monomorphisms in $Alg(T)$ are regular.

Hints. For the 'if' part note that if $E(\vec{p},\vec{q}), E(\vec{p},\vec{q}\,') \vdash u(\vec{p},\vec{q}) = u(\vec{p},\vec{q}\,')$ then, by the previous exercise, the coregular factorization of the homomorphism $f : \mathcal{F}(\vec{p}) \to \mathcal{F}(\vec{p},\vec{q})/E(\vec{p},\vec{q})$ has the following form

$$\begin{array}{ccccc} \mathcal{F}(\vec{p}) & \xrightarrow{\ f\ } & \mathcal{F}(\vec{p},\vec{q})/E(\vec{p},\vec{q}) & \overset{p_0}{\underset{p_1}{\rightrightarrows}} & \mathcal{F}(\vec{p},\vec{q},\vec{q}\,')/E(\vec{p},\vec{q}), E(\vec{p},\vec{q}\,') \\ & {\scriptstyle e}\searrow & \nearrow{\scriptstyle m} & & \\ & & \mathcal{F}(\vec{p})/\Delta(\vec{p}) & & \end{array}$$

where all the morphisms indicated send generators to the same generators except for p_1 which sends $\vec{q}$ to $\vec{q}\,'$, and $\Delta(\vec{p})$ is a (possibly infinite) set of equations of terms with variables from $\vec{p}$. Then use the fact that e is a regular epi to conclude that $f(t(\vec{p})) = u(\vec{p},\vec{q})$ for some $t(\vec{p})$ in $\mathcal{F}(\vec{p})$.

For the 'only if' part present an arbitrary monomorphism in $Alg(T)$ as

$$\mathcal{F}(\vec{P})/\Delta(\vec{P}) \xrightarrow{\ f\ } \mathcal{F}(\vec{P},\vec{Q})/\Gamma(\vec{P},\vec{Q})$$

and its cokernel as

$$\mathcal{F}(\vec{P},\vec{Q})/\Gamma(\vec{P},\vec{Q}) \overset{p_0}{\underset{p_1}{\rightrightarrows}} \mathcal{F}(\vec{P},\vec{Q},\vec{Q}')/\Gamma(\vec{P},\vec{Q}),\Gamma(\vec{P},\vec{Q}')$$

where where f and p_0 send generators to the same generators, p_1 sends $\vec{P}$ to $\vec{P}$ and $\vec{Q}$ to $\vec{Q}'$, and $\Delta(\vec{P})$ is a (possibly infinite) set of equations of terms with variables from $\vec{P}$.

Now using compactness and SBP show that if $p_0(\psi(\vec{p},\vec{q})) = p_1(\psi(\vec{p},\vec{q}))$ (here $\vec{p},\vec{q}$ are finite sets of variables from $\vec{P},\vec{Q}$), then there is $\theta(\vec{p})$ in ... such that $f(\theta(\vec{p})) = \psi(\vec{p},\vec{q})$.

19 Let T be an equational theory[16].

Using exercise 17b, show that if all monos are regular in $Alg(T)$ then all epis are regular in $Alg(T)$ and in $Alg(T)_{fp}$ as well. *Hint.* For $Alg(T)_{fp}$ keep in mind exercise 15 and Proposition 2.13.

20 Assume that L is a propositional logic which has among other connectives $\top$, $\wedge$ and $\leftrightarrow$[17]. Let $\vdash_L$ denote the deducibility relation in L. By $Alg(L)$ we denote the variety of algebras in the language having operations corresponding to the connectives of L, so the formulas of L can be identified with terms of $Alg(L)$ and $\varphi = \psi$ is an axiom of $Alg(L)$ whenever $\vdash_L \varphi \leftrightarrow \psi$. Moreover, assume that for any terms $\varphi_1, \ldots, \varphi_n$, φ, ψ and any e-formula E of $Alg(L)$, we have

$$\varphi_1, \ldots, \varphi_n \vdash_L \varphi \quad \text{iff} \quad \varphi_1 = \top, \ldots, \varphi_n = \top \models_{Alg(L)} \varphi = \top$$

$$E \models_{Alg(L)} (\varphi \leftrightarrow \psi) = \top \quad \text{iff} \quad E \models_{Alg(L)} \varphi = \psi$$

and

$$E \models_{Alg(L)} (\varphi \wedge \psi) = \top \quad \text{iff}$$
$$E \models_{Alg(L)} \varphi = \top \text{ and } E \models_{Alg(L)} \psi = \top$$

We say that the logic L has the *Beth property for deducibility*, (BPD) for short, iff for any formula $\varphi(\vec{p}, x)$ of L, if $\varphi(\vec{p}, x), \varphi(\vec{p}, y) \vdash_L x \leftrightarrow y$ then there is a formulas $\psi(\vec{p})$ of L such that $\varphi(\vec{p}, x) \vdash_L x \leftrightarrow \psi(\vec{p})$.

Show that L has BPD iff $Alg(L)$ has BP[18].

21 Assume that L is a propositional logic as in the previous exercise 20.

We say that the logic L has the *strong Beth property for deducibility*, (SBPD) for short, iff for any formulas $\varphi(\vec{p}, \vec{q})$ and $\psi(\vec{p}, \vec{q})$ of L, if $\varphi(\vec{p}, \vec{q}), \varphi(\vec{p}, \vec{q}') \vdash_L \psi(\vec{p}, \vec{q}) \leftrightarrow \psi(\vec{p}, \vec{q}')$ then there is a formula $\theta(\vec{p})$ of L such that $\varphi(\vec{p}, \vec{q}) \vdash_L \psi(\vec{p}, \vec{q}) \leftrightarrow \theta(\vec{p})$.

Show that L has SBPD iff $Alg(L)$ has SBP.

---

[16]This exercise shows that the algebraic characterization of the strong Beth property SBP implies the algebraic characterization of the Beth property BP. The two properties are indeed distinct: for instance, SBP is quite rare in varieties of Heyting algebras [Ma6], whereas BP holds for all of them.

[17]We don't want to enter into the discussion of what an algebraizable logic is, since it can be a subject of another book, due to many borderline cases which should be taken into account in such considerations. However, if a correspondence between a logic L and a variety of algebras $Alg(L)$ is 'reasonable', which can mean in different contexts different things, we can transfer back and forth some properties of logics into properties of varieties and vice versa. The above exercise specifies some 'reasonable' assumptions about such correspondence which is good enough to make a link between (BPD) and (BP). These assumptions are clearly satisfied by all superintuitionistic logics and by all modal logics extending the system **K4**.

[18]See also Proposition 2.14.

Chapter 3

MODEL COMPLETIONS

1. r-Heyting categories

In this section we introduce the notions of r-regular and r-Heyting categories and study some of their basic properties. Roughly speaking, these notions are obtained from the extensively studied notions of regular and Heyting category (see e.g. [MR1], [MR2]]) 'by replacing monos with regular monos and regular epis by epis'. In case all subobjects are regular, the two notions coincide (this is evident from Proposition 3.3 below), so, for instance, any topos is r-regular and also r-Heyting. In case not all monos are regular, the two concepts are quite distinct: posets and order-preserving maps, for instance, form an r-Heyting category which is not even regular. As we saw in Proposition 2.14, regularity of monos in the opposite of the category of finitely presented algebras follows from some appropriate version of Beth theorem, which is often true (e.g. it holds in all varieties of Heyting and of **K4**-algebras, see [Ma5] and Section 5.6 below). Up to some extent, the theory of r-regular and r-Heyting categories goes parallel to that of regular and Heyting categories: some of the properties established in this section, for instance, are obtained through adaptations of standard arguments.

The best way to introduce r-regular categories is probably through stable factorization systems. Given a category **C**, a pair of classes of arrows $\langle \mathcal{E}, \mathcal{M} \rangle$ is said to be a *stable factorization system* for **C** iff the following four conditions are satisfied (see [FK], but we follow the equivalent formulation of [CJKP]):

(i) both $\mathcal{E}$ and $\mathcal{M}$ contain identities and are closed under left and right composition with isomorphisms;

(ii) each map f in **C** can be written as $m \circ e$ with $m \in \mathcal{M}$ and $e \in \mathcal{E}$;

(iii) whenever we have a commutative square,

$$\begin{array}{ccc} A & \xrightarrow{e} & B \\ {\scriptstyle u}\downarrow & & \downarrow{\scriptstyle v} \\ C & \xrightarrow[m]{} & D \end{array}$$

with $m \in \mathcal{M}$ and $e \in \mathcal{E}$, there is a unique $w : B \longrightarrow C$ such that $w \circ e = u$ and $m \circ w = v$;

(iv) whenever we have a pullback square

$$\begin{array}{ccc} A & \xrightarrow{u'} & B \\ {\scriptstyle e'}\downarrow & & \downarrow{\scriptstyle e} \\ C & \xrightarrow[u]{} & D \end{array}$$

the fact that $e \in \mathcal{E}$ implies that $e' \in \mathcal{E}$.

The decomposition in (ii) is said to be a *factorization* for f; this factorization is unique in the sense that if $f = m \circ e$ can be factored as well as $m' \circ e'$, for $m' \in \mathcal{M}$ and $e' \in \mathcal{E}$, then using (iii), it can be shown that there is an invertible map w such that $w \circ e = e'$ and $m' \circ w = m$. In a factorization system, both $\mathcal{E}$ and $\mathcal{M}$ are closed under composition [CJKP]: let us recall how to show it for $\mathcal{M}$ (for $\mathcal{E}$ the proof is analogous). It is sufficient to have a characterization of arrows in $\mathcal{M}$, from which the desired property easily follows. So let us prove that for all $f : C \to D$, f belongs to $\mathcal{M}$ iff f is *orthogonal* to $\mathcal{E}$, i.e. iff the following condition is satisfied:

- 'for every $e \in \mathcal{E}$ and for every commutative square

$$\begin{array}{ccc} A & \xrightarrow{e} & B \\ {\scriptstyle u}\downarrow & & \downarrow{\scriptstyle v} \\ C & \xrightarrow[f]{} & D \end{array}$$

there is a unique $w : B \longrightarrow C$ such that $w \circ e = u$ and $f \circ w = v$.'

One side is just (iii); for the other side, take a factorization $m \circ e$ of f and consider the square

$$\begin{array}{ccc} C & \xrightarrow{e} & D' \\ {\scriptstyle id_C}\downarrow & & \downarrow{\scriptstyle m} \\ C & \xrightarrow[f]{} & D \end{array}$$

to get w such that $w \circ e = id_C$. As $e \circ w$ and $id_{D'}$, both fills the diagonal of the square

$$\begin{array}{ccc} C & \xrightarrow{e} & D' \\ {\scriptstyle e}\downarrow & & \downarrow{\scriptstyle m} \\ D' & \xrightarrow[m]{} & D \end{array}$$

they are equal by (iii), so e is iso and $f \in \mathcal{M}$ by (i).

Having established that an arrow belongs to $\mathcal{M}$ iff it is orthogonal to $\mathcal{E}$, it is not difficult to see that if $C \xrightarrow{m_1} D \xrightarrow{m_2} E$ are both orthogonal to $\mathcal{E}$, so is $m_2 \circ m_1$.

We say that a category **C** is *r-regular* iff it has finite limits and moreover taking $\mathcal{E}$ =all epis and $\mathcal{M}$ =all regular monos, we get a stable factorization system for **C**. As conditions (i) and (iii) are trivially true in this case (by the definition of epi and regular mono), **C** is r-regular iff it has finite limits, each arrow has an epi/regular mono factorization and epis are stable under pullbacks.

PROPOSITION 3.1 *If* **C** *is r-regular, then the pullback functors operating on regular subobjects have left adjoints satisfying the Beck-Chevalley condition.*

Proof. The statement of the Proposition says that for every arrow $f : B \to A$ in **C**, for every regular subobject $S \hookrightarrow B$, there is a regular subobject $\exists_f(S) \hookrightarrow A$ satisfying the condition

$$\exists_f(S) \leq T \quad \text{iff} \quad S \leq f^*(T)$$

for every regular subobject $T \hookrightarrow A$. The Beck-Chevalley condition says that for every pullback square

$$\begin{array}{ccc} C & \xrightarrow{p_1} & B_1 \\ {\scriptstyle p_2}\downarrow & & \downarrow{\scriptstyle f_1} \\ B_2 & \xrightarrow[f_2]{} & A \end{array} \tag{3.1}$$

and for every regular subobject $S \hookrightarrow B_1$, the equation

$$f_2^*(\exists_{f_1}(S)) = \exists_{p_2}(p_1^*(S)) \tag{3.2}$$

holds in $Sub_r(B_2)$.

Let us fix a morphism $f : B \to A$ and a regular mono $s : S \hookrightarrow B$ in $\mathbf{C}$. We take as $\exists_f(S)$ the second component of the factorization of $f \circ s$. We can form a commuting diagram

(3.3)

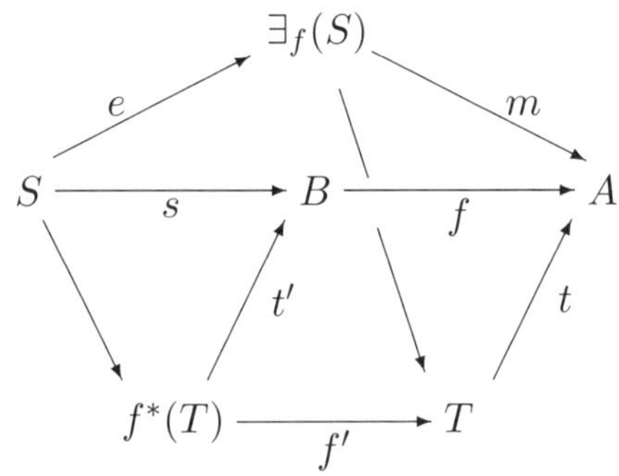

of all the named arrows; the arrows without names might not exist, but if they do, they are unique making the obvious shapes to commute, ($f^*(T)$ is a pullback of t along f). Now if $\exists_f(S) \le T$, i.e. if $\exists_f(S) \to T$ exists in (3.3), then the two arrows

$$S \xrightarrow{s} B \xrightarrow{f} A \qquad S \xrightarrow{e} \exists_f(S) \longrightarrow T \xrightarrow{t} A$$

are equal, hence by the universal property of the pullback, $S \to f^*(T)$ exists in (3.3), showing that $S \le f^*(T)$.

On the other hand, if $S \le f^*(T)$ i.e. $S \to f^*(T)$ exists in (3.3), then the outer penthagon in (3.3) commutes, and $\exists_f(S) \to T$ exists, by the property (iii) of the definition of a stable factorization system, as e is an epi, and t is a regular mono.

It remains to show the Beck-Chevalley condition. Let us consider a pullback square (3.1) and let $S \overset{s}{\hookrightarrow} B_1$ be a regular mono. Take the further pullback

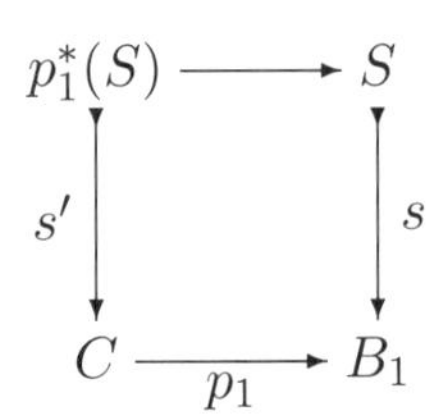

What we have to show is that the factorization of $p_2 \circ s'$ is just (up to an isomorphism) the factorization of $f_1 \circ s$ pulled back along f_2. But it is a general property of stable factorization systems that in a pullback square

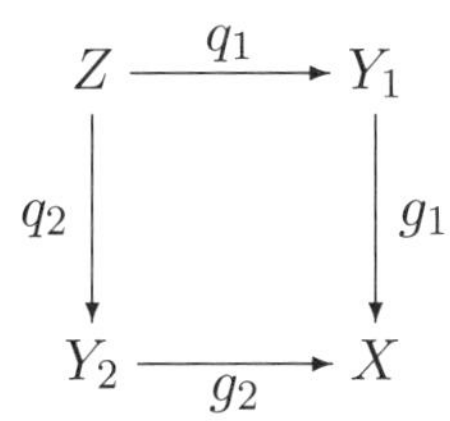

the factorization of q_2 is obtained by taking the factorization $m \circ e$ of g_1 and by successively pulling back m and e: this property is essentially due to condition (iv) ensuring that members of $\mathcal{E}$ are pullback-stable, whereas pullback-stability of members of $\mathcal{M}$ follows from conditions (i)-(iii) [CJKP] (in our case, anyway, stability of regular monos under pullbacks is a general fact). $\square$

We shall reverse Proposition 3.1 in order to get an alternative definition of r-regular category. First, we prove a Lemma:

LEMMA 3.2 *Let* $\mathbf{C}$ *a category with finite limits, and* $f : B \to A$ *a morphism in* $\mathbf{C}$. *Then*

(i) f *is epi iff for every regular subobject* $S \hookrightarrow A$, *we have that* $id_B \leq f^*(S)$ *implies* $id_A \leq S$;

(ii) *if, moreover, the pullback functor* $f^* : Sub_r(A) \to Sub_r(B)$ *has a left adjoint* $\exists_f : Sub_r(B) \to Sub_r(A)$, *then* f *is epi iff* $id_A \leq \exists_f(id_B)$.

Proof. Ad (i). Let $f : B \to A$ be an epi and let $S \hookrightarrow A$ be a regular mono such that $id_B \leq f^*(S)$; this means that we have a pullback square

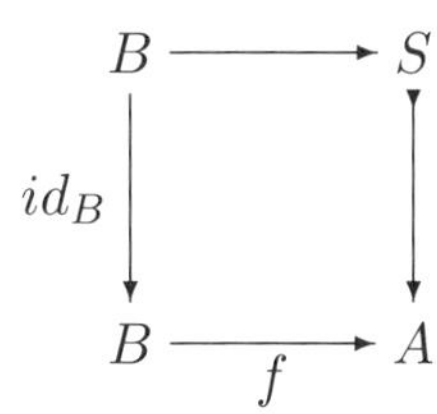

so $S \hookrightarrow A$ is epi as a second component of an epic arrow. Being also a regular mono, it is an isomorphism. Vice versa, suppose that $id_B \leq f^*(S)$ implies $id_A \leq S$ for all $S \in Sub_r(A)$ and suppose that $g_1 \circ f = g_2 \circ f$ for some parallel arrows of domain A. Let $S \overset{s}{\hookrightarrow} A$ be the equalizer of g_1, g_2; by the universal property of equalizers, we have a unique map $f' : B \to S$ such that $s \circ f' = f$. As s is mono, it turns out that the diagram

$$\begin{array}{ccc} B & \xrightarrow{f'} & S \\ {\scriptstyle id_B}\downarrow & & \downarrow{\scriptstyle s} \\ B & \xrightarrow[f]{} & A \end{array}$$

is a pullback, so that id_B is equal to $f^*(S)$; this implies by our assumption that s is an isomorphism and consequently $g_1 = g_2$.

We deduce (ii) from (i). For any $f : B \to A$ such that $\exists_f$ exists, we have $id_B \leq f^*(\exists_f(id_B))$. Then, by (i), putting $S = \exists_f(id_B)$, we get $id_A \leq \exists_f(id_B)$. Conversely, if $id_A \leq \exists_f(id_B)$ and $S \in Sub_r(A)$ is such that $id_B \leq f^*(S)$, then

$$id_A \leq \exists_f(id_B) \leq \exists_f(f^*(S)) \leq S$$

and hence, by (i), f is epi, as required. □

PROPOSITION 3.3 *A category with finite limits* **C** *is r-regular iff the pullback functors operating on regular subobjects have left adjoints satisfying the Beck-Chevalley condition.*

Proof. The necessity of the condition is just Proposition 3.1. For the sufficiency, we have to show that epis are pullback-stable and that each arrow has an epi/regular mono factorization.

Using Lemma 3.2(ii) and the Beck-Chevalley condition, we shall show that epis are pullback-stable. Consider a pullback

$$\begin{array}{ccc} D & \xrightarrow{k} & B \\ {\scriptstyle h}\downarrow & & \downarrow{\scriptstyle f} \\ C & \xrightarrow[g]{} & A \end{array}$$

and suppose that f is epic, that is $id_A \leq \exists_f(id_B)$. It follows that

$$id_C = g^*(id_A) \leq g^*(\exists_f(id_B)) = \exists_h(k^*(id_B)) = \exists_h(id_D)$$

which implies, by Lemma 3.2(ii) that h is an epi.

Let us now look for a factorization of $f : B \to A$. Take the regular mono $\exists_f(id_B) \overset{m}{\hookrightarrow} A$. As, by adjointness, $id_B \leq f^*(\exists_f(id_B))$ we have a pullback

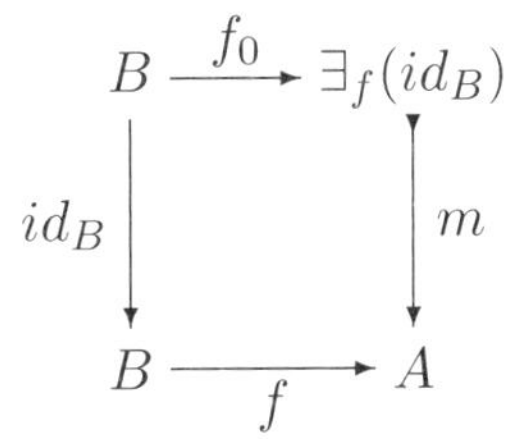

Using the Beck-Chevalley condition, we have in $Sub_r(\exists_f(id_B))$

$$id_{\exists_f(id_B)} = m^*(\exists_f(id_B)) = \exists_{f_0}(id_B^*(id_B)) = \exists_{f_0}(id_B)$$

(the first equality is because the pullback of m against itself is the identity, as m is mono). Thus, by Lemma 3.2(ii), f_0 is epi. $\Box$

Notice that from the above proof it follows once again that, in an r-regular category, for $f : A \longrightarrow B$ and $S \overset{s}{\hookrightarrow} A$, the regular subobject $\exists_f(S)$ is just the second component of the factorization of $f \circ s$.

LEMMA 3.4 *In an r-regular category,*

(i) *if* $T \overset{t}{\hookrightarrow} A$ *and* $S \in Sub_r(T)$, *then* $t^*(\exists_t(S)) = S$;

(ii) *regular monos compose;*

(iii) *given two regular subobjects* $S \overset{s}{\hookrightarrow} A$ *and* $T \overset{t}{\hookrightarrow} A$, *we have that* $\exists_s(s^*(T)) = S \wedge T$;

Proof. Let $S \overset{s}{\hookrightarrow} T \overset{t}{\hookrightarrow} A$ be regular monos. We have $S = t^*(\exists_t S)$ because of the Beck-Chevalley condition applied to the pullback square

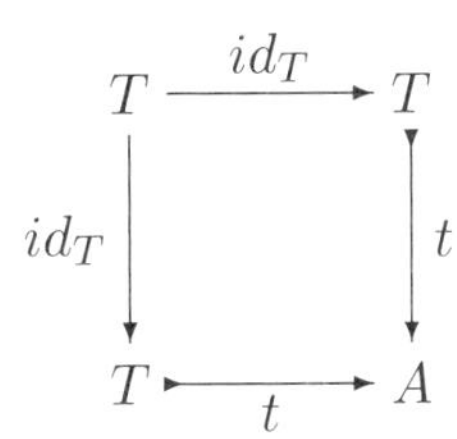

This shows (i) and moreover that we have a pullback

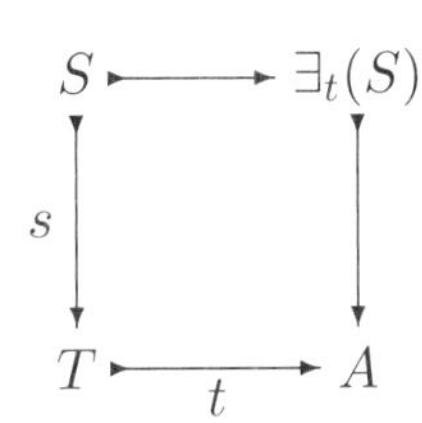

proving (ii), namely that $S \overset{tos}{\hookrightarrow} A$ is equal to the regular mono $T \wedge \exists_t(S)$.

It remains to show (iii); consider the pullback

$$\begin{array}{ccc} S \wedge T & \rightarrowtail & T \\ \downarrow & & \downarrow t \\ S & \underset{s}{\rightarrowtail} & A \end{array}$$

and notice that $S \wedge T \hookrightarrow S \hookrightarrow A$ has the identity followed by itself as factorization (because it is a regular mono). $\square$

PROPOSITION 3.5 *In an r-regular category, given an arrow $f : B \to A$ and regular subobjects $S \overset{s}{\hookrightarrow} A$ and $T \overset{t}{\hookrightarrow} B$, the Frobenius reciprocity*

$$(F) \qquad S \wedge \exists_f(T) = \exists_f(f^*(S) \wedge T)$$

holds.

Proof. Consider the pullback

$$\begin{array}{ccc} f^*(S) & \overset{s_0}{\rightarrowtail} & B \\ f_0 \downarrow & & \downarrow f \\ S & \underset{s}{\rightarrowtail} & A \end{array} \qquad (3.4)$$

Using the Beck-Chevalley condition for (3.4) and Lemma 3.4, we get

$$S \wedge \exists_f(T) = \exists_s(s^*(\exists_f(T))) = \exists_s(\exists_{f_0}(s_0^*(T))) =$$

$$\exists_f(\exists_{s_0}(s_0^*(T))) = \exists_f(f^*(S) \wedge T)$$

$\square$

C is said to be an *r-Heyting* category iff

(i) **C** is r-regular;

(ii) for every object A, $Sub_r(A)$ is a lattice with 0 and 1;

(iii) for every arrow $f : A \to B$ in **C**, the pullback functor f^* has also a right adjoint $\forall_f : Sub_r(A) \to Sub_r(B)$, that is for every regular subobject $S \hookrightarrow A$, there is a regular subobject $\forall_f(S) \hookrightarrow B$ satisfying the condition

$$T \leq \forall_f(S) \quad \text{iff} \quad f^*(T) \leq S$$

for every regular subobject $T \hookrightarrow B$.

Notice that in an r-Heyting category, the Beck-Chevalley condition for universal quantification

$$\forall_{p_2}(p_1^*(S)) = f_2^*(\forall_{f_1}(S))$$

relative to a pullback square

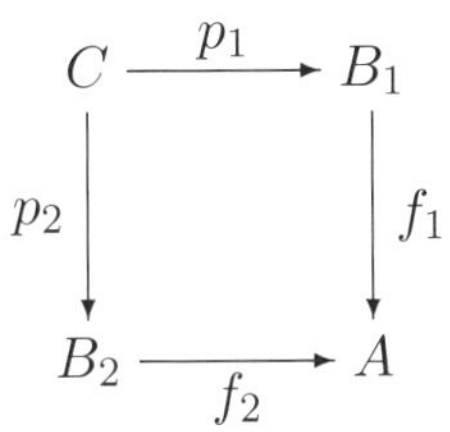

also holds for every $S \hookrightarrow B_1$ (it is actually equivalent to (3.2)[1]). In an r-Heyting category, for every arrow $f : B \longrightarrow A$, taking pullback $f^* : Sub_r(A) \to Sub_r(B)$ is a lattice morphism because of the presence of the two adjoints.

PROPOSITION 3.6 *In an r-Heyting category, the regular monos functor takes values into the category of Heyting algebras.*

Proof. We have to show that for every object A, $Sub_r(A)$ has implications (i.e. relative pseudocomplements) and that for every arrow $f : B \to A$, f^* preserves them.

Let $S \overset{s}{\hookrightarrow} A$ and $T \overset{t}{\hookrightarrow} A$ be regular subobjects of A; define $S \to T$ as $\forall_s(s^*(T))$. For $U \in Sub_r(A)$, we have the following sequence of equivalences

$$\frac{\frac{\frac{\frac{S \wedge U \leq T}{\exists_s(s^*(U)) \leq T}}{s^*(U) \leq s^*(T)}}{U \leq \forall_s(s^*(T))}}{U \leq S \to T}$$

i.e. $S \to T$ is indeed a relative pseudocomplement. In order to show that $f^* : Sub_r(A) \to Sub_r(B)$ preserves relative pseudocomplements, we use

[1]This is seen as follows. The functor $\forall_{p_2} \circ p_1^*$ is the right adjoint to $\exists_{p_1} \circ p_2^*$, and the functor $f_2^* \circ \forall_{f_1}$ is the right adjoint to $f_1^* \circ \exists_{f_2}$. Thus, by uniqueness of the adjoints, $\exists_{p_1} \circ p_2^* = f_1^* \circ \exists_{f_2}$ iff $\forall_{p_2} \circ p_1^* = f_2^* \circ \forall_{f_1}$.

Frobenius reciprocity, proved in Proposition 3.5. For $U \in Sub_r(B)$, we have

$$\begin{array}{c} U \leq f^*(S \to T) \\ \hline \exists_f(U) \leq S \to T \\ \hline S \wedge \exists_f(U) \leq T \\ \hline \exists_f(f^*(S) \wedge U) \leq T \\ \hline f^*(S) \wedge U \leq f^*(T) \\ \hline U \leq f^*(S) \to f^*(T) \end{array}$$

Since U is arbitrary, we obtain

$$f^*(S \to T) = f^*(S) \to f^*(T)$$

as required. □

The following Lemma will be needed in next section:

LEMMA 3.7 *Let* **C** *be an r-regular category in which* $Sub_r(A)$ *has implications for every object* A*;*

(i) *suppose we are given a commutative square*

$$\begin{array}{ccc} B & \xrightarrow{f} & A \\ {\scriptstyle b}\downarrow & & \downarrow{\scriptstyle a} \\ B_0 & \xrightarrow[f_0]{} & A_0 \end{array} \tag{3.5}$$

such that a *and* b *are regular monos. If* f_0^* *has a right adjoint, so does* f^**;*

(ii) *if for a projection* $p_A : A \times B \longrightarrow A$*,* p_A^* *has a right adjoint, so does any arrow* $f : B \to A$ *with domain* B *and codomain* A*.*

Proof. Ad (i). Let $S \in Sub_r(B)$. We put

$$\forall_f(S) = a^*(\forall_{f_0}(\exists_b(B) \to \exists_b(S))) \tag{3.6}$$

where B in (3.6), is the top element in $Sub_r(B)$. In order to check that $f^* \vdash \forall_f$, using Lemma 3.4, Frobenius reciprocity and commutativity of (3.5), we have

$$\begin{array}{c} T \leq \forall_f(S) \\ \hline T \leq a^*(\forall_{f_0}(\exists_b(B) \to \exists_b(S))) \\ \hline f_0^*(\exists_a(T)) \leq \exists_b(B) \to \exists_b(S) \\ \hline \exists_b(B) \wedge f_0^*(\exists_a(T)) \leq \exists_b(S) \\ \hline \exists_b(B \wedge b^*(f_0^*(\exists_a(T)))) \leq \exists_b(S) \\ \hline B \wedge b^*(f_0^*(\exists_a(T))) \leq b^*(\exists_b(S)) \\ \hline b^*(f_0^*(\exists_a(T))) \leq S \\ \hline f^*(a^*(\exists_a(T))) \leq S \\ \hline f^*(T) \leq S \end{array}$$

Thus $f^* \dashv \forall_f$, as required.

Ad (ii). For any $f : B \to A$ in **C**, the square

$$\begin{array}{ccc} B & \xrightarrow{\ f\ } & A \\ {\scriptstyle < f, id_B >}\big\downarrow & & \big\downarrow{\scriptstyle id_A} \\ A \times B & \xrightarrow[p_A]{} & A \end{array}$$

commutes. Thus, by (i), it is enough to show that $< f, id_B >: B \to A \times B$ is a regular mono. But this is indeed the case since

$$B \xrightarrow{< f, id_B >} A \times B \overset{p_A}{\underset{f \circ p_B}{\rightrightarrows}} A$$

is an equalizer in **C** (for any $c : C \to A \times B$, $p_B \circ c : C \to B$ is the unique morphism such that $c =< f, id_B > \circ p_B \circ c$, when ever c equalizes p_A and $f \circ p_B$). $\square$

Before closing this section, we make some little remarks (not to be used in the sequel) about joins of regular subobjects. In important cases (including the case of the opposite category of finitely presented algebras, in the one-sorted or also in the finite-sorted case), the existence of joins of regular subobjects follows from conditions (i) and (iii) of the definition of r-Heyting category. This is shown as follows. Notice that, in a category with finite limits, any regular mono $S \hookrightarrow A$ is, by definition, the equalizer of a pair of arrows $f, g : A \longrightarrow X$, so it is equal to $\langle f, g\rangle^*(\Delta_X)$, where we recall that Δ_X is the regular mono given by the diagonal map $\langle id_X, id_X\rangle : X \hookrightarrow X \times X$. We say in this case that $S \hookrightarrow A$ is X*-definable*. More generally, given a set $\mathcal{S}$ of objects, $S \hookrightarrow A$ is said to be $\mathcal{S}$-definable iff S can be represented as a finite meet $\bigwedge_i \langle f_i, g_i\rangle^*(\Delta_{X_i})$, for $X_i \in \mathcal{S}$. A category with finite limits is said to be *of finite type* iff all regular subobjects are $\mathcal{S}$-definable for a finite set $\mathcal{S}$ of objects. Now it turns out that a category with finite limits which is of finite type is r-Heyting iff only conditions (i) and (iii) hold. Joins among regular subobjects can in fact be introduced, for every $S_1 \hookrightarrow A, \ldots, S_n \hookrightarrow A$, through the formula

$$S_1 \vee \cdots \vee S_n \ =$$

$$= \ \bigwedge_{X \in \mathcal{S}} \forall_{p_A}((\bigwedge_{i=1}^{n}(p_A^*(S_i) \to p_{X\times X}^*(\Delta_X))) \to p_{X\times X}^*(\Delta_X)),$$

where p_A and $p_{X\times X}$ are the projections of domain $A \times (X \times X)$ and codomain A and $X \times X$, respectively.[2] We also mention that in an r-Heyting category of finite type, also $\exists_f(S)$ (for $f : B \longrightarrow A$ and $S \in Sub_r(B)$) can be defined in terms of dual images by

$$\exists_f(S) \;=\; \bigwedge_{X\in\mathcal{S}} \forall_{p_A}(\forall_{f\times p_{id_{X\times X}}}(p_B^*(S) \to p_{X\times X}^*(\Delta_X)) \to p_{X\times X}^*(\Delta_X)).$$

Notice that these characterizations of $\vee$ and $\exists_f$ in terms of $\forall, \to$ have evident analogies with the way disjunction and existential quantification are defined in second order intuitionistic logic.

2. Model completions and fp algebras

We turn to our universal algebra context; as usual, let T be an equational theory and let $Alg(T)$ be the category of algebras which are models of T. By $\mathbf{T}$ we still mean the opposite of the category of finitely presented algebras in $Alg(T)$. We shall relate in this Section the categorical properties of $\mathbf{T}$ with the existence of a model completion T^* for T.

THEOREM 3.8 *If $\mathbf{T}$ is an r-Heyting category, then T admits a model completion.*

Proof. During the proof, we shall make extensive use of notational conventions established at the beginning and at the end of section 2.4. We need some preliminary facts. Let us fix a model A of T and e-formulas $E(x, \vec{a}/\vec{x})$ and $E_1(\vec{a}/\vec{x}), E_2(\vec{a}/\vec{x})$ with parameters $\vec{a}$ from A. Recall that at the same time $[\![E(x, \vec{x})]\!]$ and $[\![E_1(\vec{x})]\!], [\![E_2(\vec{x})]\!]$ are regular monos in $\mathbf{T}$ of codomains $\mathcal{F}(x, \vec{x})$ and $\mathcal{F}(\vec{x})$, respectively. Let p be the projection in $\mathbf{T}$ from $\mathcal{F}(x, \vec{x})$ into $\mathcal{F}(\vec{x})$ (it is the morphism corresponding to the inclusion at the level of the free generators). By $A[\mathbf{x}]$ we mean the algebra of polynomials with coefficients in A, which is defined as the coproduct in $Alg(T)$ of A and of the free algebra on one generator. We have that:[3]

(i) $A \models [\exists_p[\![E]\!]](\vec{a})$ iff the arrow

$$A \longrightarrow A[\mathbf{x}] \longrightarrow A[\mathbf{x}]/E(\mathbf{x}, \vec{a}),$$

[2] In the general case, i.e. for categories which may not be of finite type, these formulas define a weak kind of joins, which we call $\mathcal{S}$-joins. Formally, the definition is as follows. We say that a category with finite limits has *stable $\mathcal{S}$-joins* (where $\mathcal{S}$ is a finite set of objects) iff for every A, for every $S_1, \ldots, S_n \in Sub_r(A)$ ($n \geq 0$), we are given a regular subobject of A, to be called $S_1 \vee_{\mathcal{S}} \cdots \vee_{\mathcal{S}} S_n$, in such a way that the following conditions are satisfied: i) for every $i = 1, \ldots, n$, $S_i \leq S_1 \vee_{\mathcal{S}} \cdots \vee_{\mathcal{S}} S_n$; ii) for every $\mathcal{S}$-definable regular mono $T \hookrightarrow A$ such that $S_i \leq T$ (for every $i = 1, \ldots, n$), we have that $S_1 \vee_{\mathcal{S}} \cdots \vee_{\mathcal{S}} S_n \leq T$; iii) for every arrow $f : B \longrightarrow A$, we have that $f^*(S_1 \vee_{\mathcal{S}} \cdots \vee_{\mathcal{S}} S_n) = f^*(S_1) \vee_{\mathcal{S}} \cdots \vee_{\mathcal{S}} f^*(S_n)$.

[3] Recall that for every regular subobject S of $\mathcal{F}(\vec{x})$, the notation $[S](\vec{x})$ is used in order to indicate any formula representing it (that is, $[S](\vec{x})$ is any $E(\vec{x})$ such that $[\![E]\!] = S$.

obtained by composing the inclusion into the algebra of polynomials with the obvious quotient map, is a monomorphism;

(ii) $A \models [\forall_p \llbracket E \rrbracket](\vec{a})$ iff $A[\mathbf{x}] \models E(\mathbf{x}, \vec{a})$;

(iii) $A \models [\llbracket E_1 \rrbracket \to \llbracket E_2 \rrbracket](\vec{a})$ iff $A/E_1(\vec{a}) \models E_2(\vec{a})$.

(iii) is Lemma 2.18. Recall that $\mathcal{F}(A)/\Delta$, where Δ is the positive diagram of A, is a presentation for A.

Proof of (i). As $\mathcal{F}(A)/\Delta$ is a presentation for A, $A[\mathbf{x}]/E(\mathbf{x}, \vec{a})$ has the presentation $\mathcal{F}(A \cup \{\mathbf{x}\})/(\Delta \cup \{E(\mathbf{x}, \vec{a})\})$. Suppose first that $A \models [\exists_p \llbracket E \rrbracket](\vec{a})$; in order to prove that the map $A \to A[\mathbf{x}]/E(\mathbf{x}, \vec{a})$ is a monomorphism, it is sufficient to show that for every e-formula E' with parameters from A, we have that

$$A[\mathbf{x}]/E(\mathbf{x}, \vec{a}) \models E' \text{ implies } A \models E'$$

(note that E' could be for instance $b_1 = b_2$ for $b_1, b_2 \in A$). So take any such e-formula E' and suppose that $A[\mathbf{x}]/E(\mathbf{x}, \vec{a}) \models E'$, i.e. that for some e-formula Δ_0 obtained by taking a finite conjunction of members of Δ, we have that

$$\vdash_T \Delta_0(\vec{a}, \vec{b}) \;\&\; E(\mathbf{x}, \vec{a}) \;\Rightarrow\; E'(\vec{a}, \vec{b}),$$

where the list $\vec{a}, \vec{b}$ contains all parameters from A involved in Δ_0 and E'. We shall simply prove that

$$\vdash_T \Delta_0(\vec{a}, \vec{b}) \;\&[\exists_p \llbracket E \rrbracket](\vec{a}) \;\Rightarrow\; E'(\vec{a}, \vec{b}).$$

Taking variables instead of constants, what we shall show is that the condition

$$\vdash_T \Delta_0(\vec{x}, \vec{y}) \;\&\; E(x, \vec{x}) \;\Rightarrow\; E'(\vec{x}, \vec{y}) \tag{3.7}$$

implies the condition

$$\vdash_T \Delta_0(\vec{x}, \vec{y}) \;\&[\exists_p \llbracket E \rrbracket](\vec{x}) \;\Rightarrow\; E'(\vec{x}, \vec{y}). \tag{3.8}$$

Let us now interpret this internally in $\mathbf{T}$: we have regular subobjects $\llbracket \Delta_0 \rrbracket \hookrightarrow \mathcal{F}(\vec{x}, \vec{y})$, $\llbracket E' \rrbracket \hookrightarrow \mathcal{F}(\vec{x}, \vec{y})$ and $\llbracket E \rrbracket \hookrightarrow \mathcal{F}(x, \vec{x})$; we also have a pullback square in $\mathbf{T}$ entirely formed by projections

$$\begin{array}{ccc}
\mathcal{F}(x, \vec{x}, \vec{y}) & \xrightarrow{\;p_{x,\vec{x}}\;} & \mathcal{F}(x, \vec{x}) \\
{\scriptstyle p_{\vec{x},\vec{y}}}\downarrow & & \downarrow{\scriptstyle p} \\
\mathcal{F}(\vec{x}, \vec{y}) & \xrightarrow[\;p_{\vec{x}}\;]{} & \mathcal{F}(\vec{x})
\end{array}$$

Consequently, '(3.7) implies (3.8)' says [4] that the inequality

$$p^*_{\vec{x},\vec{y}}([\![\Delta_0]\!]) \wedge p^*_{x,\vec{x}}([\![E]\!]) \leq p^*_{\vec{x},\vec{y}}([\![E']\!])$$

implies the inequality

$$[\![\Delta_0]\!] \wedge p^*_{\vec{x}}(\exists_p[\![E]\!]) \leq [\![E']\!],$$

which is true in any r-regular category by the Beck-Chevalley and the Frobenius conditions. Conversely, suppose that the map $A \to A[\mathbf{x}]/E(\mathbf{x}, \vec{a})$ is a monomorphism; as the inequality

$$[\![E]\!] \leq p^*(\exists_p[\![E]\!]),$$

holds in all r-regular categories, it follows that

$$\vdash_T E(x, \vec{x}) \Rightarrow [\exists_p[\![E]\!]](\vec{x}).$$

Consequently, we have that $A[\mathbf{x}]/E(\mathbf{x}, \vec{a}) \models [\exists_p[\![E]\!]](\vec{a})$ and then by assumption that $A \models [\exists_p[\![E]\!]](\vec{a})$.

Proof of (ii). As $\mathcal{F}(A)/\Delta$ is a presentation for A, $A[\mathbf{x}]$ has the presentation $\mathcal{F}(A \cup \{\mathbf{x}\})/\Delta$. Now $A \models [\forall_p[\![E]\!]](\vec{a})$ holds iff for some $\Delta_0(\vec{a}/\vec{x}, \vec{b}/\vec{y})$ obtained by taking a finite conjunction of members of Δ, we have

$$\vdash_T \Delta_0(\vec{x}, \vec{y}) \Rightarrow [\forall_p[\![E]\!]](\vec{x}).$$

This condition is equivalent to the inequation in $\mathbf{T}$

$$[\![\Delta_0]\!] \leq p^*_{\vec{x}}(\forall_p[\![E]\!]);$$

applying Beck-Chevalley condition to the above pullback square, we equivalently get that

$$[\![\Delta_0]\!] \leq \forall_{p_{\vec{x},\vec{y}}}(p^*_{x,\vec{x}}[\![E]\!]),$$

and, by adjointness, that

$$p^*_{\vec{x},\vec{y}}[\![\Delta_0]\!] \leq p^*_{x,\vec{x}}[\![E]\!].$$

[4] In order to understand properly this and similar passages, keep in mind the observations at the end of Section 2.4. In particular recall that for a couple of e-formulas $E_1(\vec{x}), E_2(\vec{x}), \vdash_T E_1 \Rightarrow E_2$ is equivalent to the inequality $[\![E_1]\!] \leq [\![E_2]\!]$ among regular subobjects of $\mathcal{F}(\vec{x})$. Moreover, for a couple of e-formulas $E_1(\vec{x}), E_2(\vec{x})$, we have that $[\![E_1]\!] \wedge [\![E_2]\!] = [\![E_1 \& E_2]\!]$. Finally, recall that taking pullback in $\mathbf{T}$ is like taking substitution from a logical point of view. Projections like $p_{\vec{z}} : \mathcal{F}(\vec{z}, \vec{w}) \longrightarrow \mathcal{F}(\vec{z})$ in $\mathbf{T}$ corresponds to the canonical embedding of $\mathcal{F}(\vec{z})$ into $\mathcal{F}(\vec{z}, \vec{w})$, i.e. to substitutions that don't do anything, except from the fact that they simply 'consider a formula in at most the variables $\vec{z}$ as a formula in at most the variables $\vec{z}, \vec{w}$'; so for any e-formula $E(\vec{z})$, we have $p^*_{\vec{z}}([\![E]\!]) = [\![E]\!]$ (with a more cumbersome notation, this is written as $p^*_{\vec{z}}([\![E]\!]_{\mathcal{F}(\vec{z})}) = [\![E]\!]_{\mathcal{F}(\vec{z},\vec{w})}$).

This relation, from a logical point of view, means that

$$\vdash_T \Delta_0(\vec{x}, \vec{y}) \Rightarrow E(x, \vec{x}).$$

In conclusion, $A \models [\forall_p[\![E]\!]](\vec{a})$ holds iff for some $\Delta_0(\vec{a}, \vec{b})$ obtained by taking a finite conjunction of members of Δ, we have $\vdash_T \Delta_0(\vec{a}, \vec{b}) \Rightarrow E(\mathbf{x}, \vec{a})$, i.e. iff $A[\mathbf{x}] \models E(\mathbf{x}, \vec{a})$, as claimed.

We define the model completion T^* of T as T augmented by the axioms $\varphi' \Rightarrow \varphi$, where if $\varphi(\vec{x})$ is the primitive formula

$$\exists x(E(x, \vec{x}) \& \sim E_1(x, \vec{x}) \& \cdots \& \sim E_n(x, \vec{x})),$$

$\varphi'(\vec{x})$ is the quantifier-free formula

$$[\exists_p[\![E]\!]](\vec{x}) \;\&\; \sim [\forall_p([\![E]\!] \to [\![E_1]\!])](\vec{x}) \;\&\cdots$$
$$\cdots \& \;\sim [\forall_p([\![E]\!] \to [\![E_n]\!])](\vec{x}).$$

Recalling the definition of a model completion from Section 2.2, we have to show that:

(iv) T^* admits quantifier-elimination;

(v) each model of T is embeddable in a model of T^*.

Proof of (iv). It is sufficient to show that $\vdash_T \varphi \Rightarrow \varphi'$ for every primitive formula $\varphi(\vec{x})$. Suppose that φ is $\exists x(E(x, \vec{x}) \& \sim E_1(x, \vec{x}) \& \cdots \& \sim E_n(x, \vec{x}))$; the claim follows if we show that

$$\vdash_T E(x, \vec{x}) \Rightarrow [\exists_p[\![E]\!]](\vec{x}) \tag{3.9}$$

and that

$$\vdash_T E(x, \vec{x}) \;\&\; \sim E_i(x, \vec{x}) \Rightarrow \sim [\forall_p([\![E]\!] \to [\![E_i]\!])](\vec{x}), \tag{3.10}$$

for every $i = 1, \ldots, n$. But (3.9) and (3.10) are equivalent to the inequations

$$[\![E]\!] \leq p^*(\exists_p[\![E]\!]) \quad \text{and} \quad [\![E]\!] \wedge p^*(\forall_p([\![E]\!] \to [\![E_i]\!])) \leq [\![E_i]\!],$$

which are easily seen to hold in any r-Heyting category.

Proof of (v). Here it is sufficient to show that, given an algebra A, given $\vec{a} \in A$ and given a primitive formula $\varphi(\vec{x})$ such that $A \models \varphi'(\vec{a})$, there exists an extension A' of A such that $A' \models \varphi(\vec{a})$. If this holds, we can take a well-ordering of the pairs $\langle \vec{a}, \varphi(\vec{x}) \rangle$, where $\vec{a} \in A$ and $\varphi(\vec{x})$ is primitive and $A \models \varphi'(\vec{a})$. Then we build at each step an expansion in which $\varphi(\vec{a})$ holds (if $\langle \vec{a}, \varphi(\vec{x}) \rangle$ is the currently examined pair) and take union at limits ordinals. In this way, we produce a first expansion A_1 of A in which all formulas $\varphi'(\vec{a}) \Rightarrow \varphi(\vec{a})$ are true, *provided the parameters $\vec{a}$ are taken from A* (in fact, $A_1 \models \varphi'(\vec{a})$ holds iff $A \models \varphi'(\vec{a})$ holds because φ' is quantifier free, notice also that $\varphi(\vec{a})$ is existential, hence preserved in extensions). We *repeat ω-times* this passage from A to A_1 and take union, in order to produce a genuine model of T^* which is an extension of A.

So let us show that if $A \models \varphi'(\vec{a})$, then there exists an extension A' of A such that $A' \models \varphi(\vec{a})$. If, as usual $\varphi(\vec{a})$ is $\exists x(E(x,\vec{a}) \& \sim E_1(x,\vec{a}) \& \cdots \& \sim E_n(x,\vec{a}))$, take the algebra $A[\mathbf{x}]/E(\mathbf{x},\vec{a})$. As $A \models \varphi'(\vec{a})$, we have in particular that $A \models [\exists_p [\![E]\!]](\vec{a})$, so by (i), $A[\mathbf{x}]/E(\mathbf{x},\vec{a})$ is an extension of A. Moreover, for $i = 1,\ldots,n$, we cannot have $A[\mathbf{x}]/E(\mathbf{x},\vec{a}) \models E_i(\mathbf{x},\vec{a})$, otherwise, by (iii) we get $A[\mathbf{x}] \models [[\![E]\!] \to [\![E_i]\!]](\mathbf{x},\vec{a})$ and by (ii) we get $A \models [\forall_p([\![E]\!] \to [\![E_i]\!])](\vec{a})$, in contrast to the fact that $A \models \varphi'(\vec{a})$. □

The statement of the above Theorem cannot be reversed, without additional assumptions (for instance, the theory of abelian groups does admit a model completion but the opposite category of finitely presented abelian groups is not r-Heyting). Moreover, the procedure for quantifier-eliminations in T^* we used in the proof of the Theorem is too peculiar to be completely general. Notice for instance that, according to such a procedure, a formula of the kind $\forall x E(x,\vec{x})$ (for any e-formula E) is always equivalent in T^* to a single positive formula, namely $[\forall_p [\![E]\!]]$, where $p : \mathcal{F}(x,\vec{x}) \longrightarrow \mathcal{F}(\vec{x})$ is the projection in $\mathbf{T}$. As a special case, we have that $\forall x_1 \forall x_2 (x_1 = x_2)$ is equivalent in T^* to any variable-free e-formula representing the regular subobject $\forall_p([\![x_1 = x_2]\!])$, where now $p : \mathcal{F}(x_1,x_2) \longrightarrow \mathcal{F}(\emptyset)$ is the unique map into the terminal object of $\mathbf{T}$. From the formulas defining finite joins in terms of $\forall, \to$ in r-Heyting categories of finite type (see the end of the previous section), it follows that $\forall_p([\![x_1 = x_2]\!])$ is the minimum regular subobject of the terminal object. This minimum regular subobject expresses a degeneracy condition (like $0 = 1$ in the case of Heyting algebras), because the relation $p^*(\forall_p([\![x_1 = x_2]\!])) \leq [\![x_1 = x_2]\!]$ once read in T gives that $\vdash_T [\forall_p([\![x_1 = x_2]\!])] \Rightarrow x_1 = x_2$. In conclusion, according to the quantifier-elimination procedure given in the proof of Theorem 3.8, $\forall x_1 \forall x_2 (x_1 = x_2)$ is equivalent in T^* to a degeneracy condition, which may not even be expressible in the general case (see e.g. the example of abelian groups) by using a variable-free e-formula.

In [Wh] a necessary and sufficient condition for a Horn theory to admit a model completion is given, but such a condition is not very manageable. We shall assume something about $\mathbf{T}$ in order to be able to prove the converse of Theorem 3.8. The assumptions we shall require are rather strong, but they are often satisfied in varieties arising from algebraic logic: we already met and discussed them in Section 2.5 (see that Section and the Exercises at the end of Chapter 2 for relevant examples). We shall assume that

$(*)$ **T** *has stable finite joins and stable implications among regular subobjects.*

This means that for every object A in $\mathbf{T}$, $Sub_r(A)$ is not only a semilattice but it is in fact a Heyting algebra and that such a Heyting algebra structure is preserved by taking pullback along any arrow. In Proposition 2.16 we have shown that $(*)$ is equivalent to some well-known conditions from universal algebra, namely to the fact that $Alg(T)$ has (EDO) (i.e. equationally definable principal congruences and equationally definable principal meets).

The following Lemma is well-known:

LEMMA 3.9 *Suppose that* $\mathbf{T}$ *has stable implications among regular subobjects. Then* $Alg(T)$ *has the congruence extension property.*

Proof. We have to show that for every monomorphism $m : A \longrightarrow B$ and for every quotient $q : A \longrightarrow C$, we can fill the square

$$\begin{array}{ccc} A & \stackrel{m}{\rightarrowtail} & B \\ {\scriptstyle q}\downarrow & & \downarrow{\scriptstyle q'} \\ C & \underset{m'}{\rightarrowtail} & D \end{array}$$

in such a way that m' is mono and q' is also a quotient. We can assume that m is just inclusion, so that if $\mathcal{F}(A)/\Delta$ is the canonical presentation of A and if $\mathcal{F}(B)/\Theta$ is the canonical presentation of B, then we have $A \subseteq B$ and $\Delta \subseteq \Theta$. The fact that m is mono says also that for every e-formula $E(\vec{a})$ with parameters from A

$$\vdash_T \Theta \Rightarrow E \text{ implies } \vdash_T \Delta \Rightarrow E, \tag{3.11}$$

where $\vdash_T \Theta \Rightarrow E$ (and similarly for $\vdash_T \Delta \Rightarrow E$) means that E is provable from some finite conjunction of members of Θ. We can present C as $\mathcal{F}(A)/(\Delta \cup \Delta')$ for some Δ'. We take $\mathcal{F}(B)/(\Theta \cup \Delta')$ as D. What we have to show is that for every e-formula E with parameters from A,

$$\vdash_T (\Theta \cup \Delta') \Rightarrow E \text{ implies } \vdash_T (\Delta \cup \Delta') \Rightarrow E.$$

Suppose that we have

$$\vdash_T \Theta_0(\vec{a}, \vec{b}) \ \& \ \Delta'_0(\vec{a}) \Rightarrow E(\vec{a}),$$

where Θ_0, Δ'_0 are finite conjunctions from Θ and Δ', respectively, and $\vec{a}, \vec{b}$ are all parameters involved from A and $B \backslash A$, respectively. Taking variables instead of constants and passing to the internal meaning in $\mathbf{T}$, we get

$$[\![\Theta_0(\vec{x}, \vec{y})]\!] \wedge p^*_{\vec{x}}([\![\Delta'_0(\vec{x})]\!]) \leq p^*_{\vec{x}}([\![E(\vec{x})]\!]).$$

As $\mathbf{T}$ has stable implications, we can pass to

$$[\![\Theta_0(\vec{x},\vec{y})]\!] \leq p^*_{\vec{x}}([\![\Delta'_0(\vec{x})]\!] \to [\![E(\vec{x})]\!]);$$

this means

$$\vdash_T \Theta_0(\vec{a},\vec{b}) \Rightarrow [[\![\Delta'_0]\!] \to [\![E]\!]](\vec{a});$$

by (3.11), there is a finite conjunction $\Delta_0(\vec{a}/\vec{x}, \vec{a}'/\vec{z})$ from Δ such that

$$\vdash_T \Delta_0(\vec{a},\vec{a}') \Rightarrow [[\![\Delta'_0]\!] \to [\![E]\!]](\vec{a}).$$

In $\mathbf{T}$, this means

$$[\![\Delta_0]\!] \leq p^*_{\vec{x}}([\![\Delta'_0]\!] \to [\![E]\!])$$

and also

$$[\![\Delta_0]\!] \wedge p^*_{\vec{x}}([\![\Delta'_0]\!]) \leq p^*_{\vec{x}}([\![E]\!]);$$

this inequality shows the claim as it says that

$$\vdash_T \Delta_0(\vec{a},\vec{a}') \;\&\; \Delta'_0(\vec{a}) \Rightarrow E(\vec{a}). \quad \square$$

PROPOSITION 3.10 *Suppose that* $\mathbf{T}$ *has stable implications among regular subobjects. Then, if* T *admits a model completion,* $\mathbf{T}$ *is an r-regular category.*

Proof. Suppose that T admits a model completion T^*. By Proposition 2.8 we know that $Alg(T)$ has the amalgamation property, hence by the previous Lemma it follows that injections are transferable in $Alg(T)$ (see condition (IT) from section 2.4). Keeping in mind the existence of pushouts in $Alg(T)$ and the fact that the first component of a mono is mono, it immediately follows that epis are pullback-stable in $\mathbf{T}$ (recall that 'epi' in $\mathbf{T}$ is equivalent to 'mono' in $Alg(T)_{fp}$ and 'mono' in $Alg(T)_{fp}$ is the same as 'set-theoretically injective', by an easy argument based on the use of finitely generated free algebras).

Before showing the existence of epi/regular mono factorizations in $\mathbf{T}$[5] (which will complete the proof of the fact that $\mathbf{T}$ is r-regular), we need the following preliminary information:

(i) for every e-formula $E(\vec{x},\vec{y})$, there is an e-formula $E_0(\vec{y})$ such that $\vdash_{T^*} \exists\vec{x}E(\vec{x},\vec{y}) \Leftrightarrow E_0(\vec{y})$.

Essentially, (i) (whose proof below depends on the fact that (IT) holds under our current hypotheses), says that in T^* an existential quantifier applied to an e-formula is eliminated by an e-formula as well.

[5]Essentially, this amounts to show that any finitely generated subalgebra of a finitely presented algebra is also finitely presented. In [Wh] it is actually shown that this so-called 'coherence' property follows from the existence of a model completion. We give a different proof of this result, which is less general because it uses the hypothesis of the Proposition, but which is on the other hand slightly more informative.

Proof of (i). Using disjunctive normal forms and the fact that T^* eliminates quantifiers, we have that

$$\vdash_{T^*} \exists \vec{x} E(\vec{x}, \vec{y}) \Leftrightarrow D_1(\vec{y}) \mathrm{v} \cdots \mathrm{v} D_n(\vec{y}), \tag{3.12}$$

where for each $i = 1, \ldots, n$

$$D_i \;=\; E_i(\vec{y}) \;\&\; \sim F_{i1}(\vec{y}) \;\&\cdots\&\; \sim F_{ik_i}(\vec{y})$$

and E_i, F_{ij} are e-formulas. We first show that

$$\vdash_{T^*} E_i(\vec{y}) \Rightarrow \exists \vec{x} E(\vec{x}, \vec{y}) \;\text{ or }\; \vdash_T \sim D_i(\vec{y}), \tag{3.13}$$

for every $i = 1, \ldots, n$. In order to prove (3.13), we suppose that $\not\vdash_T \sim D_i(\vec{y})$ and we show that

$$\vdash_{T^*} E_i(\vec{y}) \Rightarrow \exists \vec{x} E(\vec{x}, \vec{y}). \tag{3.14}$$

Take the algebra $\mathcal{F}(\vec{y})/E_i(\vec{y})$; we have

$$\mathcal{F}(\vec{y})/E_i(\vec{y}) \not\models F_{i1}(\vec{y}) \;\mathrm{v}\cdots\mathrm{v}\; F_{ik_i}(\vec{y})$$

(otherwise $\vdash_T E_i \Rightarrow F_{i1}(\vec{y}) \;\mathrm{v}\cdots\mathrm{v}\; F_{ik_i}(\vec{y})$ and $\vdash_T \sim D_i$); consequently $\mathcal{F}(\vec{y})/E_i(\vec{y}) \models D_i(\vec{y})$ holds and by (3.12), there is an extension A of $\mathcal{F}(\vec{y})/E_i(\vec{y})$ to a model of T^* such that $A \models \exists \vec{x} E(\vec{x}, \vec{y})$ holds. We show that for every model B of T^*, for every $\vec{b} \in B$, if $B \models E_i(\vec{b})$ then $B \models \exists \vec{x} E(\vec{x}, \vec{b})$ (this will prove (3.14)). For such a B, there is a morphism $\mathcal{F}(\vec{y})/E_i(\vec{y})$ mapping the $\vec{y}$'s into the $\vec{b}$'s and so, since injections are transferable in $Alg(T)$, we can complete the square

$$\begin{array}{ccc} \mathcal{F}(\vec{y})/E_i(\vec{y}) & \rightarrowtail & A \\ \downarrow & & \downarrow \\ B & \rightarrowtail & D \end{array}$$

in such a way that the bottom morphism is mono (because A is an extension of $\mathcal{F}(\vec{y})/E_i(\vec{y})$). We can also suppose that D is a model of T^* (because every model of T embeds into a model of T^*). But, as $A \models \exists \vec{x} E(\vec{x}, \vec{y})$, we have that $D \models \exists \vec{x} E(\vec{x}, \vec{b})$ and also that $B \models \exists \vec{x} E(\vec{x}, \vec{b})$ because T^* is model complete, hence a monomorphism between two models of its is elementary.

Having shown (3.13), by (3.12) and elementary logical passages, we get

$$\vdash_{T^*} \exists \vec{x} E(\vec{x}, \vec{y}) \Leftrightarrow E'_1(\vec{y}) \mathrm{v} \cdots \mathrm{v} E'_k(\vec{y}), \tag{3.15}$$

where $E'_1, \ldots, E'_k$ are some of the $E_1, \ldots, E_n$. But T and T^* prove the same quantifier-free formulas, hence we get $\vdash_T E(\vec{x}, \vec{y}) \Rightarrow E'_1(\vec{y}) \mathrm{v} \cdots \mathrm{v} E'_k(\vec{y})$; as T

is equational, by Proposition 2.12, there is $i = 1, \ldots, k$ such that $\vdash_T E(\vec{x}, \vec{y}) \Rightarrow E_i'(\vec{y})$ (this shows also that k cannot be 0). From this and (3.15), it follows that $\vdash_{T^*} \exists\vec{x}E(\vec{x}, \vec{y}) \Leftrightarrow E_i'(\vec{y})$, which is precisely the claim (i).

Let us look for an epi/regular mono factorization of an arrow in $\mathbf{T}$. Given an arrow in $Alg(T)_{fp}$

$$\mathcal{F}(\vec{y})/E(\vec{y}) \longrightarrow \mathcal{F}(\vec{z})/E'(\vec{z}),$$

we must be able to factor it into a quotient followed by a monomorphism (recall Proposition 2.13). We shall take the usual image factorization for algebras and show that the kernel of the image is a finitely generated congruence. Suppose the morphism is determined by the correspondence $\vec{y} \mapsto \vec{t}(\vec{z})$, for some tuple of terms $\vec{t}$. For any terms $u_1(\vec{y}), u_2(\vec{y})$, the pair given by the equivalence classes of u_1 and u_2 is in the kernel iff

$$\vdash_T E'(\vec{z}) \Rightarrow u_1(\vec{t}(\vec{z})) = u_2(\vec{t}(\vec{z})),$$

i.e. iff (we write $\vec{y} = \vec{t}(\vec{z})$ for the obvious componentwise conjunction)

$$\vdash_T E'(\vec{z}) \ \& \ \vec{y} = \vec{t}(\vec{z}) \Rightarrow u_1(\vec{y}) = u_2(\vec{y})$$

(to see it, on one side use substitutivity of equality and on the other side just replace $\vec{y}$ with $\vec{t}(\vec{z})$). This is the same as

$$\vdash_{T^*} \exists\vec{z}(E'(\vec{z}) \ \& \ \vec{y} = \vec{t}(\vec{z})) \Rightarrow u_1(\vec{y}) = u_2(\vec{y}); \qquad (3.16)$$

By (i), there is an e-formula $E_0(\vec{y})$ such that $\vdash_{T^*} \exists\vec{z}(E'(\vec{z}) \ \& \ \vec{y} = \vec{t}(\vec{z})) \Leftrightarrow E_0(\vec{y})$. Thus (3.16) is equivalent to

$$\vdash_{T^*} E_0(\vec{y}) \Rightarrow u_1(\vec{y}) = u_2(\vec{y}) \qquad (3.17)$$

and, since T and T^* prove the same quantifier-free formulas, (3.17) is equivalent to

$$\vdash_T E_0(\vec{y}) \Rightarrow u_1(\vec{y}) = u_2(\vec{y}),$$

which shows that the image of our morphism admits the finite presentation $\mathcal{F}(\vec{y})/E_0(\vec{y})$. □

The following result is a key point for the applications to propositional logics we shall meet in next Chapters:

THEOREM 3.11 *Suppose that* $\mathbf{T}$ *has stable finite joins*[6] *and stable implications among regular subobjects. Then* T *admits a model completion iff* $\mathbf{T}$ *is an r-Heyting category.*

[6]We shall not deal here with the many-sorted case. However, we mention that in this case a sufficient hypothesis is that $\mathbf{T}$ has stable $\mathcal{S}$-joins for every finite set of sorts $\mathcal{S}$.

Proof. The 'if' part of the claim is covered by Theorem 3.8. So let us assume that $\mathbf{T}$ has stable joins and implications and that T admits a model completion T^*. As in the proof of Proposition 3.10, we need a preliminary information concerning this time the way universal quantifiers applied to e-formulas are eliminated in T^*. We claim that:

(ii) for every e-formula $E(\vec{x}, \vec{y})$, there is an e-formula $E_0(\vec{y})$ such that $\vdash_{T^*} \forall \vec{x} E(\vec{x}, \vec{y}) \Leftrightarrow E_0(\vec{y})$.

Proof of (ii). By conjunctive normal forms and the fact that T^* eliminates quantifiers, we have that

$$\vdash_{T^*} \exists \vec{x} \sim E(\vec{x}, \vec{y}) \Leftrightarrow C_1(\vec{y}) \ \& \ \cdots \ \& C_n(\vec{y}), \tag{3.18}$$

where for each $i = 1, \ldots, n$

$$C_i(\vec{y}) \quad = \quad \sim E_i(\vec{y}) \ \mathrm{v} \ F_{i1}(\vec{y}) \ \mathrm{v} \cdots \ \mathrm{v} F_{ik_i}(\vec{y})$$

and E_i, F_{ij} are e-formulas. We prove that the F_{ij} can be removed; in fact, for every $i = 1, \ldots, n$ we have (recall that T and T^* agree on quantifier-free formulas)

$$\vdash_T \sim E(\vec{x}, \vec{y}) \Rightarrow C_i(\vec{y}),$$

that is

$$\vdash_T E_i(\vec{y}) \Rightarrow E(\vec{x}, \vec{y}) \ \mathrm{v} \ F_{i1}(\vec{y}) \ \mathrm{v} \cdots \ \mathrm{v} F_{ik_i}(\vec{y}).$$

By Proposition 2.12, this implies that either $\vdash_T E_i(\vec{y}) \Rightarrow E(\vec{x}, \vec{y})$ or $\vdash_T E_i(\vec{y}) \Rightarrow F_{i1}(\vec{y}) \ \mathrm{v} \cdots \ \mathrm{v} F_{ik_i}(\vec{y})$. In the second case the i-th conjunct of (3.18) is a provable formula in T, hence it can be removed. In the first case, as we have $\vdash_T \sim E(\vec{x}, \vec{y}) \Rightarrow \sim E_i(\vec{y})$, we can remove the $F_{ij}(\vec{y})$ from the i-th conjunct of (3.18). This means that we have shown that

$$\vdash_{T^*} \forall \vec{x} E(\vec{x}, \vec{y}) \Leftrightarrow E'_1(\vec{y}) \mathrm{v} \cdots \mathrm{v} E'_k(\vec{y}) \tag{3.19}$$

(the E'_i's are some of the e-formulas $E_1, \ldots, E_n$). Let us now consider $\bigvee_i [\![E'_i]\!] \hookrightarrow \mathcal{F}(\vec{y})$ in $\mathbf{T}$; we claim that

$$\vdash_{T^*} \forall \vec{x} E(\vec{x}, \vec{y}) \Leftrightarrow [[\![E'_1(\vec{y})]\!] \vee \cdots \vee [\![E'_k(\vec{y})]\!]].$$

In fact, we have for every $j = 1, \ldots, n$ that $[\![E'_j]\!] \leq \bigvee_i [\![E'_i]\!]$, which means that

$$\vdash_T E'_j(\vec{y}) \Rightarrow [[\![E'_1(\vec{y})]\!] \vee \cdots \vee [\![E'_k(\vec{y})]\!]].$$

From this, we can pass to

$$\vdash_T E'_1(\vec{y}) \mathrm{v} \cdots \mathrm{v} E'_k(\vec{y}) \Rightarrow [[\![E'_1(\vec{y})]\!] \vee \cdots \vee [\![E'_k(\vec{y})]\!]]$$

which, together with (3.19), shows one half of the claim, namely that

$$\vdash_{T^*} \forall \vec{x} E(\vec{x}, \vec{y}) \Rightarrow [\llbracket E'_1(\vec{y}) \rrbracket \vee \cdots \vee \llbracket E'_k(\vec{y}) \rrbracket].$$

For the remaining half, notice that (by (3.19) and the fact that T and T^* agree on quantifier-free formulas)

$$\vdash_T E'_1(\vec{y}) \mathrm{v} \cdots \mathrm{v} E'_k(\vec{y}) \Rightarrow E(\vec{x}, \vec{y}),$$

which means that, for every $i = 1, \ldots, k$, $\vdash_T E'_i(\vec{y}) \Rightarrow E(\vec{x}, \vec{y})$. If we read this relation in $\mathbf{T}$, we get

$$p^*_{\vec{y}}(\llbracket E'_i(\vec{y}) \rrbracket) \leq \llbracket E(\vec{x}, \vec{y}) \rrbracket.$$

As joins are pullback-stable, we obtain

$$p^*_{\vec{y}}(\bigvee_i \llbracket E'_i(\vec{y}) \rrbracket) \leq \llbracket E(\vec{x}, \vec{y}) \rrbracket.$$

We finally get

$$\vdash_T [\llbracket E'_1(\vec{y}) \rrbracket \vee \cdots \vee \llbracket E'_k(\vec{y}) \rrbracket] \Rightarrow E(\vec{x}, \vec{y}),$$

which completes the proof of (ii).[7]

We know from Proposition 3.10 that $\mathbf{T}$ is r-regular. In order to show that $\mathbf{T}$ is r-Heyting, it is sufficient to show that universal quantification exists along any projection arrow whose domain and codomain are finitely generated free algebras: this is enough by Lemma 3.7 (in fact, according to condition (ii) of that Lemma, from this we can draw existence of universal quantification along any arrow among finitely generated free algebras and according to condition (i) of that Lemma, we finally get existence of universal quantification along any arrow whatsoever, because objects of $\mathbf{T}$ are regular subobjects of the objects corresponding to finitely generated free algebras). Let us consequently take a projection $p_{\vec{y}} : \mathcal{F}(\vec{x}, \vec{y}) \longrightarrow \mathcal{F}(\vec{y})$, and a regular subobject $\llbracket E(\vec{x}, \vec{y}) \rrbracket \hookrightarrow \mathcal{F}(\vec{x}, \vec{y})$; we must find $\llbracket E_0(\vec{y}) \rrbracket \hookrightarrow \mathcal{F}(\vec{y})$ such that for all e-formulas $G(\vec{y})$ we have

$$\vdash_T G(\vec{y}) \Rightarrow E_0(\vec{y}) \text{ iff } \vdash_T G(\vec{y}) \Rightarrow E(\vec{x}, \vec{y}).$$

We take the $E_0(\vec{y})$ suggested by (ii). The left-to-right side follows from the fact that $\vdash_{T^*} E_0(\vec{y}) \Rightarrow E(\vec{x}, \vec{y})$ and the fact that T and T^* agree on quantifier free formulas. For the right-to-left side, notice that $\vdash_T G(\vec{y}) \Rightarrow E(\vec{x}, \vec{y})$ implies $\vdash_T G(\vec{y}) \Rightarrow \forall \vec{x} E(\vec{x}, \vec{y})$ and also (by (ii)) $\vdash_{T^*} G(\vec{y}) \Rightarrow E_0(\vec{y})$. Therefore $\vdash_T G(\vec{y}) \Rightarrow E_0(\vec{y})$ holds, as well. □

[7] Notice that the argument works also in the case $k = 0$: recall that the empty disjunction is $\perp$ and the empty join is the minimum (pullback-stable by hypothesis) regular subobject.

Notice that conditions (i) and (ii) proved in Proposition 3.10 and Theorem 3.11 agree with the quantifier-elimination axiomatization for T^* given in Theorem 3.8.

3. Exercises

1 Let $(\mathcal{E}, \mathcal{M})$ be a stable factorization system.

(a) Show that if $f \in \mathcal{M}$ and if $f \in \mathcal{E}$ then f is iso.

(b) Show that morphisms in $\mathcal{M}$ are pullback stable.

2 Let $Alg(T)$ be a locally finite variety with EDO. Show that $\mathbf{T}$ is an r-Heyting category iff AP^* holds in $\mathbf{T}$.

3 Verify that all the categories

(a) Top - of topological spaces and continuous functions;

(b) $Graph$ - of graphs, i.e. the category of sets equipped with a binary relation and functions preserving these relations;

are r-Heyting categories, in spite of the fact that the first two are not even regular.

4 Show that a subcategory A of an r-Heyting category B closed in B under finite limits and regular subobjects is again an r-Heyting category. Conclude from this and the previous exercise that the categories:

(a) $Trans$ - of sets equipped with a binary transitive relation and functions preserving these relations;

(b) $Poset$ - of partially ordered sets and monotone functions;

(c) $Quasi$ - of pre-ordered sets and monotone functions;

are r-Heyting categories.

5 Prove that if for the projection $p_A : A \times B \to A$ there is a right adjoint $\forall_{p_A}$ to the pullback functor $p_A^* : Sub_r(A) \to Sub_r(A \times B)$, then for any $f : B \to A$ there is a right adjoint $\forall_f : Sub_r(A) \to Sub_r(B)$ to the pullback functor f^*, which is given by for $S \in Sub_r(A)$ by the formula:

$$\forall_f(S) = \forall_{p_A}((id_A \times f)^*(\Delta_A) \to p_B^*(S))$$

where Δ_A be the internal equality, i.e. the regular subobject of $A \times A$ given by the diagonal map $\langle id_A, id_A \rangle : A \hookrightarrow A \times A$.

Chapter 4

HEYTING ALGEBRAS

We shall develop a duality for the category finitely presented Heyting algebras HA_{fp}. Using a combinatorial description $\mathbf{M_H}$ of HA_{fp}^{op} we shall prove that it is a Heyting category and hence, according to Theorem 3.8, the theory of Heyting algebras T_H admits a model completion T_H^*. Then we shall study some further properties of HA_{fp}^{op} and we shall derive some conclusions from these studies for intuitionistic propositional logic IpC. We introduce a sheaf semantics for second order logic and show how to use it to eliminate quantifiers in T_H^*.

1. Basic definitions

In this Chapter, we shall mainly deal with finite posets, to be indicated with the letters $P, Q, \ldots$; their elements will be usually written as $p, q, \ldots$ and their ordering (reflexive, transitive and antisymmetric) relations will be written simply as $\leq$, leaving the more explicit notation $\leq_P, \leq_Q, \ldots$ for contexts requiring such further specification. A poset P is called *rooted* iff it has a greatest element $\rho(P)$ (sometimes we indicate it simply as ρ instead of $\rho(P)$). If a basic finite poset L (the poset of 'labels') is fixed, we call an *L-evaluation* or simply an *evaluation* a pair $\langle P, u\rangle$, where P is a rooted finite poset and $u : P \to L$ is an order-preserving map. This notion has a strict relation with finite Kripke models. In fact if $\langle L, \leq\rangle$ is $\langle \mathcal{P}(\vec{p}), \supseteq\rangle$ (where $\vec{p}$ is a finite list of propositional letters), then an L-evaluation $u : P \to L$ is the same as a Kripke model for the propositional intuitionistic language built up from $\vec{p}$.[1]

We define for every $n \in \omega$ and for every pair of L-evaluations $u : P \to L$ and $v : Q \to L$, the notions of being *n-equivalent* (written $u \sim_n v$) and of

[1]According to our conventions, we have that (for $p, q \in P$) if $p \leq q$ then $u(p) \supseteq u(q)$, that is we use $\leq$ where standard literature uses $\geq$.

being *n-less than or equal to* (written $u \leq_n v$). These notions are motivated by the fact (implicit in what is proved in the next section) that in the case of Kripke models being n-equivalent means exactly to satisfy the same formulas up to implicational degree n, see Exercise 1. Similarly, $u \leq_n v$ means that u satisfies all formulas up to implicational degree n that v satisfies. We define also for two L-evaluations u, v the notions of being *infinitely equivalent* (written $u \sim_\infty v$) and *infinitely less than or equal to* (written $u \leq_\infty v$). All these notions are parallel to the analogous definitions introduced in [Fi1], [Fi2] for modal logic, see Chapter 5. We prefer to introduce them by means of Ehrenfeucht-Fraissé games.

Let $u : P \to L$ and $v : Q \to L$ be two L-evaluations. The *game* we are interested in has two players, player I and player II. Player I can choose either a point in P or a point in Q and player II must answer by choosing a point in the other poset and the only rule is that, if $\langle p \in P, q \in Q \rangle$ is the last move played, then in the successive move the two players can only choose points $\langle p', q' \rangle$ such that $p' \leq p$ and $q' \leq q$. If $\langle p_1, q_1 \rangle, \ldots, \langle p_i, q_i \rangle, \ldots$ are the points chosen at the end of the game, after infinitely many moves, player II wins iff for every $i = 1, 2, \ldots$, we have that $u(p_i) = v(q_i)$. We say that

- $u \sim_\infty v$ iff *player II has a winning strategy*;
- $u \sim_n v$ (for $n > 0$) iff *player II has a winning strategy for the first n-moves*, i.e. he has a winning strategy provided we stipulate that the game terminates after n moves;
- $u \sim_0 v$ iff $u(\rho(P)) = v(\rho(Q))$;
- $u \leq_\infty v$ iff *player II has a winning strategy in the modified game*, where the word 'modified' refers to the fact that we have an additional rule forcing player I to play in the domain of u the first move;
- $u \leq_n v$ (for $n > 0$) iff *player II has a winning strategy for the first n moves in the modified game*;
- $u \leq_0 v$ iff $u(\rho(P)) \leq v(\rho(Q))$.

The relations $\sim_n$ and $\sim_\infty$ are clearly equivalence relations, whereas the relations $\leq_n$ and $\leq_\infty$ are only reflexive and transitive. Notice also that for every n, $u \sim_n v$ implies $u \sim_0 v$ because $\leq_L$ is antisymmetric. The straightforward proposition below (to be often used without explicit mention in the following) provides equivalent definitions. We fix a notation: if $u : P \to L$ is an L-evaluation and $p \in P$, u_p is u restricted in the domain to the downward closed subset $\downarrow p = \{p' : p' \leq p\}$.

PROPOSITION 4.1 *Given two L-evaluations $u : P \to L, v : Q \to L$, and $n \in \omega$, we have*

(i) $u \sim_{n+1} v$ *iff* $\forall p \in P\ \exists q \in Q\ (u_p \sim_n v_q)$ *and vice versa;*

(ii) $u \leq_{n+1} v$ *iff* $\forall p \in P\ \exists q \in Q\ (u_p \sim_n v_q)$*;*

(iii) $u \leq_\infty v$ *iff* $(\forall p \in P\, \exists q \in Q$ *such that* $u_p \sim_\infty u_q)$ *iff* $(\exists q \in Q\, (u \sim_\infty v_q))$*;*

(iv) $u \sim_\infty v$ *iff* $(u \leq_\infty v$ *and* $v \leq_\infty u)$. $\square$

Example Let L be a three element partial order with set of nodes $\{a, b, c\}$ given by the following Hasse diagram

L :

Consider the following three L-evaluations on three different four element posets:

u :

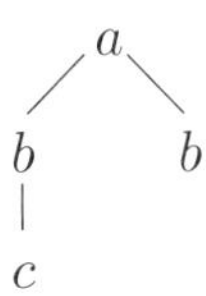

v :

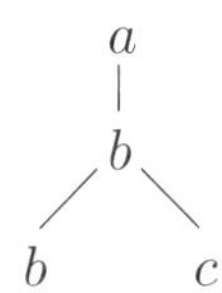

w :

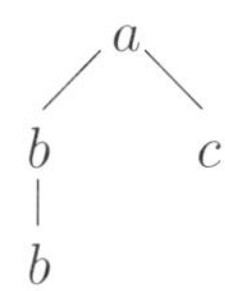

In the above picture, we put in place of points of the poset the values of evaluations at these points.

Then $u \sim_2 v$ but $u \not\sim_3 v$, and $u \sim_1 w$ but $v \not\sim_2 w$.

Notice that saying that '$u \sim_\infty v$' is equivalent to say that '$u \sim_n v$ holds for every n'. In fact, on one side it is evident that if there exists an infinite strategy, then there are also strategies for n games for every n. Vice versa, suppose that there are all such finite strategies and suppose that player I chooses a point $p \in P$. Our posets are *finite* so that, as for every $n \geq 0$ there exists $q_n \in Q$ such that $u_p \sim_n v_{q_n}$, there is also $q \in Q$ (independent of n) such that $u_p \sim_n v_q$ for every n. Player II answers this q and continuing this way, it is clear that we can define the desired winning infinite strategy.

With each L-evaluation $u : P \to L$ we associate for every $n \in \omega$ a set $Type_n(u)$ of $\sim_n$-equivalence classes by: $Type_n(u) = \{[u_p]_n : p \in P\}$ (where, of course, $[u_p]_n$ is the $\sim_n$-equivalence class of u_p). An important, although simple, fact is given by the following proposition:

PROPOSITION 4.2 *Fix a finite poset L and $n \in \omega$; then there are only finitely many equivalence classes of L-evaluations with respect to $\sim_n$.*

Proof. This is evident for $n = 0$. For $n > 0$, we argue by induction as follows. By Proposition 4.1(i), we have that $u \sim_n v$ iff $Type_{n-1}(u) = Type_{n-1}(v)$, hence there cannot be more non $\sim_n$-equivalent L-evaluations than sets of $\sim_{n-1}$ equivalence classes.[2] $\square$

For infinite equivalence the result of the previous Proposition is not true, but on the other hand infinite equivalence can be characterized in terms of open maps (see Proposition 4.4 below). Recall from Chapter 2 that an order-preserving map $h : P \to Q$ is called *open* iff for all $p \in P, q \in Q$, if $q \leq h(p)$ then there exists $p' \leq p$ such that $h(p') = q$. Notice that if Q is rooted and h open, h is surjective iff the inverse image of the root of Q is non-empty. As shown in Chapter 2, open maps are exactly Birkhoff duals of Heyting algebras homomorphisms.

LEMMA 4.3 *Given an L-evaluation $u : P \to L$ and an open map $h : Q \to P$, we have that $(u \circ h) \leq_\infty u$. Moreover, if h is also surjective, then $(u \circ h) \sim_\infty u$.*

Proof. Clearly it is sufficient to prove the second part of the claim (for $(u \circ h) \sim_\infty u_{h(\rho)}$, hence $(u \circ h) \leq_\infty u$ follows). So suppose that h is surjective. We have the following infinite strategy for player II: if player I plays $q \in Q$, the answer is $h(q)$, if he plays in P the answer is suggested by the openness of h (or by the surjectivity of h in the first move), so that we reach only positions of the kind $\langle q, h(q) \rangle$. It is evident that in this way player II wins. $\square$

PROPOSITION 4.4 *For two L-evaluations $u : P \to L$, $v : Q \to L$, we have that $u \sim_\infty v$ iff there is a commutative square*

$$\begin{array}{ccc} R & \xrightarrow{\;h\;} & P \\ {\scriptstyle k}\downarrow & & \downarrow{\scriptstyle u} \\ Q & \xrightarrow[\;v\;]{} & L \end{array}$$

such that R is a finite rooted poset and h, k are open surjective maps. Moreover $u \leq_\infty v$ iff there is a commutative square like the above one, with the only difference that now k is not required to be surjective.

Proof. Suppose that $u \sim_\infty v$. Take $R = \{\langle p, q \rangle : u_p \sim_\infty v_q\}$. Order is the restriction of the product order on $P \times Q$ and h, k are the two projections, restricted in their domains. R is clearly rooted, the square commutes and h, k are surjective and open. The latter can be shown as follows (e.g. for h).

[2]This is only an exponential strict upper bound, because not all sets of $\sim_{n-1}$-equivalence classes are legal, i.e. are of the form $Type_{n-1}(u)$ for some u (for instance, for $n = 1$ only subsets of L having a greatest element are legal and the situation becomes more involved for larger n, see [Ur] , [Gh1]).

Suppose that $\langle p, q\rangle \in R$ and that $p' \leq p$; we have that there exists $q' \leq q$ such that $u_{p'} \sim_\infty v_{q'}$. For such q' we have that $\langle p', q'\rangle \in R$, $\langle p', q'\rangle \leq \langle p, q\rangle$ and $h(p', q') = p'$.

Conversely, suppose that there is a commutative square with the required properties. Now $u \sim_\infty (u \circ h)$, by Lemma 4.3 and similarly $v \sim_\infty (v \circ k)$, hence $u \sim_\infty v$.

The characterization of $u \leq_\infty v$ follows from the above characterization of $\sim_\infty$ and Proposition 4.1 (iii). □

It is interesting to know that, given an L-evaluation u, for sufficiently large n (strictly depending on u) infinite equivalence to u is the same as n-equivalence to u. By $ht(P)$, we denote the height of the finite poset P, i.e. the length of a maximal chain in P. We have

PROPOSITION 4.5 *Let $u : P \to L$ be an L-evaluation. For $n(P) = 2\,ht(P) - 1$ we have that for every $v : Q \to L$, $u \sim_{n(P)} v$ iff $u \sim_\infty v$.*

Proof. The claim is clear for $ht(P) = 1$, because in this case P is a one-point poset and $u \sim_1 v$ means that v is constant. Suppose that $ht(P) > 1$ and that $u \sim_{n(P)} v$: we define an infinite strategy for player II. We recall that, as $u \sim_{n(P)} v$, for every $q \in Q$ there must exist a point $p \in P$ such that $u_p \sim_{n(P)-1} v_q$. Such p may or may not be the root of P. Player II behaves as follows: as long as player I plays a point $q \in Q$, such that $u \sim_{n(P)-1} v_q$, player II answers the root of u. After an initial (possibly empty) sequence of such moves we reach a position $\langle \rho(P), q\rangle$ with $u \sim_{n(P)-1} v_q$. If now player I tries with $p \in P$ (different from $\rho(P)$, otherwise answer is obviously q), then there exists $q' \leq q$ such that $u_p \sim_{n(P)-2} u_{q'}$: player II answers such q' and wins by induction hypothesis. If player I tries with $q' \in Q$ such that there exists $p \neq \rho(P)$ so that $u_p \sim_{n(P)-1} v_{q'}$, then player II answers such p and wins again by induction hypothesis. □

The next Lemma will complete the list of basic properties of the relations $\sim_n$ and $\leq_n$. Before stating the Lemma, we introduce a useful construction on L-evaluations. Suppose that $u : P \to L$, $v : Q \to L$ are L-evaluations such that $v \leq_0 u$. The *grafting* of v below the root of P is an L-evaluation $v \triangleleft u : P' \to L$ defined as follows. P' is $P + Q$ (disjoint union as sets) with the following order $\leq'$: $q \leq_{P'} p$ iff either $(p = \rho(P))$ or $(p, q \in P$ and $q \leq_P p)$ or $(p, q \in Q$ and $q \leq_Q p)$. Moreover, $v \triangleleft u$ acts as u on P and as v on Q (this is order-preserving because $v \leq_0 u$). It is immediately seen that:

LEMMA 4.6 *$(v \triangleleft u) \circ i_Q = v$ (where i_Q is the open inclusion of Q into P'), hence $v \leq_\infty (v \triangleleft u)$. Moreover, if for some $n \in \omega$, $v \leq_n u$, then $u \sim_n (v \triangleleft u)$*

□

2. Finitely presented Heyting algebras

Let $\mathbf{P_0}$ be the category of finite rooted posets and open maps. We introduce a Grothendieck topology $\mathbf{J_0}$ on it, which will turn out to be the canonical one, i.e. the finest (that is, having most covers) that makes representable functors sheaves. In the following we use the names 'products', 'pullbacks', etc. for the related constructions performed in the category of posets and order-preserving maps. Such constructions are not outside $\mathbf{P_0}$ (because projections are open and because the pullback of an open map along any order-preserving map is open), however they do not represent products and pullbacks in $\mathbf{P_0}$: the unique map required by the universal property needs not be open.

The topology $\mathbf{J_0}$ is introduced by saying that *a sieve of P covers P iff it contains an open surjective map.* Here we make some comments about the choice of $\mathbf{P_0}$: one may think that the limitation to rooted posets is rather arbitrary. A more natural choice would be that of considering the category $\mathbf{P}$ of finite posets and open maps (dual to the category of finite Heyting algebras) and the canonical topology $\mathbf{J_1}$ on it (which is characterized as the topology consisting of jointly-surjective families). This approach would give equivalent categories of sheaves. In fact the inclusion of $\mathbf{P_0}$ into $\mathbf{P}$ is a full and faithful functor, $\mathbf{J_0}$ is the topology induced by $\mathbf{J_1}$ (i.e. it is the strongest one that makes the inclusion functor continuous) and moreover each finite posets is $\mathbf{J_1}$-covered by rooted ones, hence Grothendieck's Comparison Lemma applies (see [SGA4], Exposé III, p.288 or the Appendix): the two sheaf toposes are equivalent. The reason why we make the choice of limiting ourselves to rooted posets is that $\mathbf{J_0}$ is a much simpler topology, for instance it makes unions of subsheaves easy to compute.

PROPOSITION 4.7 *$\mathbf{J_0}$ is the canonical topology on $\mathbf{P_0}$.*

Proof. We begin by showing that for every P, the representable functor

$$Y_P = \mathbf{P_0}(-, P) : \mathbf{P_0}^{op} \longrightarrow Set$$

is a sheaf for our topology. Let $S \subseteq Y_Q$ be a $\mathbf{J_0}$-cover of Q and let $u : S \to Y_P$ be a compatible family in Y_P. By the definition of $\mathbf{J_0}$, S contains an open surjection $h : Q' \to Q$. Taking the kernel pair of h we get a pullback

$$\begin{array}{ccc} Q'' & \xrightarrow{k_1} & Q' \\ {\scriptstyle k_2}\downarrow & & \downarrow{\scriptstyle h} \\ Q' & \xrightarrow[h]{} & Q \end{array}$$

which, as it can be easily verified, is a pushout as well. The fact that u is compatible entails that $u(h) \circ k_1 = u(h) \circ k_2$, hence there exists a unique $u' : Q \to P$ such that $u' \circ h = u(h)$. As u' is open, it can be easily checked to be the desired extension.

Let us take now a topology $\mathbf{J}$ such that representable functors are $\mathbf{J}$-sheaves. Let $S \subseteq Y_P$ be a $\mathbf{J}$-cover of P. We show that it contains a map for which the inverse image of the root of P is not empty. Suppose the contrary: then build P' as P with the addition of a top element t. If $k : P' \to P$ is the open map that acts identically except for t which is mapped onto the root of P, then $k^{-1}(S) = \{h : k \circ h \in S\}$ is a cover of P', whose corresponding compatible family has two extensions $Y_{P'} \to Y_{P'}$, namely the identity $P' \to P'$ and k followed by the inclusion $P \subseteq P'$. Thus $\mathbf{J} \subseteq \mathbf{J_0}$. $\square$

We shall be interested in a larger class of sheaves than the representable ones: for every finite poset L, let h_L be the presheaf so defined

$$\begin{aligned} h_L(P) &= \{u : P \to L \,:\, u \text{ is an order preserving map}\} \\ h_L(k)(u) &= u \circ k. \end{aligned}$$

h_L is actually a sheaf: one simply repeats the proof of the first part of Proposition 4.7. For $L = \mathbf{1}$ (the one-element poset $\{0\}$), h_L is the terminal object in $Sh(\mathbf{P_0}, \mathbf{J_0})$. Notice also that for $L = \mathbf{2}$ (the two-element poset $\{0, 1\}$ with $0 \leq 1$), h_L is the subobject classifier Ω in $Sh(\mathbf{P_0}, \mathbf{J_0})$: the classifying map $\chi_S : X \to \Omega$ of a subobject $S \subseteq X$ is defined by $(\chi_S)_P(x)(p) = 0$ iff $X(i_p)(x) \in S_{\downarrow p}$, for $x \in X_P$ (here for $p \in P$, $i_p :\downarrow p \to P$ is the open inclusion). The map $true : h_\mathbf{1} \to h_\mathbf{2}$ is an inclusion.

It is a standard fact that for every sheaf X, the set of subsheaves $Sub(X)$ of X, ordered with pointwise inclusion, is a complete Heyting algebra. As our topology is peculiar, it turns out that not only meets but also joins are pointwise. Implication is given by (we abbreviate $X(h)(x)$ as $h(x)$):

$$(S_1 \to S_2)_P = \tag{4.1}$$

$$\{x \in X_P \,:\, \forall h : Q \to P \; (h(x) \in (S_1)_Q \;\Rightarrow\; h(x) \in (S_2)_Q)\}$$

or, equivalently, by:[3]

$$(S_1 \to S_2)_P = \tag{4.2}$$

$$= \{x \in X_P \,:\, \forall p \in P \; (i_p(x) \in (S_1)_{\downarrow p} \;\Rightarrow\; i_p(x) \in (S_2)_{\downarrow p})\}$$

[3] For the proof of the equivalence, notice that any $h : Q \to P$ factors as $Q \xrightarrow{h'} \downarrow p \xrightarrow{i_p} P$, where h' is open and surjective. Moreover, because of the sheaf condition, given a subsheaf $S \subseteq X$, an element $u \in X_{P'}$ and a map $h' : Q \to P'$ open and surjective, we have that $u \in S_{P'}$ iff $h'(u) \in S_Q$.

Using the notions of the previous section, we can define interesting subalgebras of Heyting algebras of the kind $Sub(h_L)$. Let $\mathcal{S}(h_L)$ be the set of subsheaves S of h_L satisfying the following condition for some $n \geq 0$

$$\forall u : P \to L,\ \forall v : Q \to L\ (u \in S_P\ \&\ u \sim_n v\ \Rightarrow\ v \in S_Q) \tag{4.3}$$

When this happens, we say that n is *an index* for S. $\mathcal{S}(h_L)$ is a sub-Heyting algebra of $Sub(h_L)$ (which is in general not complete): in fact, it can be easily seen for instance that if S_1, S_2 have index n, then $S_1 \to S_2$ has index $n+1$. Important elements of $\mathcal{S}(h_L)$ are given by the minimum subsheaves $\downarrow_\infty u$ containing a given $u : P \to L$. In fact, by Proposition 4.4, we have that

$$(\downarrow_\infty u)_Q = \{v : Q \to L\ :\ v \leq_\infty u\} \tag{4.4}$$

To find an index, we use Propositions 4.1 (iii) and 4.5: for $n \geq n(P)$, if $v \leq_\infty u$ and $v' \sim_n v$, then $v' \sim_\infty u_p$ for some $p \in P$, hence $v' \leq_\infty u$. From the fact that such single-generator subsheaves $\downarrow_\infty u$ have an index, it clearly follows that $Sub(h_L)$ is a completion of $\mathcal{S}(h_L)$ under infinite joins.

Condition 4.3 can be expressed in a slightly different way by

$$\forall u : P \to L,\ \forall v : Q \to L\ (u \in S_P\ \&\ v \leq_n u\ \Rightarrow\ v \in S_Q). \tag{4.5}$$

Clearly (4.5) implies (4.3). For the converse, suppose that $v \leq_n u$ and that $u \in S$ (sometimes we use this abbreviated notation for $u \in S_P$); we want to show that $v \in S$. By the 'grafting Lemma' 4.6, $u \sim_n v \lhd u$, hence $v \lhd u \in S$ by (4.3), so $v \in S$ by taking the restriction with respect to the open inclusion i_Q. We will use (4.3) or (4.5) indifferently.

Recall from section 2.1 that $\mathcal{D}(L)$ is the Birkhoff dual of L, i.e. the distributive lattice of downward closed subsets of L. Notice that there is a lattice embedding $\iota_L : \mathcal{D}(L) \to \mathcal{S}(h_L)$ associating with a downward closed subset d of L, the subsheaf $\iota_L(d)_P = \{u : P \to L\ :\ u(\rho(P)) \in d\}$. We shall prove that $\mathcal{S}(h_L)$ is the free Heyting algebra generated by the finite distributive lattice $\mathcal{D}(L)$ with ι_L as the canonical embedding. This construction is similar to well-known ones. In fact, 'canonical' representative elements are considered in the literature (within a different terminology) both for ∞-equivalence classes [Be], [Ry], [Gh2] and for the n-equivalence classes [Ur], [Gh1]. We prefer however not to deal with them, in order to be less demanding when performing certain constructions on L-evaluations, without being forced at each step to reduce the models we build into such 'canonical' forms. Moreover, the use of canonical representatives (interesting in itself for other purposes) would destroy the sheaf structure which is one of the main points of this approach.

LEMMA 4.8 *The image of ι_L generates $\mathcal{S}(h_L)$ as a Heyting algebra.*

Proof. Clearly the elements of $\mathcal{S}(h_L)$ having index 0 are exactly the elements of the image of ι_L. Now consider an element having index $n+1$. Because of Proposition 4.2, it is a finite union of elements of the kind $(\downarrow_{n+1} u)$, where $(\downarrow_{n+1} u)_Q = \{v : Q \to L \ :\ v \leq_{n+1} u\}$. We can express such elements in terms of elements having index n by:

$$(\downarrow_{n+1} u) = \bigcap_{\{v : \forall p \in dom(u)\ v \not\sim_n u_p\}} ((\downarrow_n v) \to \bigcup_{\{w : v \not\leq_n w\}} (\downarrow_n w)) \tag{4.6}$$

(notice that all intersections and unions involved are finite).[4] In fact, suppose that $z \leq_{n+1} u$ and take v such that for every point p in the domain of u, $v \not\sim_n u_p$. We show that $z \in (\downarrow_n v) \to \bigcup_{\{w : v \not\leq_n w\}} (\downarrow_n w)$ by using (4.1).

To check it, take a point q in the domain of z and suppose that $z_q \leq_n v$. Now from $z \leq_{n+1} u$ we conclude that there exists p such that $z_q \sim_n u_p$. Consequently, $v \not\leq_n z_q$ (otherwise $z_q \sim_n v$ and so $v \sim_n u_p$, which contradicts the choice of v) and so $z_q \in \bigcup_{\{w : v \not\leq_n w\}} (\downarrow_n w)$. Vice versa, suppose that $z \not\leq_{n+1} u$. It follows that there is a point q in the domain of z such that for every point p in the domain of u, $z_q \not\sim_n u_p$. We check that $z \notin (\downarrow_n z_q) \to \bigcup_{\{w : z_q \not\leq_n w\}} \downarrow_n (w)$. This is clear as $z_q \in (\downarrow_n z_q)$ and $z_q \notin \bigcup_{\{w : z_q \not\leq_n w\}} (\downarrow_n w)$. This ends the proof of (4.6) and of the Lemma. □

PROPOSITION 4.9 *$\mathcal{S}(h_L)$ is the free Heyting algebra generated by the finite distributive lattice $\mathcal{D}(L)$ with ι_L as the canonical embedding.*

Proof. Let us write F (for the time being) for this free algebra and let $\eta : \mathcal{D}(L) \to F$ be the related universal map. We have a commutative triangle:

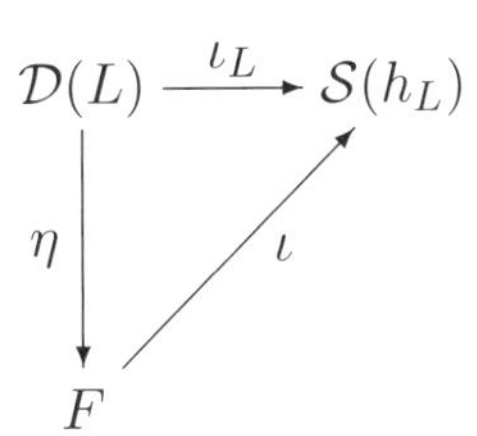

where the Heyting morphism ι is surjective by the previous Lemma. We prove that it is also injective. As F is a finitely presented Heyting algebra, by Theorem 2.3 and Lemma 2.19, there is an embedding $\langle h_i \rangle_i : F \longrightarrow \prod_i \mathcal{D}(P_i)$ where the P_i are finite rooted posets. The claim is proved if we show that for every i there is a Heyting algebra morphism $h'_i : \mathcal{S}(h_L) \longrightarrow \mathcal{D}(P_i)$ such that the square

[4]First of all notice that there are only finitely many elements of the kind $\downarrow_n w$, because there are only finitely many non n-equivalent w. Moreover, if $v_1 \sim_n v_2$, we have that $(\downarrow_n v_1) \to \bigcup_{\{w : v_1 \not\leq_n w\}} (\downarrow_n w)$ is equal to $(\downarrow_n v_2) \to \bigcup_{\{w : v_2 \not\leq_n w\}} (\downarrow_n w)$. Formulas 4.6, although quite involved at a first glance, arise naturally from the construction of [Gh1].

$$\begin{array}{ccc} \mathcal{D}(L) & \xrightarrow{\iota_L} & \mathcal{S}(h_L) \\ \eta \downarrow & & \downarrow h'_i \\ F & \xrightarrow[h_i]{} & \mathcal{D}(P_i) \end{array}$$

commutes. In fact if this holds, $\langle h_i \rangle_i \circ \eta = \langle h'_i \rangle_i \circ \iota_L = \langle h'_i \rangle_i \circ \iota \circ \eta$ which implies (by uniqueness) $\langle h_i \rangle_i = \langle h'_i \rangle_i \circ \iota$ and now ι is injective because it is the first component of an injection. We only have to find the h'_i. Notice that, by the duality theorem between finite distributive lattices and finite posets, the lattice morphism $h_i \circ \eta : \mathcal{D}(L) \to \mathcal{D}(P_i)$ is the inverse image morphism $\mathcal{D}(u)$ for an order-preserving map $u : P_i \to L$, so we can simply put $h'_i(S) = \{p \in P_i : u_p \in S_{\downarrow p}\}$ for every $S \in \mathcal{S}(h_L)$. Preservation of Heyting operations and commutativity of the triangle

$$\begin{array}{ccc} \mathcal{D}(L) & \xrightarrow{\iota_L} & \mathcal{S}(h_L) \\ & \mathcal{D}(u) \searrow & \downarrow h'_i \\ & & \mathcal{D}(P_i) \end{array}$$

are easy. □

COROLLARY 4.10 *A Heyting algebra is finitely presented iff it is isomorphic to one of the kind $\mathcal{S}(C)$, where $C \in \mathcal{S}(h_L)$ and $\mathcal{S}(C)$ is the Heyting algebra of all $C' \in \mathcal{S}(h_L)$ such that $C' \subseteq C$ (the order relation is again inclusion and consequently in order to perform the Heyting operations one first performs them in $\mathcal{S}(h_L)$ and then takes intersection with C).*

Proof. Just notice that finitely presented Heyting algebras can be described as the Heyting algebras freely generated by a finite distributive lattice divided by a single equation. □

3. Duality for Heyting algebras

We begin with the definition of the category $\mathbf{M_H}$. This category can be thought of as the category whose objects are collections (=sheaves equipped with an additional structure) of all the finite models of a finitely axiomatized theory in intuitionistic propositional logic. An object of $\mathbf{M_H}$ is a pair $\langle L, C \rangle$ such that L is a finite poset and $C \in \mathcal{S}(h_L)$. We usually write C^L for $\langle L, C \rangle$; sometimes we write simply C, meaning that L (which is an essential datum for our objects) is understood. So, every object of $\mathbf{M_H}$ is equipped with relations

$\sim_n$ and $\leq_n$ for $n \in \omega \cup \{\infty\}$, inherited from h_L. We write $v \in C$ if $v \in C(P)$, where P is the domain of v. A morphism $f : C \to C'$ in $\mathbf{M_H}$ is a natural transformation such that for some natural number k, if $v, v' \in C$ and $v \sim_k v'$ then $f(v) \sim_0 f(v')$. Any k satisfying the above condition is called an *index* of f.

The following Lemma gives some equivalent conditions for being a morphism in $\mathbf{M_H}$. In particular it ensures that the morphisms in $\mathbf{M_H}$ do compose.

LEMMA 4.11 *Let $f : C \to C'$ be a natural transformation between objects of $\mathbf{M_H}$. The following conditions are equivalent for $k \in \omega$:*

(i) $\forall v, v' \in C$ *if* $v \leq_k v'$ *then* $f(v) \leq_0 f(v')$

(ii) $\forall v, v' \in C$ *if* $v \sim_k v'$ *then* $f(v) \sim_0 f(v')$

(iii) $\forall n\, \forall v, v' \in C$ *if* $v \sim_{n+k} v'$ *then* $f(v) \sim_n f(v')$

(iv) $\forall n\, \forall v, v' \in C$ *if* $v \leq_{n+k} v'$ *then* $f(v) \leq_n f(v')$

Proof. Clearly (i) $\Rightarrow$ (ii) and (iv) $\Rightarrow$ (i).

(ii) $\Rightarrow$ (iii). We prove by induction on n that

$$\forall v, v' \in C \;\; \textit{if} \;\; v \sim_{n+k} v' \;\; \textit{then} \;\; f(v) \sim_n f(v') \qquad (*)_\mathrm{n}$$

For $n = 0$, $(*)_\mathrm{n}$ is just (ii). Suppose that $(*)_\mathrm{n}$ holds for some n. Let $v \in C(P)$, $v' \in C(P')$ and $v \sim_{n+k+1} v'$. We shall prove that $f(v) \leq_{n+1} f(v')$. The other inequality is similar. To this aim we show that

$$\forall p \in P\, \exists p' \in P' \;\; f(v)_p \sim_n f(v')_{p'}$$

Fix $p \in P$. Since $v \sim_{n+k+1} v'$, there is $p' \in P'$ such that $v_p \sim_{n+k} v'_{p'}$. Using the inductive assumption and the naturality of f, we obtain:

$$f(v)_p = f(v_p) \sim_n f(v'_{p'}) = f(v')_{p'}$$

(iii) $\Rightarrow$ (iv). Assume (iii). Fix n and $v, v' \in C$, for $v : P \to L$, $v' : P' \to L$, such that $v \leq_{n+k} v'$. By Lemma 4.6, $v \triangleleft v' \sim_{n+k} v'$ and $(v \triangleleft v') \circ i_P = v$. Hence, by (iii), $f(v') \sim_n f(v \triangleleft v')$

and

$$f(v) = f((v \triangleleft v') \circ i_P) = f(v \triangleleft v') \circ i_P.$$

Therefore $f(v) \leq_\infty f(v \triangleleft v')$ and so $f(v) \leq_n f(v')$. $\square$

We define the following functor $\mathbf{T_H}$:

$$\begin{array}{ccc} \mathbf{M_H} & \xrightarrow{\mathbf{T_H}} & HA^{op}_{fp} \\ C & & \mathcal{S}(C) \\ f \downarrow & \longmapsto & \uparrow f^{-1} = \mathbf{T_H}(f) \\ C' & & \mathcal{S}(C') \end{array}$$

where $\mathcal{S}(C)$ is as in Corollary 4.10 and f^{-1} is the inverse image function.

LEMMA 4.12 (i) $\mathbf{T_H}$ *is a well defined functor.*

(ii) $\mathbf{T_H}$ *is essentially surjective.*

Proof. By Corollary 4.10, $\mathcal{S}(C)$ is a finitely presented Heyting algebra and every Heyting algebra is isomorphic to one of that form. Hence $\mathbf{T_H}$ is well defined on objects and essentially surjective. Clearly $\mathbf{T_H}$ preserves compositions and identities. We need to show that for any subset D of C' with an index, $f^{-1}(D)$ has an index and that f^{-1} is a Heyting algebra morphism. The latter follows from the fact that $f^{-1} : Sub(C') \longrightarrow Sub(C)$ is a Heyting algebra morphism and that $\mathcal{S}(C), \mathcal{S}(C')$ are sub-Heyting algebras of $Sub(C), Sub(C')$, respectively.

Let n be an index of D and m an index of f. We shall show that $f^{-1}(D)$ has index $n+m$. Let $v \in f^{-1}(D)$, $v' \in C$ and $v \sim_{n+m} v'$. By Lemma 4.11, $f(v) \sim_n f(v')$. Since $f(v) \in D$ and D is $\sim_n$-closed, it follows that $f(v') \in D$ and then $v' \in f^{-1}(D)$. □

Let $C^L \in \mathbf{M_H}$ and let $v : P \to L$ be an L-evaluation such that $v \in C^L$; we define a Heyting algebras morphism

$$ev_v : \mathcal{S}(C^L) \longrightarrow \mathcal{D}(P)$$

by putting, for $X \in \mathcal{S}(C^L)$,

$$ev_v(X) = \{p \in P : v_p \in X\}.$$

Conversely, given a Heyting algebra morphism, $\alpha : \mathcal{S}(C^L) \longrightarrow \mathcal{D}(P)$, we define an L-evaluation

$$\overline{\alpha} : P \to L$$

as the dual of the distributive lattice morphism

$$\mathcal{D}(L) \xrightarrow{\iota_L} \mathcal{S}(h_L) \xrightarrow{q_{C^L}} \mathcal{S}(C^L) \xrightarrow{\alpha} \mathcal{D}(P)$$

where $q_{C^L}(X) = X \cap C^L$, for $X \in \mathcal{S}(h_L)$. Taking in mind that the dual $\mathcal{D}(\overline{\alpha})$ of $\overline{\alpha}$ is the inverse image function $(\overline{\alpha})^{-1}$, we have $(\overline{\alpha})^{-1} = \alpha \circ q_{C^L} \circ \iota_L$.

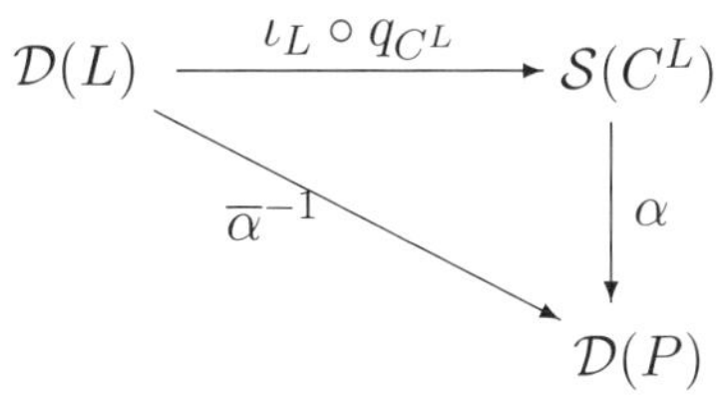

The following proposition shows that C^L is the sheaf of Heyting algebra morphisms from $\mathcal{S}(C^L)$ to all finite Heyting algebras.

PROPOSITION 4.13 *For L, C^L, and P as above, there is a bijective correspondence between the Heyting algebra morphisms $\alpha : \mathcal{S}(C^L) \to \mathcal{D}(P)$ and the L-evaluations $v : P \to L \in C^L$, given by the mutually inverse maps*

$$\alpha \mapsto \overline{\alpha} \qquad v \mapsto ev_v.$$

The correspondence is natural in P.

Proof. First we show the correspondence for the special case $C^L = h_L$. For $p \in P$ and $d \in \mathcal{D}(P)$, we have

$$\begin{array}{c} \overline{ev_v}(p) \in d \\ \hline p \in (\overline{ev_v})^{-1}(d) \\ \hline p \in ev_v \circ \iota_L(d) \\ \hline v_p \in \iota_L(d) \\ \hline v(p) \in d \end{array}$$

i.e. $v = \overline{ev_v}$, and

$$\begin{array}{c} p \in ev_{\overline{\alpha}} \circ \iota_L(d) \\ \hline \overline{\alpha}_p \in \iota_L(d) \\ \hline \overline{\alpha}(p) \in d \\ \hline p \in (\overline{\alpha})^{-1}(d) \\ \hline p \in \alpha \circ \iota_L(d) \end{array}$$

hence, by Lemma 4.8, $\alpha = ev_{\overline{\alpha}}$.

To finish the proof, let us note the following bijective correspondences

$$\begin{array}{c} \alpha : \mathcal{S}(C^L) \longrightarrow \mathcal{D}(P) \\ \hline \alpha' : \mathcal{S}(h_L) \longrightarrow \mathcal{D}(P) \text{ and } \alpha'(C^L) = P \\ \hline v : P \to L \in h_L \text{ and } ev_v(C^L) = P \\ \hline v : P \to L \in C^L \end{array}$$

where α and α' are Heyting algebra morphisms and the second equivalence follows from the above special case.

The naturality of this correspondence means that for any morphism $f : Q \rightarrow P$ in $\mathbf{P_0}$ we have $\overline{\alpha} \circ f = \overline{f^{-1} \circ \alpha}$ and $ev_{u \circ f} = f^{-1} \circ ev_u$. The verifications of these equations are left to the reader. □

Let C^L, D^M be objects of $\mathbf{M_H}$, $\mu : \mathcal{S}(C^L) \rightarrow \mathcal{S}(D^M)$ be a morphism of Heyting algebras. We define, for $P \in \mathbf{P_0}$, a function

$$\mu_P^* : D^M(P) \longrightarrow C^L(P)$$

such that, for $u \in D^M(P)$

$$\mu_P^*(u) = \overline{ev_u \circ \mu}.$$

Note that by the above Proposition, it is immediate that $\mu_P^*(u) \in C^L(P)$ and that $\mu^* : D^M \rightarrow C^L$ is a natural transformation. Moreover

PROPOSITION 4.14 *With the notation as above, we have*

(i) $ev_{\mu^*(u)} = ev_u \circ \mu$ *for* $u \in D^M$.

(ii) *For* $X \in \mathcal{S}(C^L)$, *and* $u \in D^M$, *we have* $\mu^*(u) \in X$ *iff* $u \in \mu(X)$.

(iii) $\mu^* : D^M \rightarrow C^L$ *is a morphism in* $\mathbf{M_H}$.

(iv) $\mu = (\mu^*)^{-1}$.

(v) *For any morphism* $f : D^M \longrightarrow C^L$ *in* $\mathbf{M_H}$ *we have* $f = (f^{-1})^*$.

Proof. Ad (i). We have $ev_{\mu^*(u)} = ev_{\overline{ev_u \circ \mu}} = ev_u \circ \mu$ by the bijective correspondence of Proposition 4.13.

Ad (ii). By (i), for $X \in \mathcal{S}(C^L)$, and $u : P \rightarrow M \in D^M$ $ev_{\mu^*(u)}(X) = ev_u \circ \mu(X)$. Moreover $ev_{\mu^*(u)}(X) = P$ iff $\mu^*(u) \in X$ and $ev_u \circ \mu(X) = P$ iff $u \in \mu(X)$. Hence $\mu^*(u) \in X$ iff $u \in \mu(X)$.

Ad (iii). We need to show that the transformation μ^* has an index. Let $n \in \omega$ be the maximum of the indices of sets of $\mu(X)$ where X is of the kind $C^L \cap \iota_L(d)$ for some $d \in \mathcal{D}(L)$. Notice that there are only finitely many such X's, moreover for $w, w' \in C^L$ we have $w \sim_0 w'$ iff w, w' belong to the same such X's. Then for any $u, v \in D^M$, such that $u \sim_n v$ and for $X = C^L \cap \iota_L(d)$, we have

$$\frac{\frac{\frac{\mu^*(v) \in X}{v \in \mu(X)}}{u \in \mu(X)}}{\mu^*(u) \in X}$$

Thus $\mu^*(v) \sim_0 \mu^*(u)$ and μ^* has index n.

Ad (iv). For any $X \in \mathcal{S}(C^L)$ and $v \in D^M$ we have

$$\frac{\dfrac{v \in (\mu^*)^{-1}(X)}{\mu^*(v) \in X}}{v \in \mu(X)}$$

i.e. $\mu = (\mu^*)^{-1}$.

Ad (v). Let $v : P \to M \in D^M$, $p \in P$ and $d \in \mathcal{D}(L)$. Then, we have

$$\begin{array}{c} (f^{-1})^*(v)(p) \in d \\ \hline (f^{-1})^*(v_p) \in \iota_L(d) \\ \hline v_p \in (f^{-1})(\iota_L(d)) \\ \hline f(v_p) \in \iota_L(d) \\ \hline f(v)(p) \in d \end{array}$$

Since v, p and d were arbitrary $f = (f^{-1})^*$. □

Thus we have

THEOREM 4.15 (DUALITY THEOREM) *The functor* $\mathbf{T_H} : \mathbf{M_H} \longrightarrow HA^{op}_{fp}$ *is an equivalence of categories.*

Proof. Lemma 4.12 shows that $\mathbf{T_H}$ is a functor which is essentially surjective and by Lemma 4.14(iv-v) $\mathbf{T_H}$ is full and faithful, i.e. $\mathbf{T_H}$ is an equivalence of categories. □

We can define a functor

$$\mathbf{\Phi_H} : HA^{op}_{fp} \longrightarrow Sh(\mathbf{P_0}, \mathbf{J_0})$$

sending a finitely presented Heyting algebra H to a sheaf

$$\mathbf{\Phi_H}(H) = [P \mapsto HA(H, \mathcal{D}(P))]$$

(i.e. $\mathbf{\Phi_H}(H)$ associates to every finite rooted poset P the set of all Heyting morphisms from H to the Heyting algebra $\mathcal{D}(P)$ of downward closed subsets of P). Both $\mathbf{\Phi_H}(H)$ and $\mathbf{\Phi_H}$ act on morphisms in the obvious way, by composition.

We shall prove that the 'forgetful' functor sending $\langle L, C\rangle$ to C and acting identically on morphisms

$$\mathbf{\Psi_H} : \mathbf{M_H} \longrightarrow Sh(\mathbf{P_0}, \mathbf{J_0})$$

is isomorphic to the functor $\mathbf{\Phi_H}$ composed with the equivalence of categories $\mathbf{T_H}$.

Therefore we will be able to deduce all categorical properties of HA^{op}_{fp} and $\mathbf{\Phi_H}$ from those we will establish for $\mathbf{M_H}$ and $\mathbf{\Psi_H}$.

PROPOSITION 4.16 *The functors $\mathbf{\Psi_H}$ and $\mathbf{\Phi_H} \circ \mathbf{T_H}$ are naturally isomorphic.*

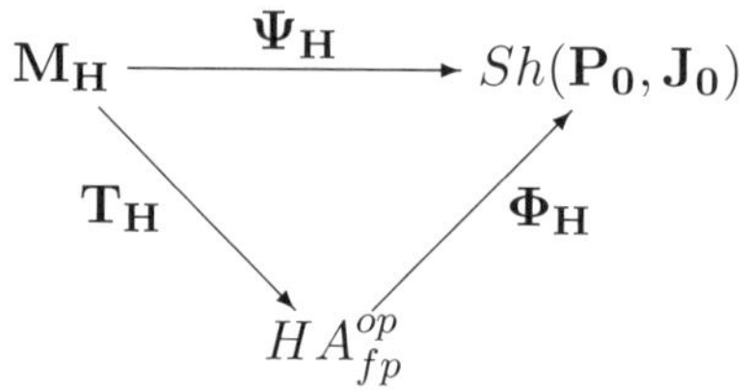

Proof. Let C^L be in $\mathbf{M_H}$, P in $\mathbf{P_0}$. Then, we have

$$\mathbf{\Psi_H}(C^L)(P) = C^L(P) \cong HA(\mathcal{S}(C^L), \mathcal{D}(P)) = (\mathbf{\Phi_H} \circ \mathbf{T_H})(C^L)(P)$$

(the natural isomorphism in the middle comes from Proposition 4.13). □

4. A combinatorial result

This Section contains the main technical argument of the Chapter which is needed in order to solve a combinatorial problem about our games. The solution to this problem will give us as an immediate consequence that $\mathbf{M_H}$ *is closed under the sheaf operation of taking images*. From this, it will be easy in the next Section to show that $\mathbf{M_H}$ is a Heyting category.

For $n > 0$ and $u : P \to L$, the *n-rank of u*, written $rk_n(u)$, is the cardinality of the finite set $Type_{n-1}(u)$ introduced on page 75. A point $p \in P$ is said to have *full n-rank* iff $rk_n(u) = rk_n(u_p)$. An obvious property of this notion is the following:

LEMMA 4.17 *Let $u : P \to L$ and $p, q \in P$ be given. If p has full n-rank, then there exists $p' \leq p$ such that $u_{p'} \sim_{n-1} u_q$. Consequently, if two points $p_1, p_2 \in P$ both have full n-rank, then $u_{p_1} \sim_n u_{p_2}$.*

Proof. Clearly $\{[u_{p'}]_{n-1} : p' \leq p\} \subseteq \{[u_{p'}]_{n-1} : p' \in P\}$ and the assumption that p has full n-rank means that these two finite sets have the same number of elements, so the claim follows immediately. □

LEMMA 4.18 *Let $f : C^L \to D^M$ be a morphism in $\mathbf{M_H}$ and consider $n > 0$ big enough to be an index both for C and f. Take $u : P \to L$ in C_P and $v : Q \to M$ in D_Q such that $v \sim_{2r-1} f(u)$, where r is the n-rank of u. Then there exist $u' : P' \to L$ such that $u' \sim_n u$ (hence $u' \in C$) and an open surjection $h : P' \to Q$ such that the triangle*

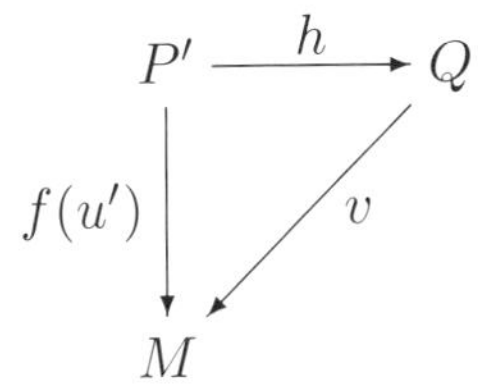

commutes.

Proof. P' is built as a subset of the cartesian product $P \times Q$ (the partial order on P' is the restriction of the product partial order). For a pair $\langle p_0, q_1 \rangle$ to be in P' a 'justification' is needed. The justification is another pair $\langle p_1, q_0 \rangle \in P \times Q$ that satisfies the following requirements: 1) $p_1 \leq_P p_0$ and $q_1 \leq_Q q_0$; 2) u_{p_0} and u_{p_1} have the same n-rank (let s be this rank); 3) $v_{q_0} \sim_{2s-1} f(u_{p_0})$ and $v_{q_1} \sim_{2s-2} f(u_{p_1})$. Let us repeat this key definition in a formal way:

$$\begin{aligned} P' &= \{\langle p_0, q_1 \rangle \in P \times Q \, : \, \exists p_1 \in P, \exists q_0 \in Q, \exists s \\ &\quad (p_1 \leq p_0 \;\&\; q_1 \leq q_0 \;\&\; rk_n(u_{p_0}) = s = rk_n(u_{p_1}) \;\& \\ &\quad \& \; v_{q_0} \sim_{2s-1} f(u_{p_0}) \;\&\; v_{q_1} \sim_{2s-2} f(u_{p_1}))\}. \end{aligned}$$

P' is a finite rooted poset: the root is clearly $\langle \rho(P), \rho(Q) \rangle$, because $f(u) \sim_{2r-1} v$ is our hypothesis (the p_1, q_0 needed for a justification are again $\rho(P), \rho(Q)$, respectively). We put $u' = u \circ \pi_0$, where π_0 is the first projection, restricted in its domain.

We show that for every $\langle p_0, q_1 \rangle \in P'$ we have that

$$(*) \qquad u'_{\langle p_0, q_1 \rangle} \sim_n u_{p_0}$$

(notice that, by taking the roots, $(*)$ implies that $u' \sim_n u$, which is the first part of the claim). We show by induction on $k = 0, \ldots, n$ that if $\langle p_0, q_1 \rangle \in P'$, then $u'_{\langle p_0, q_1 \rangle} \sim_k u_{p_0}$. For $k = 0$, this follows immediately from the definition of u'. For $k > 0$, let s be the rank of p_0 and let us fix a pair p_1, q_0 that justifies the membership of $\langle p_0, q_1 \rangle$ in P'. We define a k-strategy for player II: there are two cases.

Case (i) If player I plays in P' a point $\langle p', q' \rangle \leq \langle p_0, q_1 \rangle$, then player II answers p' and has a $(k-1)$-strategy by induction hypothesis.

Case (ii) Suppose that player I plays $p'_0 \leq p_0$ in P. As $p_1 \leq p_0$ and $rk_n(p_1) = rk_n(p_0)$, it follows by Lemma 4.17 applied to u_{p_0}, that there exists $p_2 \leq p_1$ such that $u_{p'_0} \sim_{n-1} u_{p_2}$. We distinguish two subcases.

Subcase (ii1) If $rk_n(p_2) = s$, then $u_{p_0} \sim_n u_{p_2}$ by Lemma 4.17 again. The induction hypothesis tells us that $u_{p_0} \sim_{k-1} u'_{\langle p_0, q_1 \rangle}$. Putting all these data together, we get

$$u_{p'_0} \sim_{n-1} u_{p_2} \sim_n u_{p_0} \sim_{k-1} u'_{\langle p_0, q_1 \rangle},$$

hence player II wins by answering $\langle p_0, q_1 \rangle$ (recall that $k - 1 \leq n - 1$).

Subcase (ii2) If $rk_n(p_2) < s$, then from $f(u_{p_1}) \sim_{2s-2} v_{q_1}$ it follows that there exists $q_2 \leq q_1$ such that $f(u_{p_2}) \sim_{2s-3} v_{q_2}$, hence $\langle p_2, q_2 \rangle \in P'$ (the justification is the pair p_2, q_2 itself, because now equivalence up to $2(s-1) - 1 = 2s - 3$ is enough). Player II wins by answering $\langle p_2, q_2 \rangle$ by induction hypothesis and because $u_{p'_0} \sim_{n-1} u_{p_2}$.

Let $h = \pi_1$ (i.e. h is the second projection, restricted in its domain). We claim that h *is open* (hence also surjective because it maps the root of P' onto the root of Q). In fact let $\langle p_0, q_1 \rangle \in P'$ and let $q_2 \leq q_1$. We look for $\langle p'_0, q_2 \rangle \in P'$ such that $p'_0 \leq p_0$. Let p_1, q_0 be a pair that justifies the membership of $\langle p_0, q_1 \rangle$ in P' and let s be the rank of p_0, p_1. As $f(u_{p_0}) \sim_{2s-1} v_{q_0}$, there exists $p_2 \leq p_0$ such that $v_{q_2} \sim_{2s-2} f(u)_{p_2} = f(u_{p_2})$. If p_2 has again rank equal to s, then $\langle p_0, q_2 \rangle \in P'$ (the justification is now the pair p_2, q_0). If p_2 has rank less than s, then $\langle p_2, q_2 \rangle \in P'$: the justification is the pair p_2, q_2 itself (in fact, $v_{q_2} \sim_{2s-2} f(u_{p_2})$ is more than what we need, equivalence up to $2(s-1) - 1 = 2s - 3$ would be sufficient).

We claim that the triangle

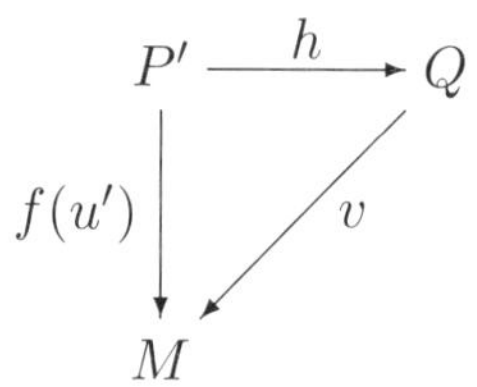

commutes. As n is an index for f, for every $\langle p_0, q_1 \rangle \in P'$ it follows from $(*)$ that $f(u'_{\langle p_0, q_1 \rangle}) \sim_0 f(u_{p_0})$. Let p_1, q_0 be a pair that justifies the membership of $\langle p_0, q_1 \rangle$ in P'. As $p_1 \leq p_0$ and $rk_n(u_{p_0}) = rk_n(u_{p_1})$, it follows by Lemma 4.17 applied to the evaluation u_{p_0} that $u_{p_0} \sim_n u_{p_1}$, hence $f(u_{p_0}) \sim_0 f(u_{p_1})$; moreover $f(u_{p_1}) \sim_0 v_{q_1}$ (we have $v_{q_1} \sim_{2s-2} f(u_{p_1})$ from the definition of P'). We conclude that $f(u')_{\langle p_0, q_1 \rangle} \sim_0 v_{q_1}$, i.e. the claim. $\square$

PROPOSITION 4.19 *For every finite poset L and $n > 0$, there is a number $G(n, L)$, such that the following holds. Let $f : C^L \to D^M$ be a morphism in $\mathbf{M_H}$ so that n is big enough to be an index both for C and f; take $u : P \to L$ in C and $v : Q \to M$ in D, such that $v \sim_{G(n,L)} f(u)$; then there exists $u' : P' \to L$ such that $u' \sim_n u$ (hence $u' \in C$) and an open surjective map $h : P' \to Q$ such that $v \circ h = f(u')$.*

Proof. It is sufficient to put $G(n, L) = 2r - 1$, where r is the maximum n-rank that an L-evaluation can have (recall that there are in general only finitely many $(n-1)$-equivalence classes of L-evaluations). The claim follows immediately from Lemma 4.18. □

COROLLARY 4.20 *Suppose that $f : C^L \to D^M$ is a morphism in $\mathbf{M_H}$ and that $C' \subseteq C$ has an index. Then the sheaf-theoretic image $\exists_f(C')$ of C' and the sheaf-theoretic dual image $\forall_f(C')$ of C' both have an index.*

Proof. We recall that $\exists_f(C')$ is computed locally, i.e.

$$v \in \exists_f(C')_Q \quad \text{iff} \quad \exists P,\ \exists u \in C'_P\ \exists h : P \to Q \text{ s.t.}$$
$$h \text{ is an open surjection and } v \circ h = f(u).$$

We show that $\exists_f(C')$ has index $G(n, L)$ if $n > 0$ is an index for both C' and f. In fact, if $v \in \exists_f(C')_Q$ and $v' \sim_{G(n,L)} v$, then there exist P, $u \in C'_P$ and $h : P \to Q$ such that h is open and surjective and $v \circ h = f(u)$. As $v \sim_\infty v \circ h$ this implies that $v' \sim_{G(n,L)} f(u)$, hence by Proposition 4.19, there exist u' and h' open and surjective such that $u' \sim_n u$ and $v' \circ h' = f(u')$. As C' has index n, $u' \in C'$ and consequently $v' \in \exists_f(C')$.

We recall that $\forall_f(C')$ is computed universally, i.e.

$$v \in \forall_f(C')_Q \quad \text{iff}$$
$$\forall P,\ \forall u \in C_P\ \forall \text{ open } h : P \to Q\ (v \circ h = f(u)\ \Rightarrow u \in C')$$

We show that $\forall_f(C')$ has index $G(n, L)+1$, if $n > 0$ is an index for C, C', and f. In fact, take $v \in \forall_f(C')_Q$, $v' \sim_{G(n,L)+1} v$, $u' \in C_{P'}$ and $h' : P' \to Q'$ open (Q' is the domain of v') such that $f(u') = v' \circ h'$. We show that $u' \in C'$. As $v' \circ h'$ is infinitely equivalent to $v'_{h'(\rho)}$, it is $G(n, L)$-equivalent to v_q for some $q \in Q$. Hence, $v_q \sim_{G(n,L)} f(u')$ and we can apply Proposition 4.19. We get $u \sim_n u'$ and h open surjective such that $v_q \circ h = f(u)$. Then $u \in C'$ because $v \in \forall_f(C')$. As n is an index for C' too, $u' \in C'$ as required. □

5. Properties of $\mathbf{M_H}$ and $\mathbf{\Psi_H}$

In this section we analyze the properties of the category $\mathbf{M_H}$ and of the embedding functor $\mathbf{\Psi_H} : \mathbf{M_H} \longrightarrow Sh(\mathbf{P_0}, \mathbf{J_0})$.

PROPOSITION 4.21 *$\mathbf{M_H}$ has finite limits and $\mathbf{\Psi_H}$ preserves them.*

Proof. The terminal object is clearly h_L, where L is the one-point poset. The product of $C_1^{L_1}$ and $C_2^{L_2}$ is $(C_1 \times C_2)^{L_1 \times L_2}$: it has as an index any n which is an index for both $C_1^{L_1}$ and $C_2^{L_2}$ (this is due to the fact that for every $n \geq 0$, for every $\langle v_1, v_2 \rangle : P \to L_1 \times L_2$, $\langle v_1', v_2' \rangle : P' \to L_1 \times L_2$, $\langle v_1, v_2 \rangle \sim_n \langle v_1', v_2' \rangle$ implies $v_1 \sim_n v_1'$ and $v_2 \sim_n v_2'$). The projection maps are the obvious ones (0 is an index for them) and the unique maps required by the universal property of product have the obvious index. Thus Ψ preserves products.

Let us now consider a pair of parallel arrows $f_1 : C_1^{L_1} \to C_2^{L_2}$ and $f_2 : C_1^{L_1} \to C_2^{L_2}$. We show that their equalizer is E^{L_1}, where E is the equalizer of f_1, f_2 in $Sh(\mathbf{P_0}, \mathbf{J_0})$. If n is an index for C_1, f_1, f_2, then $n+1$ is an index for E. In fact suppose that $v : Q \to L_1 \in E_Q$ and that $u : P \to L_1$ is such that $u \sim_{n+1} v$. We want to prove that $f_1(u) = f_2(u)$, i.e. that for every $p \in P$, $f_1(u_p) \sim_0 f_2(u_p)$. Take any $p \in P$: we have that $u_p \sim_n v_q$ for some $q \in Q$. Hence $v_q \in C_1$ and $f_1(u_p) \sim_0 f_1(v_q) = f_2(v_q) \sim_0 f_2(u_p)$, thus showing the claim. Finally, it is trivial to give an index to the unique maps required by the universal property of equalizers. □

By Proposition 4.21, an arrow $f : C^L \to D^M$ is monic in $\mathbf{M_H}$ iff it is pointwise injective. Our next step consists in proving that the inclusion functor $\mathbf{\Psi_H}$ is conservative, in the sense that for every monic $m : C_1^{L_1} \to C_2^{L_2}$ in $\mathbf{M_H}$, if $\mathbf{\Psi_H}(m)$ is iso then m is also iso. Conservativity of $\mathbf{\Psi_H}$ corresponds to a kind of conceptual completeness: it says that if m is pointwise bijective and has an index, then m^{-1} has an index too. This means that if a morphism between finitely presented Heyting algebras induces an isomorphism between their finite models, then it is an isomorphism as well.

PROPOSITION 4.22 *The functor* $\mathbf{\Psi_H} : \mathbf{M_H} \longrightarrow Sh(\mathbf{P_0}, \mathbf{J_0})$ *is conservative.*

Proof. Assume that $f : C^L \to D^M$ is an arrow in $\mathbf{M_H}$ which is a pointwise bijective natural transformation. We must find an index for f^{-1}. For all $u \in C$, the image along f of the subsheaf $(\downarrow_0 u) \cap C$ (which, we recall, at P contains the $u' : P \longrightarrow L$ such that $u' \in C_P$ and $u' \leq_0 u$) is the subsheaf $\exists_f(C \cap \downarrow_0 u)$ given at P by $\exists_f(C \cap \downarrow_0 u)_P = \{f(u') : u' \in C_P \ \&\ u' \leq_0 u\}$ (this is because f is monic and local closure is not needed). It follows from Corollary 4.20, that $\exists_f(C \cap \downarrow_0 u)$ has an index N. N depends on the indices of f and $C \cap \downarrow_0 u$. The latter always has index equal to the index of C, hence N does not depend on u. We show that N is an index for f^{-1}. In fact, suppose that $v_1, v_2 \in D$ and $v_1 \leq_N v_2$. Then, because f is onto, $v_2 = f(u_2)$ for some $u_2 \in C$. As $\exists_f(C \cap \downarrow_0 u_2)$ has index N, $v_1 \in \exists_f(C \cap \downarrow_0 u_2)$ and hence there exists $u_1 \in C$ such that $u_1 \leq_0 u_2$ and $f(u_1) = v_1$. This shows that $f^{-1}(v_1) = u_1 \leq_0 u_2 = f^{-1}(v_2)$. □

Conservativity of $\mathbf{\Psi_H}$ has important consequences, which will be used below. We say that a monic $f : C^L \longrightarrow D^M$ in $\mathbf{M_H}$ is a canonical monic iff $L = M$ and f is an inclusion. Canonical monics are exactly regular monics: in fact, we saw above that computation of equalizers gives canonical monics and, on the other hand, every canonical monic is easily seen to be an equalizer.[5] By conservativity, *every monic is isomorphic to a canonical monic in* $\mathbf{M_H}$: in fact take $\exists_f(C) \subseteq D$ (which has an index because of Corollary 4.4), then observe that f restricted in its codomain is bijective (recall that we do not need to take local closure of the image because f is injective) and has the same index as f, so that, by conservativity, it is an isomorphism of $f : C^L \longrightarrow D^M$ onto the canonical monic $(\exists_f C)^M \subseteq D^M$. The use of canonical monics will be a constant practice in this Section. We summarize the content of these remarks in the following Proposition (which is nothing but Beth's definability theorem for intuitionistic propositional logic, cf. Proposition 2.14):

PROPOSITION 4.23 *Monics are all regular in* $\mathbf{M_H}$. □

G. Kreisel in [K] gave a simple syntactic argument proving the above proposition for any superintuitionistic logic. In exercise 12 the reader will find some hints on how to reconstruct this argument.

$\mathbf{C}$ is a Heyting category iff [MR2]: (i) it has finite limits; (ii) subobjects form lattices; (iii) the pullback functors operating on subobjects have left and right adjoints; (iv) the pullback functors operating on subobjects preserve the lattice structure and satisfy the Beck-Chevalley condition. Notice that when monics are all regular, being a Heyting category is obviously the same as being an r-Heyting category.

THEOREM 4.24 $\mathbf{M_H}$ *is a Heyting category and* $\mathbf{\Psi_H}$ *preserves the Heyting category structure.*

Proof. $Sh(\mathbf{P_0}, \mathbf{J_0})$ is a Heyting category, since it is a topos. Moreover $\mathbf{M_H}$ is closed under the operations on subobjects listed in (ii)-(iii) (see Corollary 4.20). The remaining properties follow from the fact that Ψ is conservative, hence it preserves and reflects the partial order on subobjects. □

We can apply the method of Theorem 4.24 to other categorical properties: the method consists simply in showing that $\mathbf{M_H}$ is closed under certain standard constructions of $Sh(\mathbf{P_0}, \mathbf{J_0})$ (which in practice means to 'find an index' for such constructions) and then in using the conservativity of $\mathbf{\Psi_H}$ to transfer the exactness properties of such constructions back to $\mathbf{M_H}$.

Let us consider the difference operation (the dual to the intuitionistic implication). In a lattice, this operation is uniquely characterized (if it exists) by the adjointness requirement:

$$C_1 \leq C_2 \vee C \iff (C_1 \backslash C_2) \leq C.$$

[5] $S^L \subseteq D^L$ is an equalizer of the characteristic map $\chi_S : D^L \longrightarrow \Omega$ defined in Section 4.2 and of the map $true \circ ! : D^L \longrightarrow \Omega$ (such maps clearly have an index).

As joins and meets are pointwise, all infinitary distributive laws hold in the subobject lattices of $Sh(\mathbf{P_0}, \mathbf{J_0})$ and consequently such lattices are bi-Heyting algebras (i.e. Heyting algebras with difference). We transfer this property to $\mathbf{M_H}$, by finding an index for the difference. In order to characterize directly the difference of subobjects in $Sh(\mathbf{P_0}, \mathbf{J_0})$, we extend our notion of $\leq_\infty$ by applying it to arbitrary sheaves: if X is an arbitrary sheaf and $u \in X_P, v \in X_Q$ (P, Q are rooted posets), we say that $u \leq_\infty v$ iff for all subsheaf C of X,

$$v \in C_Q \implies u \in C_P.$$

It is not difficult to see that this is the same as saying that there are $h : R \to P$ and $k : R \to Q$, both open and h also surjective, such that $h(u) = k(v)$. So we have really extended our previous notion of $\leq_\infty$ for L-evaluations (see Proposition 4.4). Notice also that $\{u : u \leq_\infty v\}$ is the minimum subsheaf containing v.

It can be easily checked that if C_1, C_2 are subsheaves of X, then

$$C_1 \backslash C_2 = \{v \in X : \exists v' \in C_1 \; (v \leq_\infty v' \;\&\; v' \notin C_2)\}.$$

A *bi-Heyting category* [MR2] is a Heyting category in which subobject lattices have difference. If the difference is preserved by pullback functors along projections, we say that *the Lawvere condition* [La] is satisfied.

THEOREM 4.25 *$\mathbf{M_H}$ is a bi-Heyting category and $\mathbf{\Psi_H}$ preserves this structure; moreover $Sh(\mathbf{P_0}, \mathbf{J_0})$ (and consequently also $\mathbf{M_H}$) satisfies the Lawvere condition.*

Proof. For the first part of the claim, we must find an index for the difference operation. Let C^L be an object in $\mathbf{M_H}$. If $C_1, C_2 \subseteq C$ are subobjects of index $n+1$, then $C_1 \backslash C_2$ has index $n+1$ too: this is due to Lemma 4.6 which ensures in particular that if $u \leq_{n+1} v \leq_\infty w$, then there exists $v' = u \lhd w$ such that $u \leq_\infty v' \sim_{n+1} w$.

For the second part of the claim, it is enough to show that in the pullback functors *along projections* preserve the difference operation. Now notice that if $f : X \to Y$ is a natural transformation satisfying the following condition, for all $v \in X, u \in Y$

$$(f(v) \leq_\infty u \Rightarrow \exists v' \in X \; (v \leq_\infty v' \;\&\; f(v') \leq_\infty u \;\&\; u \leq_\infty f(v'))) \quad (4.7)$$

then the pullback functor f^* preserves the difference operation. Suppose that f is a (first) projection, say $f : X \times Y \to X$. We show that in this case (4.7) holds. In fact for $v_1 \in X_Q, v_2 \in Y_Q, u \in X_P$, if $f(v_1, v_2) = v_1 \leq_\infty u$, then there are open maps $k : R \to P$, $h : R \to Q$ such that h is surjective and moreover $k(u) = h(v_1)$. First of all, let us graft a copy of R below the root of P, that is: put $Q' = R + P$, $q_1 \leq_{Q'} q_2$ iff ($q_1, q_2 \in R$ and $q_1 \leq_R q_2$) or ($q_1, q_2 \in P$ and $q_1 \leq_P q_2$) or ($q_1 \in R$ and $q_2 = \rho(P)$). We define two open and surjective maps $k_1 : Q' \to P$ and $k_2 : Q' \to R$. k_1 is $[1_P, k]$, i.e. acts identically on P and as k on R. k_2 is defined so: $k_2(q) = q$ if $q \in R$, $k_2(q) = \rho(R)$ if $q = \rho(P)$ and $k_2(q) = c$, if $q \in P, q \neq \rho(P)$ (c is an arbitrary fixed minimal point of R). Notice that the open inclusion $i : R \to Q'$ is a section of k_2 and that $k_1 \circ i = k$. We define $v_1', v_2' \in (X \times Y)_{Q'}$ as follows: $v_1' = k_1(u)$ and $v_2' = k_2(h(v_2))$. Clearly $f(v_1', v_2') = v_1' = k_1(u)$ is infinitely equivalent to u (in the sense that $u \leq_\infty k_1(u)$ and $k_1(u) \leq_\infty u$). To complete the proof of (4.7), we only have to show that $\langle v_1, v_2 \rangle \leq_\infty \langle v_1', v_2' \rangle$, which easily follows from $h(v_1, v_2) = i(v_1', v_2')$ (recall that $k(u) = h(v_1)$). □

PROPOSITION 4.26 $\mathbf{M_H}$ *has universal and disjoint finite sums and* $\mathbf{\Psi_H}$ *preserves them.*

Proof. An initial object in $\mathbf{M_H}$ is clearly any object of the kind $\emptyset^L$. For coproducts, first of all notice that $h_{L_1+L_2}$ is the coproduct of h_{L_1} and h_{L_2} in $Sh(\mathbf{P_0}, \mathbf{J_0})$: since any poset P in $\mathbf{P_0}$ is rooted, any order-preserving map $P \longrightarrow L_1 + L_2$ must factor through L_1 or L_2. Then if C_1 and C_2 are subsheaves of h_{L_1} and h_{L_2}, of course $C_1 + C_2$ is a subsheaf of $h_{L_1+L_2}$. If C_1, C_2 have indices, $C_1 + C_2$ has index the maximum of indices of C_1, C_2; moreover the coprojections have index 0. For every D^M and for every $f_i : C_i^{L_i} \longrightarrow D^M$ ($i = 1, 2$), the universal map $[f_1, f_2] : (C_1 + C_2)^{L_1+L_2} \longrightarrow D^M$ has index the maximum of the indices of f_1, f_2. Clearly such coproducts are preserved by $\mathbf{\Psi_H}$, so universality and disjointness again follow from the fact that $\mathbf{\Psi_H}$ is conservative. □

We shall see in Section 4.8 that $\mathbf{M_H}$ *fails to be a pretopos*, because equivalence relations are not effective. It is worth recalling however that $\mathbf{M_H}$ has a subobject classifier Ω, the same as that of $Sh(\mathbf{P_0}, \mathbf{J_0})$ ($\mathbf{\Psi_H}$ preserves it), namely h_2 which is the dual of the free Heyting algebra on one generator. This subobject classifier cannot be exponentiable: this follows from the results of Section 4.8, but can be easily checked now. Global sections of Ω^Ω (supposing it exists) should be in bijective correspondence with subobjects of Ω (which are the elements of the free Heyting algebra on one generator), hence there should be infinitely many of them. However, in general, an object C^L of $\mathbf{M_H}$ cannot have more global sections than L has elements; and L is finite.

6. Second order extensions of IpC

We give in this Section the relevant hints to understand the logical applications to second order intuitionistic propositional logic of the results we reached so far.

The formulas φ of IpC can be identified with terms of the theory of Heyting algebras, they can be described by the following grammar

$$\varphi = p \mid \perp \mid \top \mid \varphi \wedge \varphi' \mid \varphi \vee \varphi' \mid \varphi \rightarrow \varphi'$$

where p is running over a set of propositional variables. Then the logic IpC can be introduced either syntactically by a proof system giving rise to the notion of provability relation $\vdash_{IpC}$, or semantically by a forcing relation $\models$ of formulas of IpC in some kind of models, giving rise to the notion of semantical consequence. We assume that $\vdash_{IpC}$ is given by a sequent calculus such as [Dr], and the *forcing relation* $\models$ is given by the Kripke semantics in finite Kripke models. The latter is specified as follows: given a formula $\varphi(\vec{p})$, given a $\vec{p}$-Kripke model, i.e. given an L-evaluation (where $L = (\mathcal{P}(\vec{p}), \supseteq)$) $v : P \rightarrow L$, and given $x \in P$, we define $v, x \models \varphi(\vec{p})$, by the following inductive clauses:

(i) $v, x \models p$ iff $p \in v(x)$;

(ii) not $v, x \models \perp$;

(iii) $v, x \models \top$;

(iv) $v, x \models \alpha(\vec{p}) \wedge \beta(\vec{p})$ iff $v, x \models \alpha(\vec{p})$ and $v, x \models \beta(\vec{p})$;

(v) $v, x \models \alpha(\vec{p}) \vee \beta(\vec{p})$ iff $v, x \models \alpha(\vec{p})$ or $v, x \models \beta(\vec{p})$;

(vi) $v, x \models \alpha(\vec{p}) \rightarrow \beta(\vec{p})$ iff
for all $y \in P$ such that $x \geq y$, we have that $v, y \models \alpha(\vec{p})$ implies $v, y \models \beta(\vec{p})$.

We put $v \models \alpha(\vec{p})$ iff $v, \rho \models \alpha(\vec{p})$.

IpC, introduced in this way, has two natural extensions to a second order logic, i.e. to a logic which allows quantification over propositional variables. The (usual) Second Order Intuititionistic Logic IpC^2 is obtained from IpC syntactically, by adding proof rules for quantifiers. The Second Order Intuitionistic Logic IpC_2 with sheaf semantics is obtained from IpC semantically, by adding new clauses for satisfaction of quantified statements in Kripke models. We call the latter quantifiers *Pitts' quantifiers* to honor A.M. Pitts who discovered them syntactically in [Pi2].

To keep the distinction explicit, we will denote the quantifiers in the former logic in the usual way by $\exists_p$ and $\forall_p$, and in the latter by $\exists^p$ and $\forall^p$, i.e. with quantified variables in superscript. The formulas of IpC^2 are given by the following grammar extending the one for IpC

$$\varphi = p \mid \perp \mid \top \mid \varphi \wedge \varphi' \mid \varphi \vee \varphi' \mid \varphi \rightarrow \varphi' \mid \exists_p \varphi \mid \forall_p \varphi$$

In the sequent calculus for IpC^2, we add the following rules for quantifiers

$$\frac{\Gamma \vdash \varphi(p)}{\Gamma \vdash \forall_p\, \varphi(p)}\; r\forall \qquad\qquad \frac{\Gamma,\, \varphi(t) \vdash \psi}{\Gamma,\, \forall_p\, \varphi(p) \vdash \psi}\; l\forall$$

$$\frac{\Gamma \vdash \varphi(t)}{\Gamma \vdash \exists_p\, \varphi(p)}\; r\exists \qquad\qquad \frac{\Gamma,\, \varphi(p) \vdash \psi}{\Gamma,\, \exists_p\, \varphi(p) \vdash \psi}\; l\exists$$

where, in $r\,\forall$ we assume that p is not free in Γ, and in $l\,\exists$ we assume that p is not free in both Γ and ψ. In this way we obtain the notion of provability in IpC^2, denoted by $\vdash_{IpC^2}$.

Formulas of IpC_2 are given by the grammar

$$\varphi = p \,|\, \bot \,|\, \top \,|\, \varphi \wedge \varphi' \,|\, \varphi \vee \varphi' \,|\, \varphi \rightarrow \varphi' \,|\, \exists^p \varphi \,|\, \forall^p \varphi$$

In order to describe the additional clauses for quantified formulas we need to introduce additional notions.

Given Kripke models $u : (P, \leq) \longrightarrow (\mathcal{P}(\vec{p}), \supseteq)$, $q \notin \vec{p}$, $u' : (Q, \leq) \longrightarrow (\mathcal{P}(\vec{p}, q), \supseteq)$, and an open map $f : (Q, \leq) \rightarrow (P, \leq)$, we say, that u' is an *expansion of u via f* iff is the square

$$\begin{array}{ccc} (Q, \leq) & \xrightarrow{\;f\;} & (P, \leq) \\ {\scriptstyle u'}\downarrow & & \downarrow{\scriptstyle u} \\ \mathcal{P}(\vec{p}, q) & \xrightarrow[r_{\vec{p}}]{} & \mathcal{P}(\vec{p}) \end{array}$$

commutes, where $r_{\vec{p}}$, is the map of 'intersecting with $\vec{p}$'. Let $x \in P$. u' is an *expansion of u via f over $x \in P$* iff u' is an expansion of u via f and $f(\rho) \geq x$. u' is an *expansion of u via f under $x \in P$* iff u' is an expansion of u via f and $f(\rho) \leq x$.

Now, we can state the clauses for satisfaction of the Pitts' quantifiers

(vii) $v, x \models \exists^q \alpha(\vec{p}, q)$ iff
there is an expansion v' of v via f over x, and $y \in dom(v')$ such that, $f(y) = x$ and $v', y \models \alpha(\vec{p}, q)$;

(viii) $v, x \models \forall^q \alpha(\vec{p}, q)$ iff
for any expansions v' of v via f under x, and any $y \in dom(v')$, we have $v', y \models \alpha(\vec{p}, q)$.

If we denote by $|\varphi(\vec{p})|$ the set of all $\vec{p}$-models of the IpC_2 formula $\varphi(\vec{p})$, which is a sheaf (actually, a subsheaf of the sheaf of all $\vec{p}$-models) on the site $(\mathbf{P_0}, \mathbf{J_0})$, the above clauses say that $|\exists^q \alpha(\vec{p}, q)|$ and $|\forall^q \alpha(\vec{p}, q)|$ are exactly the sheaf image and dual image of $|\alpha(\vec{p}, q)|$ along the natural transformation given by 'composition with' $r_{\vec{p}}$. This is why we call the above semantics the *sheaf semantics* . A formula $\varphi(\vec{p})$ of IpC_2 is *valid* (notation: $\models_{IpC_2} \varphi$) iff any $\vec{p}$-Kripke model satisfies it.

By Theorem 4.24, the sheaf image and dual image of objects in $\mathbf{M_H}$ are in $\mathbf{M_H}$, so by Duality Theorem 4.15, the formulas $\exists^q \alpha(\vec{p}, q)$ and $\forall^q \alpha(\vec{p}, q)$ are equivalent in IpC_2 to some quantifier-free formulas in the variables $\vec{p}$. For this reason we can treat formulas IpC_2 as shorthands for the equivalent (in IpC_2) formulas of IpC. The existence and effective computations of such quantifier-free formulas was first shown in [Pi2]. Thus, in fact, IpC_2 is a definitional extension of IpC. However, we have an interpretation $\mathcal{I}$ of IpC^2 in IpC_2 which sends propositional variables to the same propositional variables and commutes with all propositional constants, connectives, and quantifiers, i.e. for a formula φ of IpC^2, $\mathcal{I}(\varphi)$ is just φ 'with quantified variables moved from subscript to the superscript'. As it was shown in [Pi2] (see also [Gh3], Section 6 for a detailed analysis of the modal case), for any formula ψ in IpC^2, we have

$$\vdash_{IpC^2} \psi \text{ implies } \models_{IpC_2} \mathcal{I}(\psi)$$

In order to see this it is enough to note that the rules of IpC^2 are valid in our semantics.

7. Quantifier-elimination

In this Section, we analyze in a more explicit way the model-theoretic applications of our results.

Let T_H be the first order theory of Heyting algebras. By Theorems 3.8, 4.24 and Proposition 4.23, we know that T_H has a model completion T_H^*, which can be axiomatized by the following infinitely many formulas, involving Heyting algebra terms $t(x, \vec{y}), t_1(x, \vec{y}), \ldots, t_n(x, \vec{y})$ and Pitts's quantifiers $\forall^x$, $\exists^x$:

$$\exists x \; (t = 1 \;\&\; t_1 \neq 1 \;\& \; \cdots \;\& \; t_n \neq 1) \;\; \Leftrightarrow \tag{4.8}$$

$$(\exists^x t) = 1 \;\&\; (\forall^x (t \to t_1)) \neq 1 \;\& \cdots \&\; (\forall^x (t \to t_n)) \neq 1.$$

These formulas directly and effectively suggest how to eliminate quantifiers from an arbitrary sentence. Since quantifier-free sentences are Boolean combinations of $1 = 0$, we immediately get that:

THEOREM 4.27 *$T_H^* \cup \{1 \neq 0\}$ is decidable and complete.* □

We recall from chapter 2 that models of $T_H^* \cup \{1 \neq 0\}$ are existentially closed non-degenerate Heyting algebras. By the above Theorem, we can get in principle any 'first order' information about such algebras we would like to have.

In practice, the decision procedure is hard to run because of the difficulties in computing the quantifier free formulas equivalent to the formulas in IpC_2. Detailed, but laborious, instructions for this computation are given in [Pi2].

We give an example of the use of the sheaf semantic introduced in the previous Section to simplify the procedure of elimination of quantifiers. Let us show that in any existentially closed Heyting algebra, for every x, z we have that

$$\exists x_1 \exists x_2 (x_1 \wedge x_2 = 0 \ \&\ x = x_1 \vee x_2 \ \&\ x \vee (x_1 \to z) = x \vee (x_2 \to z)).$$

According to (4.8), this amounts to show that the formula

$$\exists^{x_1} \exists^{x_2} \left(\neg(x_1 \wedge x_2) \wedge (x \leftrightarrow (x_1 \vee x_2)) \right.$$
$$\left. \wedge ((x \vee (x_1 \to z)) \leftrightarrow (x \vee (x_2 \to z))) \right)$$

is equal to 1, i.e. that any (x, z)-Kripke model $u : (X, \leq) \longrightarrow (\mathcal{P}(x, z), \subseteq)$ whatsoever admits an expansion (over its root) to a model of the formulas

$$\neg(x_1 \wedge x_2),\ x \leftrightarrow (x_1 \vee x_2),\ (x \vee (x_1 \to z)) \leftrightarrow (x \vee (x_2 \to z)). \quad (4.9)$$

The expansion is obtained by 'duplicating' the points forcing x. More precisely, if X' is the set of such points, we define Y to be $(X \backslash X') \cup (X' \times \{1, 2\})$. Let $f : X' \times \{1, 2\} \longrightarrow X$ be the restricted first projection and let us extend it to Y in the domain by putting $f(a) = a$ for $a \in X \backslash X'$. The partial order in Y is so defined: we have that $a \leq b$ holds in Y iff (i) either $a \in X \backslash X'$ and $a \leq f(b)$ holds in $(X, \leq)$ or (ii) $a = (a', i)$ and $b = (b', i)$ $(i = 1, 2)$ and $a' \leq b'$ holds in $(X, \leq)$. We define an expansion $u' : (Y, \leq) \longrightarrow (\mathcal{P}(x, x_1, x_2, z), \subseteq)$ of u by leave a to force x or z iff $f(a)$ forces it. Finally x_1 is forced in all the points of the kind $(a, 1)$ and x_2 is forced in all the points of the kind $(a, 2)$. A short glance shows that formulas (4.9) are all forced in u'.

In some cases one can indeed run quantifier elimination just by using deduction in second order propositional intuitionistic logic (however, as we saw in the previous Section, second order propositional deduction is sound but not complete for our purposes). The following Proposition contains a list of properties of existentially closed Heyting algebras that can be verified using the sheaf semantics. Some more properties are put as exercises at the end of the Chapter.

PROPOSITION 4.28 *In any existentially closed non degenerate Heyting algebra H we have that:*

(i) *the relation $<$ is dense;*

(ii) *for every element $y \neq 0$ there is a complemented element x such that $0 < x < y$ (consequently, regular and complemented elements of H form atomless Boolean algebras);*

(iii) *the dual lattice of H has no non-zero join-irreducible elements;*

(iv) *the difference of x and y exists iff $x \leq y \vee \neg y$.*

Proof. We shall show only (iv), leaving the remaining cases to the reader. We recall that difference is, by definition, implication in the dual lattice of H, so we need to eliminate quantifiers from the statement

$$\exists z\, \forall t\, (x \leq t \vee y \;\Leftrightarrow\; z \leq t),$$

i.e. from the statement

$$\exists z\, (\sim \exists t\, (x \to t \vee y = 1 \;\&\; z \to t \neq 1)$$
$$\& \;\sim \exists t\, (z \to t = 1 \;\&\; x \to t \vee y \neq 1)).$$

As $\exists^t(x \to t \vee y)$ and $\exists^t(z \to t)$ are both provable in second order intuitionistic logic, we have to compute the statement

$$\exists z\, (\forall^t((x \to (t \vee y)) \to (z \to t)) = 1$$
$$\& \,\forall^t((z \to t) \to (x \to (t \vee y))) = 1).$$

But in second order intuitionistic logic the formulas

$$\forall^t((x \to (t \vee y)) \to (z \to t)) \;\leftrightarrow\; (z \to (x \wedge \neg y))$$
$$\forall^t((z \to t) \to (x \to (t \vee y))) \;\leftrightarrow\; (x \to (z \vee y))$$

are theorems, so we need to eliminate quantifiers from

$$\exists z((z \to (x \wedge \neg y)) = 1 \;\&\; x \to (z \vee y) = 1).$$

For this, it is sufficient to compute

$$\exists^z(((z \to (x \wedge \neg y))) \wedge (x \to (z \vee y)) = 1,$$

which, again by second order deduction, is the same as

$$x \to ((x \wedge \neg y) \vee y) = 1,$$

yielding the claim.[6] $\square$

There are still many open problems concerning existentially closed Heyting algebras. Apart from the fact that the above axiomatization (4.8) is not quite

[6] In other words, what we have shown is that the difference of x and y exists iff $x \wedge \neg y$ is the difference of x and y.

satisfactory from the algebraic point of view, we need also models for T_H^*: general facts guarantee only that each Heyting algebra embeds into a model of T_H^*, but do not exhibit concrete models of this theory. Notice that the most obvious idea for finding such models is wrong: open subsets of a topological space are never a model of T_H^*, because condition (iii) is trivially violated by taking the topological closure of any singleton. In fact, models of T_H^* which are locales (if any) must be pointless. In Chapter 7 we will come back to this point.

We mention some further facts about *varieties* of Heyting algebras. Let T be the first order theory of an equational class of Heyting algebras. If T has a model completion, then $Alg(T)$ should have the amalgamation property. Hence T must be one of the 8 equational theories listed in [Ma1]. For 6 of such theories, the model completion exists because the related varieties are locally finite, cf. [Wh].[7] The remaining 2 cases are the case of the whole class of Heyting algebras and the case of De Morgan algebras (i.e. of Heyting algebras satisfying De Morgan law $\neg(x \wedge y) = \neg x \vee \neg y$). For the latter case, the analog of the Combinatorial Lemma 4.18 also holds (to show it, use e.g. the technique of Lemma 4.18 combined with the trick we shall use in Chapter 6 for $\mathbf{S4.2.Grz}$). Therefore, we have:

THEOREM 4.29 *There are exactly 8 varieties of Heyting algebras whose first order theory admits a model completion.* □

Thus existence of a model completion is the same as the amalgamation property for equational classes of Heyting algebras. In the modal case, we shall see in Chapter 6 that the situation is quite different.

8. $\mathbf{M_H}$ is not exact

Using Section 4.3 duality we can not only prove some properties of the category of finitely presented Heyting algebras but we can also disprove such properties. In this Section we present an example of a non-effective equivalence relation in $\mathbf{M_H}$. This shows that $\mathbf{M_H}$ is not exact even if, as we know, it is a regular category.

Let R be a binary relation on C^L in $\mathbf{M_H}$, i.e. $R \in \mathcal{S}(C \times C)$. We define another subsheaf of $C^L \times C^L$ by

$$R^{(\infty)}(P) =$$

$$= \{(v_1, v_2) : \forall p \in P\ \exists (w_1^p, w_2^p) \in R\ \ (v_1)_p \sim_\infty w_1^p\ \&\ (v_2)_p \sim_\infty w_2^p\}.$$

[7] One can also apply Theorem 3.8: notice in fact that any homomorphism between finite Heyting algebras has both adjoints and the amalgamation property is equivalent to the Beck-Chevalley condition.

We still have

PROPOSITION 4.30 *Let* $f : C^L \to D^M$ *be a morphism in* $\mathbf{M_H}$. *Then, if* $R \subseteq ker(f)$ *then* $R^{(\infty)} \subseteq ker(f)$.

Proof. Assume $R \subseteq ker(f)$. Let $(v_1, v_2) : P \to L \times L$ be an evaluation such that $(v_1, v_2) \in R^{(\infty)}$. In order to show that $f(v_1) = f(v_2)$ we shall show

$$\forall p \in P \ \ f(v_1)_p \sim_0 f(v_2)_p$$

Fix $p \in P$ and w_1^p, w_2^p such that $(w_1^p, w_2^p) \in R$, $(v_1)_p \sim_\infty w_1^p$ and $(v_2)_p \sim_\infty w_2^p$. Then

$$f(v_1)_p = f((v_1)_p) \sim_\infty f(w_1^p) = f(w_2^p) \sim_\infty f((v_2)_p) = f(v_2)_p.$$

□

COROLLARY 4.31 *Let* $f : C^L \to D^M$ *be a morphism in* $\mathbf{M_H}$. *If* $R = ker(f)$ *then* $R = R^{(\infty)}$. □

LEMMA 4.32 *Let* $v : P \to L$ *be an L-evaluation which is injective. Then* $\downarrow_\infty v = \{v \circ h : h : Q \to P \in \mathbf{P_0}\}$.

Proof. Let $v : P \to L$ be injective, $w : Q \to L$ such that $w \leq_\infty v$. We need to find an open map $h : Q \to P$ such that $w = v \circ h$. By Proposition 4.4, we have a commuting square

$$\begin{array}{ccc} Q & \xrightarrow{w} & L \\ {\scriptstyle h_1}\uparrow & & \uparrow{\scriptstyle v} \\ P' & \xrightarrow[h_0]{} & P \end{array}$$

where h_0 open and h_1 open surjective. Then, since v is injective and h_1 is surjective, there is a unique diagonal fill-in $h : Q \to P$, and h is open since h_0 is. □

Example We give an example of an equivalence relation R in $\mathbf{M_H}$ which is not effective. By Corollary 4.31, it is enough to show an equivalence relation R such that $R \neq R^{(\infty)}$.

Let L be a four element partial order with set of nodes $\{a, b, c, d\}$ given by the following Hasse diagram

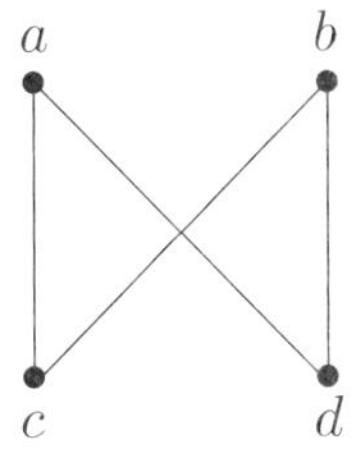

Let t_1 denote a one-element poset, t_2 a three element poset with root ρ and two leaves ρ_0 and ρ_1. We define four $L \times L$-evaluations:

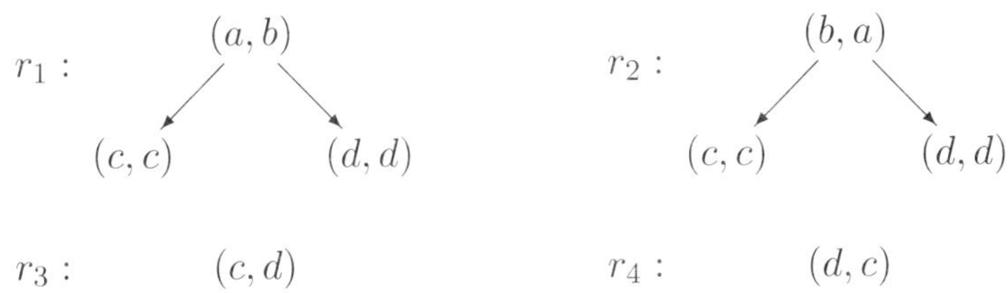

i.e. r_1 and r_2 have domain t_2, r_3 and r_4 have domain t_1 and the values of r_i's are as displayed above. Put

$$diag = \{(v, v) : v \in h_L\}$$

i.e. diag is the diagonal subobject of $h_{L \times L}$, and

$$R = diag \cup \bigcup_{i=1}^{4} (\downarrow_\infty r_i)$$

R is a subsheaf of $h_{L \times L}$ with an index, since each summand is (*diag* has index 1 and $(\downarrow_\infty r_i)$'s have indices by Proposition 4.5). We shall show that R is an equivalence relation and that $R \neq R^{(\infty)}$.

First we describe R in detail. Note that all $L \times L$-evaluations r_1, r_2, r_3, r_4 are injective and that if $h : P \to t_2$ is not surjective then $r_i \circ h \in diag$ for $i = 1, 2$. Using Lemma 4.32, we get for P in $\mathbf{P}_0$

$$R(P) = diag(P) \cup \{r_i \circ !_P : i = 3, 4\} \cup$$

$$\cup \{r_i \circ h : h : P \to t_2 \text{ is open surjective } i = 1, 2\}$$

where $!_P$ is the unique morphism from P to t_1.

Next, we show that R is an equivalence relation. Since *diag* is in R, R is reflexive. If (v_1, v_2) is in R then either $(v_1, v_2) \in diag$ and then $v_1 = v_2$ or $(v_1, v_2) = r_i \circ h$ for some $i \in \{1, 2, 3, 4\}$. If $i = 1$ or 2 then $(v_2, v_1) =$

$r_{3-i} \circ h \in R$, if $i = 3$ or 4 then $(v_2, v_1) = r_{7-i} \circ h \in R$. Thus R is symmetric. It remains to show that R is transitive. Let (v_1, v_2), $(v_2, v_3) \in R$. If (v_1, v_2) or (v_2, v_3) belongs to *diag* then clearly $(v_1, v_3) \in R$. Assume that neither (v_1, v_2) nor (v_2, v_3) belongs to *diag*. If $(v_1, v_2) = r_i \circ !_P$ then (v_2, v_3) must be $r_{7-i} \circ !_P$. Hence $(v_1, v_3) \in diag$ as a constant evaluation equal either (c, c) or (d, d). Assume $(v_1, v_2) = r_1 \circ h$ for some open surjection $h : P \to t_2$ (the case $(v_1, v_2) = r_2 \circ h$ is analogous). Since (v_2, v_3) does not belong to *diag*, it must be of form $r_2 \circ h'$, for some open surjection $h' : P \to t_2$ (consider the values at the roots of (v_1, v_2) and (v_2, v_3)). Now it is enough to show that $h = h'$, since then $v_1 = v_3$. Let $p \in P$. If $r_1 \circ h(p) = (a, b)$, then $r_2 \circ h'(p) = (b, a)$. Hence $h(p) = h'(p)$ (see definitions of r_1 and r_2). If $r_1 \circ h(p) = (c, c)$, then $r_2 \circ h'(p) = (c, c)$. So $h(p) = h'(p)$ as well. The case $r_1 \circ h(p) = (d, d)$ is similar. Therefore $h = h'$ and R is an equivalence relation.

Finally, consider the $L \times L$-evaluation

r_5 :

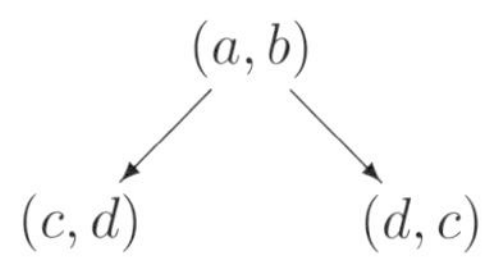

It is easy to check that $r_5 \in R^{(\infty)}$. The condition of the definition of $R^{(\infty)}$ is satisfied for leaves since r_3, $r_4 \in R$ and for the root since $r_1 \in R$ (notice the twisting!). However $r_5 \notin R$, since in any $L \times L$-evaluation w in R below the node labelled by (a, b) we can have nodes labelled by (c, c) and (d, d) only. This is not the case with r_5. This means that the equivalence relation R cannot be a kernel of any morphism in $\mathbf{M_H}$.

9. Some applications

In this Section we will draw some consequences for intuitionistic propositional calculus IpC from the negative results of the previous Section concerning HA_{fp}.

Any finitely presented Heyting algebra can be represented as a free Heyting algebra on finitely many generators, say $\vec{p} = (p_1, \ldots, p_n)$ divided by a propositional formula $\alpha(\vec{p})$ (built up from variables in $\vec{p}$ and $\bot$, $\top$, $\vee$, $\wedge$, $\to$ in the usual way). We denote such an algebra by $F(\vec{p})/\alpha(\vec{p})$ or $F(p_1, \ldots, p_n)/\alpha(p_1, \ldots, p_n)$. Elements of $F(\vec{p})/\alpha(\vec{p})$ can be regarded as intuitionistic propositional formulas modulo provable equivalence in IpC, i.e. two formulas ψ_1, ψ_2 are identified in $F(\vec{p})/\alpha(\vec{p})$ iff

$$\alpha \vdash_{IpC} \psi_1 \leftrightarrow \psi_2$$

A Heyting algebra morphism between such algebras

$$\overline{\varphi} : F(q_1, \ldots, q_m)/\beta(\vec{q}) \longrightarrow F(\vec{p})/\alpha(\vec{p})$$

is a 'substitution morphism' i.e. it is uniquely determined (up to equivalence provable in IpC from α) by formulas

$$\vec{\varphi}(\vec{p}) = (\vec{\varphi}_1(\vec{p}), \ldots, \vec{\varphi}_m(\vec{p}))$$

Then $\overline{\varphi}(\psi(\vec{q})) = \psi(\vec{\varphi}(\vec{p})/\vec{q})$.
Any sequence as above gives rise to a Heyting morphism provided

$$\alpha(\vec{p}) \vdash_{IpC} \beta(\vec{\varphi}(\vec{p})/\vec{q})$$

A *coequivalence relation* on an object $F(\vec{p})/\alpha(\vec{p})$ in HA_{fp} (i.e. an equivalence relation in HA^{op}_{fp}) is given by a formula $\rho(p_1, \ldots, p_n, p'_1, \ldots, p'_n)$ such that

$$\rho(\vec{p}, \vec{p}\,') \vdash_{IpC} \alpha(\vec{p}) \wedge \alpha(\vec{p}\,') \tag{4.10}$$

$$\alpha(\vec{p}) \vdash_{IpC} \rho(\vec{p}, \vec{p}) \tag{4.11}$$

$$\vdash_{IpC} \rho(\vec{p}, \vec{p}\,') \rightarrow \rho(\vec{p}\,', \vec{p}) \tag{4.12}$$

$$\vdash_{IpC} (\rho(\vec{p}, \vec{p}\,') \wedge \rho(\vec{p}\,', \vec{p}\,'')) \rightarrow \rho(\vec{p}, \vec{p}\,'') \tag{4.13}$$

With ρ as above, the diagram

$$F(\vec{p})/\alpha(\vec{p}) \underset{g'}{\overset{g}{\rightrightarrows}} F(\vec{p}, \vec{p}\,')/\rho(\vec{p}, \vec{p}\,')$$

is a coequivalence relation in HA_{fp}, where g sends p_i to p_i and g' sends p_i to p'_i.

On the other hand for any morphism

$$\overline{\varphi} : F(q_1, \ldots, q_m)/\beta(\vec{q}) \longrightarrow F(\vec{p})/\alpha(\vec{p})$$

in HA_{fp}, its cokernel (pushout of $\overline{\varphi}$ along itself)

$$F(\vec{p})/\alpha(\vec{p}) \underset{k'}{\overset{k}{\rightrightarrows}} F(\vec{p}, \vec{p}\,')/\sigma(\vec{p}, \vec{p}\,')$$

is a coequivalence relation (effective by definition). In this case $\sigma(\vec{p}, \vec{p}\,')$ has a special form

$$\sigma(\vec{p}, \vec{p}\,') \leftrightarrow \bigwedge_{i=1}^{m} (\varphi_i(\vec{p}) \leftrightarrow \varphi_i(\vec{p}\,')) \wedge \alpha(\vec{p}) \wedge \alpha(\vec{p}\,') \tag{4.14}$$

where φ_i's are formulas determining $\overline{\varphi}$. We say that a formula $\sigma(\vec{p}, \vec{p}')$ *separates variables* $\vec{p}$ *and* $\vec{p}'$ *relative to* $\alpha(\vec{p})$ if it is equivalent to the formula of form (4.14) for some formulas $\varphi_1, \ldots, \varphi_m$. Clearly every formula which separates variables defines a coequivalence relation on $F(\vec{p})/\alpha(\vec{p})$. We have

PROPOSITION 4.33 *A formula $\rho(\vec{p}, \vec{p}')$ defines an effective coequivalence relation on $F(\vec{p})/\alpha(\vec{p})$ iff $\rho(\vec{p}, \vec{p}')$ separates variables $\vec{p}$ and $\vec{p}'$ relative to $\alpha(\vec{p})$.*

Proof. If $\rho(\vec{p}, \vec{p}')$ defines an effective coequivalence relation on $F(\vec{p})/\alpha(\vec{p})$ then as we saw above $\rho(\vec{p}, \vec{p}')$ separates variables. On the other hand, if there are formulas $\varphi_1, \ldots, \varphi_m$ such that

$$\rho(\vec{p}, \vec{p}') \;\leftrightarrow\; \bigwedge_{i=1}^{m} (\varphi_i(\vec{p}) \leftrightarrow \varphi_i(\vec{p}')) \wedge \alpha(\vec{p}) \wedge \alpha(\vec{p}')$$

then the pair of morphisms g, g', defined as above, is the cokernel of the Heyting morphism

$$\overline{\varphi} : F(q_1, \ldots, q_m) \longrightarrow F(\vec{p})/\alpha(\vec{p})$$

sending q_i into φ_i for $i = 1, \ldots, m$. □

The example of the last Section shows that in intuitionistic propositional logic there are coequivalence relations which do not separate variables. This is in contrast with classical propositional calculus, where all coequivalence relations separate variables.

Let us note that even if a coequivalence relation in HA_{fp} is not effective it may still have a quotient. Having a quotient means that there is a smallest effective coequivalence relation containing the given one. For example, the equivalence relation in $\mathbf{M_H}$ defined in the previous section has a quotient (which is not preserved by $\mathbf{\Psi_H}$). In that case $R^{(\infty)}$ has index 2. In general we don't know whether every coequivalence relation in HA_{fp} has a quotient.

10. Projective Heyting algebras

We recall that an object P in a category $\mathbf{C}$ is said to be (regular) *projective* iff for every map $f : P \to A$ and for every regular epi $q : B \to A$ there is an arrow $\bar{f}$ such that $q \circ \bar{f} = f$. If $\mathbf{C}$ is a category of algebras or of finitely presented algebras, projective objects are just those algebras A such that if we represent them as a quotient $q : \mathcal{F}(X) \to A$ of a free algebra, then the map q has a section (namely, there exists $s : A \to \mathcal{F}(X)$ such that $q \circ s = 1_A$).

We want to characterize projective objects in our category $(\mathbf{M_H})^{\mathrm{op}}$, which is equivalent to the category of finitely presented Heyting algebras.[8] Recall that

[8] For a characterization of finite projective Heyting algebras (which can also be deduced from the results of this Section), see [BH].

free Heyting algebras correspond, through this duality, to sheaves h_L, where L is the poset obtained by taking the power set of a finite set, with reverse inclusion partial order. In this section, for simplicity, we deal only with objects of $\mathbf{M_H}$ which are of the kind C^L where L is $(\mathcal{P}(X), \supseteq)$, for a finite set X (there is no loss of generality in that, the full subcategory determined by such objects is trivially equivalent to the whole $\mathbf{M_H}$). Such an object C^L is said to be *injective* in $\mathbf{M_H}$ iff its dual Heyting algebra is projective i.e. iff (according to the above remark) there exists a map $s : h_L \to C^L$ in $\mathbf{M_H}$ such that $s(v) = v$ holds for all $v \in C^L$.

A *variant of an evaluation* $v : P \to L$ is an evaluation $v' : P \to L$ such that $v(p) = v'(p)$ holds for all p different from the root of P; we say that C^L has the *extension property* iff C^L contains a variant of every evaluation $v : P \to L$ such that $v_p \in C^L$ holds for all p different from the root of P. In other words, by iteration of the above property, if an evaluation v does not belong to C^L we must be able to change the values $v(p)$ in such a way that the resulting evaluation belongs to C^L, without however modifying the value of $v(p)$ whenever v_p is already in C^L.

Given an object C^L and given $a \in L$, let $f_a : h_L \to h_L$ be the transformation defined as follows (for $v : P \to L$, $p \in P$):

$$f_a(v)(p) \quad = \quad \begin{cases} v(p) & \text{if } v_p \in C^L, \\ a \cap \bigcap\{v(q) \,:\, q \leq p,\ v_q \in C^L\} & \text{otherwise.} \end{cases}$$

Notice that $f_a(v) = v$ in case $v \in C^L$. The main properties of f_a are summarized in the following easy

LEMMA 4.34 *For any C^L having index n and for any $a \in L$, we have that:*

(i) *f_a is a natural transformation having index $n + 1$;*[9]

(ii) *f_a is idempotent;*

(iii) *if $v : P \to L$ is an L-evaluation such that, for all $p \in P$ we have ($f_a(v_p) \in C^L$ iff $v_p \in C^L$), then for every $b \in L$, $f_b(f_a(v)) = f_b(v)$.* □

We are ready for the main result:

THEOREM 4.35 *An object C^L is injective in $\mathbf{M_H}$ iff it has the extension property.*

Proof. Suppose first that there is $f : h_L \to C^L$ such that $f(w) = w$ holds for all $w \in C^L$. In order to show that C^L has the extension property, take any

[9] If C^L corresponds to the IpC-formula $A(p_1, \ldots, p_n)$, then f_a corresponds to the substitution associating to p_i the formula $A \to p_i$, in case $p_i \in a$, and the formula $A \wedge p_i$, in case $p_i \notin a$.

$v : P \to L$ such that $v_p \in C^L$ holds for all $p \neq \rho$. Then it is immediate to see that $f(v) \in C^L$ is a variant of v.

Suppose now that C^L has the extension property; let $a_1, \ldots, a_l$ be a list of the elements of L such that

$$a_i \subseteq a_j \quad \Rightarrow \quad i \leq j \tag{4.15}$$

holds for all i, j (this is just an extension of the partial order $\subseteq$ to a total order). Let $f : h_L \to h_L$ be the composed transformation $f_{a_1} \circ \cdots \circ f_{a_l}$. Clearly f does not modify the evaluations belonging to C^L; we only have to prove that it factors through C^L. We first need the following:

Claim. Suppose that for $v : P \to L$, we have that $f(v) \in C^L$ and that $f(v)(\rho) \supseteq a_i$; then $f_{a_i} \cdots f_{a_l}(v) \in C^L$.

Proof of the Claim. If not, let $1 \leq j < i$ be the maximum index such that $f_{a_j} \cdots f_{a_i} \cdots f_{a_l}(v) \in C^L$. Then, as the $f_{a_1}, \ldots, f_{a_{j-1}}$ can't change anything whereas f_{a_j} gave the root a value smaller than a_j (by the definition of f_{a_j}), we have $f(v)(\rho) \subseteq a_j$; in conclusion $a_i \subseteq a_j$ with $j < i$, impossible by (4.15).

Let us now turn to the proof of the Theorem. Take any $v : P \to L$, the aim is to show that $f(v) \in C^L$. Suppose not; then, as our frames are finite, we can suppose without loss of generality, that $f(v_p) \in C^L$ for all $p \neq \rho$. By the extension property, it is possible to modify $f(v)$ giving the root the value, say a_i, so that the resulting evaluation belongs to C^L. Clearly $a_i \subseteq f(v)(p)$ for all $\rho > p$ (because evaluations, which are order preserving, change $\leq$ into $\supseteq$) and consequently the modified evaluation is just $f_{a_i}(f(v))$ (which, we repeat, belongs to C^L). Applying the Claim to the evaluations v_p (for $p \neq \rho$), we get that $f_{a_i} \cdots f_{a_l}(v) \in C^L$. Notice now that nothing relevant happens if we apply successively first $f_{a_{i-1}}$, then $f_{a_{i-2}}$, then ..., then finally f_{a_1} to $f_{a_i} \cdots f_{a_l}(v)$: the values of all points different from the root are not modified, the value of the root changes but still the modified evaluation is not in C^L (otherwise $f(v)$ would be in C^L which is not). Thus by repeated applications of Lemma 4.34(iii), $f_{a_i}(f_a(v))$ is equal to $f_{a_i} f_{a_i} \cdots f_{a_l}(v)$ and also to $f_{a_i} \cdots f_{a_l}(v)$, by Lemma 4.34(ii). This is impossible, as $f_{a_i}(f(v))$ belongs to C^L, whereas $f_{a_i} \cdots f_{a_l}(v)$ does not. □

COROLLARY 4.36 *Injective objects in $\mathbf{M_H}$ are closed under images.*

Proof. Suppose that for $g : C^L \to D^M$ in $\mathbf{M_H}$, we have that D^M is the sheaf image of its domain and that C^L has the extension property; we show that D^M has the extension property as well. Pick $v : P \to M$ such that $v_p \in D^M$ holds for all $p \neq \rho$. As D^M is the local closure of the presheaf image of C^L (within h_M), there are $w^p \in C^L$ such that $g(w^p) \sim_\infty v_p$. Take the disjoint union of the domains of the w^p and add it a new root; as C^L has the extension property, there is in C^L an evaluation w on the finite poset so built, which moreover restricts to the w^p's. In this way, you get an evaluation $g(w) \in D^M$ which is infinitely equivalent to a variant of v. $\square$

The dual of the above Corollary is a remarkable fact about Heyting algebras: *any finitely generated subalgebra of a finitely presented projective Heyting algebra is finitely presented and projective as well.* Applications of the above results to the E-unification theory of Heyting algebras are given in [Gh4].

11. Exercises

1 The *implicational degree* $d(\varphi)$ of an IpC formula φ is defined as follows. For atomic sentences p, $d(p) = 0$, moreover $d(\alpha \vee \beta) = d(\alpha \wedge \beta) = max(d(\alpha), d(\beta))$, $d(\alpha \rightarrow \beta) = max(d(\alpha), d(\beta)) + 1$. Show that, if $L = (\mathcal{P}(\vec{x}), \supseteq)$, $d(\varphi(\vec{x})) \leq n$, and u, v are L-evaluations such that $v \sim_n u$ then

$$v \models \varphi(\vec{x}) \text{ iff } u \models \varphi(\vec{x}).$$

Conclude that, if $d(\varphi(\vec{x})) \leq n$, then the subsheaf of h_L of those L-evaluations v that $v \models \varphi(\vec{x})$ has index n.

2 Show that the functor $\mathcal{D}(-) : \mathbf{P_0}^{op} \longrightarrow Set$ is the subobject classifier in $Sh(\mathbf{P_0}, \mathbf{J_0})$.

3 (a) Show that for any Heyting algebra H, the presheaf

$$HA(H, \mathcal{D}(-)) : \mathbf{P_0}^{op} \longrightarrow Set$$

(HA denote the category of Heyting algebras), of models of H in finite subdirectly irreducible algebras, is a $\mathbf{J_0}$-sheaf.

(b) Show that the functor

$$HA(=, \mathcal{D}(-)) : HA^{op} \longrightarrow Sh(\mathbf{P_0}, \mathbf{J_0}) \qquad (4.16)$$

given by

$$H \mapsto HA(H, \mathcal{D}(-))$$

is an extension of the essential inverse of $\mathbf{T_H}$, i.e. the diagram

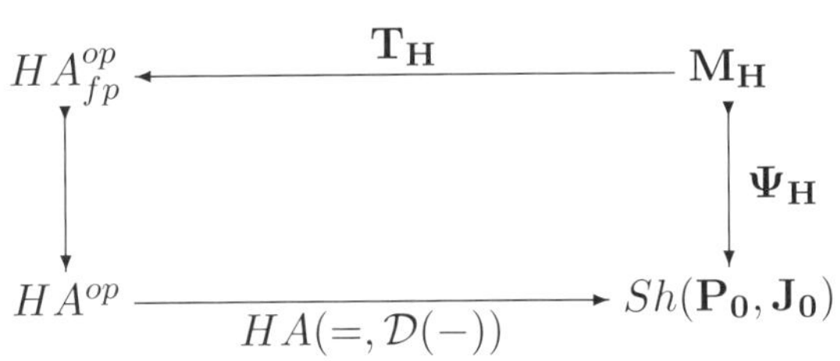

commutes.

(c) Show that the functor (4.16) has a left adjoint.

4 Let the functor

$$Y : \mathbf{P_0} \longrightarrow Sh(\mathbf{P_0}, \mathbf{J_0})$$

be the Yoneda embedding, i.e. $Y(P)(P') = \mathbf{P_0}(P', P)$, for $P, P' \in \mathbf{P_0}$. Show that

(a) $Y(P)$ is a subsheaf of h_P with index 2.

(b) For any $v,\ v' \in Y(P)$, $v \sim_0 v'$ iff $v \sim_\infty v'$.

(c) For any C in $\mathbf{M_H}$ and any natural transformation $f : Y(P) \to C$ has index 0. In particular, it is a morphism in $\mathbf{M_H}$.

(d) Any subsheaf of $Y(P)$ has index 0.

(e) Deduce from the above that the Yoneda embedding Y factors through $\mathbf{M_H}$, i.e. we have an embedding functor (also denoted by Y)

$$Y : \mathbf{P_0} \longrightarrow \mathbf{M_H}$$

(f) Show that the functors $\mathcal{D}, \mathbf{T_H} \circ Y : \mathbf{P_0} \to \mathbf{M_H}$ are naturally isomorphic, i.e. that $Y(P)$ corresponds to $\mathcal{D}(P)$ under $\mathbf{T_H}$.
Hint. Show that the map

$$\mathcal{D}(P) \xrightarrow{\iota_P} \mathcal{S}(h_P) \xrightarrow{q_{Y(P)}} \mathcal{S}(Y(P))$$

gives the required isomorphism.

5 Let us show that a coproduct of finite Heyting algebras needs not be finite and that the category of finite Heyting algebras does not have binary coproducts.

(a) Show that for any C^L in $\mathbf{M_H}$, C^L has infinitely many $\sim_\infty$ equivalence classes iff the Heyting algebra $\mathcal{S}(C^L)$ is infinite.

(b) Let $L = \{1 \geq 0\}$ and $M = \{2 \geq 1 \geq 0\}$ be posets, $Y(L)$ be the subsheaf of h_L of open L-evaluations, and C^M be the subsheaf of h_M, such that $v : P \to M \in C^M$ iff $v(p) = 0$ for every minimal point of P. Show that both $Y(L)$ and C^M are objects of $\mathbf{M_H}$, $\mathcal{S}(C^M)$ is infinite whereas $\mathcal{S}(Y(L))$ is finite.

(c) Let $\bar{f}, \bar{g} : M \to L$ be the monotone maps given by $\bar{f}(2) = \bar{f}(1) = 1$, $\bar{f}(0) = 0$, $\bar{g}(2) = 1$, $\bar{g}(1) = \bar{g}(0) = 0$; let also $f, g : C^M \to Y(L)$ be the maps in $\mathbf{M_H}$ given by $f(v) = \bar{f} \circ v$ and $g(v) = \bar{g} \circ v$. Show that the map

$$< f, g >: C^M \longrightarrow Y(L) \times Y(L)$$

is mono.

(d) Conclude that the coproduct of the finite Heyting algebras $\mathcal{S}(Y(L)) + \mathcal{S}(Y(L))$ is infinite.

(e) Conclude, using Exercise 16 from Chapter 2, that the category of finite Heyting algebras does not have binary coproducts.

6 An evaluation $v : P \longrightarrow L$ is said to be irreducible iff for all $p, q \in P$, if $v_p \sim_\infty v_q$ then $p = q$. Show that for any evaluation $v : P \longrightarrow L$ there are an irreducible evaluation $v_{irr} : P_{irr} \longrightarrow L$ and an open surjection $h : P \longrightarrow P_{irr}$ such that $v_{irr} \circ h = v$. Prove also that $v \sim_\infty w$ iff v_{irr} and w_{irr} are L-isomorphic.

7 We show here a method to generate inductively all possible irreducible L-evaluations for a finite poset L (the construction is well-known, see [Be], [Gh2], [Ry], etc.). Define for $n \geq 1$ a (non necessarily rooted) poset U_n^L and an order-preserving map $\chi_n^L : U_n^L \longrightarrow L$ as follows. U_1^L is just L with the discrete partial order and χ_1^L is identity. Suppose now that U_n^L, χ_n^L are already defined. A pair $\langle a, S \rangle$ (with $a \in L$ and $S \subseteq U_n^L$) is good iff: a) the inclusion $S \subseteq U_n^L$ is open; b) S contains an n-element chain; c) $\chi_n^L(x) \leq a$ for all $x \in S$; d) in case S has a root ρ, then $a \neq \chi_n^L(\rho)$. Let U_{n+1}^L be the union of U_n^L and of all good pairs; extend the partial order of U_n^L to U_{n+1}^L by adding it all the pairs $x \leq \langle a, S \rangle$ for $x \in S$ (add of course also the trivial reflexive pairs $\langle a, S \rangle \leq \langle a, S \rangle$). Finally extend χ_n^L by $\chi_{n+1}^L(\langle a, S \rangle) = a$.

(a) Show that for every order-preserving map $v : P \longrightarrow L$ based on a poset P having height less or equal to n, there is a unique open map $h : P \longrightarrow U_n^L$ such that $\chi_n^L \circ h = v$.

(b) Conclude that a rooted evaluation is irreducible iff it is equal (up to an L-isomorphism) to an evaluation of the kind $(\chi_n^L)_p$ for some $p \in U_n^L$.

(c) Let $U^L = \bigcup_n U_n^L$ and $\chi^L = \bigcup_n \chi_n^L$; show that the map

$$j_L : \mathcal{S}(h_L) \longrightarrow \mathcal{D}(U^L)$$

defined by $j_L(S) = \{p \,|\, \chi_p^L \in S_{U_p^L}\}$ is an embedding of the Heyting algebra freely generated by the finite distributive lattice $\mathcal{D}(L)$ into the Heyting algebra of sieves of the poset U^L. Keeping in mind that all subsheaves of the kind $\downarrow_\infty v$ have index and consequently belong to $\mathcal{S}(h_L)$, show also that all cones $\downarrow p = \{q \,|\, q \leq p\}$ are in the image of j_L.

8 Let $L = \mathbf{2}$ (the two-element chain). Using the same notation as in the previous exercise, prove that U^L can be identified in this case with the poset $\langle N, \leq \rangle$, where N is the set of natural numbers and $n \leq m$ holds iff ($n = m$ or $n \leq m-2$). The (infinitary) L-evaluation χ^L is the characteristic function of the subset $\{0\}$. Notice that, as all sieves of U^L are in this case either the total sieve or the union of at most two cones, the embedding j_L turns out to be

an isomorphism. This gives the following picture of the Rieger-Nishimura lattice, i.e. of the free Heyting algebra on one generator:

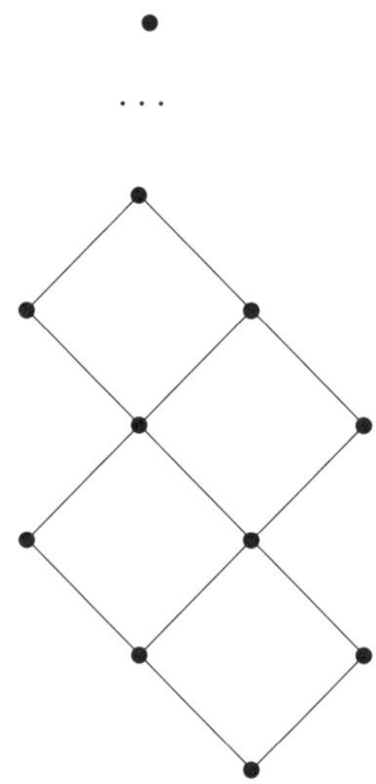

9 Prove that, for a morphism f in $\mathbf{M_H}$ or in $Sh(\mathbf{P_0}, \mathbf{J_0})$, f^{-1} preserves the difference iff (4.7) holds (use Proposition 4.5). Consequently, show that there are morphisms f such that f^{-1} does not preserve the difference (take for instance some diagonal morphism).

10 Show that the following formulas of IpC_2 are valid:

(a) $\exists^p(\varphi(\vec{q}) \to \psi(p, \vec{q})) \leftrightarrow (\varphi(\vec{q}) \to \exists^p\psi(p, \vec{q}))$
(b) $\exists^p((p \to x) \to y) \leftrightarrow (((x \to y) \to x) \to y)$
(c) $\forall^p(\varphi(p, \vec{q}) \vee \varphi(p, \vec{q})) \leftrightarrow (\forall^p\varphi(p, \vec{q}) \vee \forall^p\varphi(p, \vec{q}))$[10]
(d) $\forall^p((x \to p) \to ((y \to p) \to p)) \leftrightarrow (x \vee y)$
(e) $\forall^p(x \to (y \to p)) \leftrightarrow \neg(x \wedge y)$

11 Let $\psi(\vec{q}, p)$ be a formula in which the variable p occurs only positively. Show that there is a *least fixed point* of ψ, i.e. that there is a formula $\varphi(\vec{q})$ such that

(a) $\vdash \psi(\vec{q}, \varphi(\vec{q})) \to \varphi(\vec{q})$;
(b) for every other formula φ', if $\vdash \psi(\vec{q}, \varphi') \to \varphi'$ then $\vdash \varphi \to \varphi'$.

Notice that, from this, $\vdash \psi(\vec{q}, \varphi(\vec{q})) \leftrightarrow \varphi(\vec{q})$ follows immediately. (*Hint*: Use Pitts' quantifiers and take $\forall^p((\psi(\vec{q}, p) \to p) \to p)$ as φ).[11]

---

[10] This quite peculiar IpC_2-valid formula is first mentioned in [Pi2].
[11] For recent results on least fixed points in modal logics, see [Mar].

12 Prove (i), (ii), (iii) of the Proposition 4.28.

13 Show that in T_H^* the following holds:

(a) $\exists_p (p \to x \leq y) \;\Leftrightarrow\; ((x \to y) \to x \leq y)$

(b) $\exists_p (p \leq u \,\&\, p \to z \leq p) \;\Leftrightarrow\; (u \to z \leq u)$

14 Show that a projective Heyting algebra H is prime (namely, $0 \neq 1$ and for all $a, b \in H$, if $a \vee b = 1$ then $a = 1$ or $b = 1$), but that the converse does not hold. *Hint*: show that the algebra

$$\mathcal{F}(p, q, r)/(\neg p \to q \vee r) = 1$$

is not projective.

15 Find all projective finitely presented Heyting algebras generated by a single element.

16 (Kreisel) Let $\varphi(p, \vec{x})$ be a formula of IpC. Show that if

$$\varphi(p, \vec{x}), \varphi(q, \vec{x}) \vdash_{IpC} p \leftrightarrow q$$

then

$$\varphi(p, \vec{x}) \vdash_{IpC} p \leftrightarrow \varphi(1, \vec{x})$$

Conclude that the Beth property holds for all superintuitionistic logics and that epimorphisms between finitely presented algebras are regular in any subvariety of the variety of Heyting algebras.

Chapter 5

DUALITY FOR MODAL ALGEBRAS

In this Chapter we shall develop a duality for finitely presented modal algebras in a similar way we have developed a duality for finitely presented Heyting algebras in the previous Chapter. We point out below to some slight differences between these dualities.

The duality for modal algebras has a parameter $\mathbf{S}$, being an equational theory containing the equational theory of $\mathbf{K4}$-algebras with the finite model property and (AP) for finite algebras. In varieties of $\mathbf{K4}$-algebras, the principal congruences do not correspond to elements of algebras, but rather to elements of form $a \wedge \Box a$. This is a source of some additional technical complications. In order to define properly the dual category $\mathbf{M_S}$ we need to use games to define two kinds of relations $\sim_n$, $\approx_n$ which reflects the difference between arbitrary elements of algebras and those that are of form $a \wedge \Box a$. The first relation serves to define morphisms and the second to define objects in the dual categories. As one could see from an exercise at the end of the previous Chapter the index n in the relation $\sim_n$ was reflecting the implicational degree of the intuitionistic formulas. This time the index is related to the nesting of the necessity connective $\Box$. The site on which the sheaves are defined in the dual category differ slightly, as well. The frames we consider are the duals of all finite $\mathbf{S}$-algebras not only of those which are subdirectly irreducible.

In the exercises the reader will find some hints on how one can develop a similar duality for a theory $\mathbf{S}$ without (AP) for finite algebras.

1. Frames, evaluations and games

In this Section we define frames, evaluations and games on evaluations. Games give rise to some equivalence relations. We also study some basic properties of these notions that will be used later.

Recall from Section 2.1 that **K4** denote the first order equational theory in the language $\wedge$, $\vee$, $\rightarrow$, $\neg$, 0, 1, $\Box$ consisting of the following axioms:

(i) axioms of Boolean algebras;

(ii) $\Box 1 = 1$ and $\Box(p \wedge q) = \Box p \wedge \Box q$;

(iii) $\Box p \leq \Box\Box p$.

The connective $\Diamond$ is defined in the usual way as $\neg\Box\neg$. *Modal formulas* are terms in the above language.

If **S** is an equational theory containing **K4** then by $Alg(\mathbf{S})$ we denote the category of all **S**-algebras (i.e. algebras satisfying the axioms of **S**), and by $Alg(\mathbf{S})_{fin}$ and $Alg(\mathbf{S})_{fp}$ we denote the full subcategories of $Alg(\mathbf{S})$ of finite and finitely presented **S**-algebras, respectively.

If X, Y are sets and $f : X \rightarrow Y$ is a function then Y^X denotes the set of functions from X to Y, and $f^\circ : 2^Y \rightarrow 2^X$ is the dual map of 'composing with f'. The Boolean operations on 2^X are denoted by $\wedge_X, \vee_X, \rightarrow_X, \neg_X, 0_X, 1_X$.

We will use the same letter for Boolean algebras and their universes. If B is a boolean algebra, and together with the operation $\Box : B \rightarrow B$ satisfies the axioms of **K4**, then the corresponding **K4**-algebra will be denoted by $(B, \Box)$. Recall that by a *frame* we mean a pair (X, R) where X is a finite set and R is a binary relation on X. The relation R is called the *accessibility relation* of the frame (X, R). If R is transitive, the frame is called *transitive* as well. From now on all frames are assumed to be transitive. The *frame algebra* of (X, R) is the **K4**-algebra $(2^X, \wedge_X, \vee_X, \rightarrow_X, \neg_X, 0_X, 1_X, \Box_R)$ (to be denoted $(2^X, \Box_R)$), where for $v : X \rightarrow 2 \in 2^X$ and $x \in X$, $\Box_R(v)(x) = 1$ iff for all $y \in X$, $x\, R\, y$ implies $v(y) = 1$.

By a *morphism of frames* $f : (X, R) \rightarrow (Y, S)$ we mean a function $f : X \rightarrow Y$ which is an open frame map, i.e.

(i) $x\, R\, x'$ implies $f(x)\, S\, f(x')$, for $x, x' \in X$;

(ii) for any $x \in X$ and $y \in Y$, if $f(x)\, S\, y$ then there is $x' \in X$ such that $x\, R\, x'$ and $f(x') = y$.

It can be easily checked that $f : (X, R) \rightarrow (Y, S)$ is a frame morphism iff the dual map $f^\circ : (2^Y, \Box_S) \rightarrow (2^X, \Box_R)$ is a homomorphism of **S**-algebras.

In this way we have defined a category **F** of finite frames and morphisms of frames. The category **F**, as remarked in Section 2.1 is equivalent to the dual of the category of finite **K4**-algebras. More precisely, we have a functor

$$Alg(\mathbf{K4})_{fin} \longrightarrow \mathbf{F}^{op} \tag{5.1}$$

$$(B, \Box) \longmapsto (at(B), R_\Box)$$

where $at(B)$ is the set of atoms of the boolean algebra B and $R_\Box$ is the binary relation on $at(B)$ defined by $x\, R_\Box\, x'$ iff $x \leq \Diamond x'$. The functor acts on morphisms in the obvious way. Moreover, we have a functor

$$\mathbf{F}^{op} \longrightarrow Alg(\mathbf{K4})_{fin} \tag{5.2}$$

$$(X, R) \longmapsto (2^X, \Box_R)$$

associating with a finite frame its frame algebra and with every frame morphism its dual map. They are essential inverses one to the other establishing the mentioned duality $Alg(\mathbf{K4})_{fin} \simeq \mathbf{F}^{op}$. Note that via this duality the empty frame $(\emptyset, \emptyset)$ correspond to the **K4**-algebra in which $0 = 1$.

This duality restricts to some subcategories. Let **S** be an equational theory containing $\mathbf{K4}$, $\mathbf{F_S}$ be the full subcategory of the category **F** corresponding via the above duality to the subcategory $Alg(\mathbf{S})_{fin}$, i.e. we have $Alg(\mathbf{S})_{fin} \simeq \mathbf{F_S}^{op}$. By an **S**-*frame* we mean an object of $\mathbf{F_S}$. For some theories **S**, the objects of the category $\mathbf{F_S}$ can be described in a simple way in terms of their accessibility relations. We list below some of them (notice that any equational theory **S** containing **K4** can be axiomatized by equations of the form $t = 1$, where t is any term, hence it is possible to introduce **S** simply by specifying a set of terms, which could be viewed as a set of axioms of a logic):

Logic **S**	Axioms	Description of the accessibility relation
S4	**K4** + $\Box p \to p$	reflexive
S4.2	**S4** + $\Diamond\Box p \to \Box\Diamond p$	reflexive and locally confluent
S4.3	**S4** + $\Box(\Box p_1 \to p_2) \vee \Box(\Box p_2 \to p_1)$	reflexive and locally linear
Grz	**S4** + $(\Box(\Box(p \to \Box p) \to p) \to p)$	partial order
G	**K4** + $(\Box(\Box p \to p) \to \Box p)$	irreflexive
S5	**S4** + $\Diamond p \to \Box\Diamond p$	equivalence relation

A frame (X, R) is *locally confluent* iff for any $x, y, z \in X$, if (xRy and xRz) then there is $t \in X$ such that (yRt and zRt). A frame (X, R) is *locally linear* iff for any $x, y, z \in X$, if (xRy and xRz) then (yRz or zRy).

In all the above cases, the class of frame algebras of the finite frames indicated in the third column generates the variety $Alg(\mathbf{S})$, where **S** is axiomatized as shown in the second column. The fact that $Alg(\mathbf{S})_{fin}$ generates $Alg(\mathbf{S})$ means that **S** has *the finite model property*. The finite model property can be equivalently stated by asking that finitely generated free **S**-algebras embed into products of finite **S**-algebras. In view of Proposition 2.19, the same property, if true, extends in our case to finitely presented algebras. In order to establish the finite model property, specific techniques are needed: in Section 2.1 we saw only how to get it quickly for the basic **K4**-case; we won't even sketch in the sequel the proofs for other cases (including the cases mentioned in the above table), the reader is referred to [CZ] as an excellent textbook on such questions. In fact, there is no need for our purposes to enter in such (interesting but non trivial) field. For instance, we never need to know that **G**-algebras (also called *diagonalizable* algebras) can be axiomatized by the single equation $\Box(\Box p \to p) \to \Box p = 1$, we could simply define **G**-algebras as the algebras belonging to the variety generated by the class of finite irreflexive frame algebras. Only proof-theoretic (or at least decidability) questions would be sensitive to the existence of nice axiomatizations, but such questions are outside the scope of this book. On the other hand, we shall almost exclusively deal with systems **S** having the finite model property, so that all such systems are fully specified once the class of finite **S**-frame algebras is given.

By an *L-evaluation*, or simply *evaluation*, $v : (X, R) \to L$ we mean a function $v : X \to L$, where (X, R) is a frame and L is a finite set. Note that if $L = \mathcal{P}(p_1, \ldots, p_n)$ then an L-evaluation is nothing but a usual Kripke model which forces modal formulas in variables $p_1, \ldots, p_n$ (see the exercises for the description of the forcing relation).

We fix an arbitrary equational theory **S** containing **K4** for the rest of the section. Consequently, from now on by a frame we mean an **S**-frame. Let (X, R) be a frame, $Y \subseteq X$, $S = R \cap (Y \times Y)$. Then (Y, S) is a *generated subframe* of (X, R) iff for any $y \in Y$ and $x \in X$, if yRx then $x \in Y$. We have the following easy Lemma (for further information concerning classes of finite frames see the exercises):

LEMMA 5.1 (i) *If (Y,S) is a generated subframe of an* **S***-frame (X,R) then (Y,S) is an* **S***-frame as well;*

(ii) *If* (X_i, R_i) *are* **S***-frames for i=1,...,n then the disjoint sum* $\coprod_{i=1}^n (X_i, R_i)$ *is an* **S***-frame;*

(iii) *If* $f : (X, R) \to (Y, S)$ *is a surjective morphism of frames and* (X, R) *is an* **S***-frame then* (Y, S) *is an* **S***-frame as well.*

We shall define three relations on evaluations in terms of some games. They are variants of the Ehrenfeucht-Fraissé games adopted by K.Fine to the context of modal logic. In fact, we will describe only one game and the others will be obtained as slight modifications of its. Let $v : (X, R) \to L$ and $u : (Y, S) \to L$ be two L-evaluations, n be a natural number or ∞. The n-game on v and u is played by two players, Player I and Player II. In the first move Player I chooses one frame, either (X, R) or (Y, S) and a node in it. Player II answers by choosing a node in the other frame. After k moves the players already constructed sequences $< x_1, \ldots, x_k >, < y_1, \ldots, y_k >$ with $x_i\, R x_{i+1}$, $y_i\, S y_{i+1}$ for every $i < k$. Then in the $(k+1)$-st move Player I chooses one of the frames, say (X, R), and a node $x_{k+1} \in X$ such that $x_k\, R\, x_{k+1}$. Player II chooses a node in the other frame, say $y_{k+1} \in Y$, such that $y_k\, S\, y_{k+1}$. If Player II can't make a move[1] at his turn then he immediately loses. Otherwise, the game terminates after m moves if either Player I can't make a move or $m = n$ (if $n = \infty$ then $m = n$ means that both players plays infinitely many moves). Then Player II wins iff for all natural $k \leq m$, $v(x_k) = u(y_k)$.

We define for $n \geq 1$

- $v \approx_n u$ iff Player II has a winning strategy in n-game on v and u.
- $v \leq_n u$ iff Player II has a winning strategy in n-game on v and u modified so that Player I must play the first move in X.

For $n \geq 0$ and $x \in X$, $y \in Y$ we define

- $(v, x) \sim_n (u, y)$ iff Player II has a winning strategy in $n+1$-game on v and u modified so that the first moves of the players must be x and y.

Notice that the relations $\leq_n$ and $\approx_n$ are defined on L-evaluations whereas the relation $\sim_n$ is defined on L-evaluations with a distinguished nodes. The relations $\sim_n$ and $\approx_n$ are clearly equivalence relations, whereas the relation $\leq_n$ is only reflexive and transitive. The equivalence class of a pair (v, x) with respect to the equivalence relation $\sim_n$ will be denoted by $[(v, x)]_n$. The straightforward proposition below provides equivalent definitions of these relations.

PROPOSITION 5.2 *Given two L-evaluations $v : (X, R) \rightarrow L$, $u : (Y, S) \rightarrow L$, nodes $x \in X$, $y \in Y$, and $n \in \omega$. We have:*

(i) $(v, x) \sim_0 (u, y)$ *iff* $v(x) = u(y)$;

(ii) $(v, x) \sim_{n+1} (u, y)$ *iff* $v(x) = u(y)$, *and* $\forall x' \in X\ \exists y' \in Y$ (*if* $x\, R\, x'$ *then* $y\, S\, y'$ *and* $(v, x') \sim_n (u, y')$), *and vice versa;*

(iii) $v \leq_{n+1} u$ *iff* $\forall x \in X\ \exists y \in Y\ (v, x) \sim_n (u, y)$;

(iv) $v \approx_{n+1} u$ *iff* $v \leq_{n+1} u$ *and* $v \geq_{n+1} u$.

(v) $(v, x) \sim_\infty (u, y)$ *iff* $v(x) = u(y)$, *and* $\forall x' \in X\ \exists y' \in Y$ (*if* $x\, R\, x'$ *then* $y\, S\, y'$ *and* $(v, x') \sim_\infty (u, y')$), *and vice versa;*

[1] Since accessibility relations of our frames need not to be reflexive it may happen that there is no point accessible from a given point.

(vi) $v \leq_\infty u$ *iff* $\forall x \in X\ \exists y \in Y\ (v, x) \sim_\infty (u, y)$;

(vii) $v \approx_\infty u$ *iff* $v \leq_\infty u$ *and* $v \geq_\infty u$. □

Notice that $u \approx_\infty v$ is equivalent to $\forall_{n\in\omega}\ u \approx_n v$. The argument is similar to the one given in the previous Chapter. On one side it is evident that if there exists a strategy for the ∞-game, then there are also strategies for n-games, for all $n \in \omega$. On the other hand, suppose that there are all such finite strategies and suppose that Player I chooses a point $x \in X$. Since our frames are finite and for every $n \geq 0$ there exists $y_n \in Y$ such that $(v, x) \sim_n (u, y_n)$, there is also $y \in Y$ (independent of n) such that $(v, x) \sim_n (u, y)$ for all $n \in \omega$. If Player II answers this y and continues this way, it is clear that he will win the ∞-game. So we found the desired winning strategy for Player II in the ∞-game.

Let $v : (X, R) \to L$ be an L-evaluation, $x \in X$. We introduce the following notation:

$$X_{\uparrow x} = \{x\} \cup \{x' \in X \mid xRx'\}$$

$$R_{\uparrow x} = R \cap (X_{\uparrow x} \times X_{\uparrow x})$$

$$i_{\uparrow x} : (X_{\uparrow x}, R_{\uparrow x}) \longrightarrow (X, R)$$

$$v_{\uparrow x} = v \circ i_{\uparrow x} \quad .$$

where $i_{\uparrow x}$ is the inclusion morphism in $\mathbf{F_S}$. Thus the L-evaluation $v_{\uparrow x}$ is the restriction of v to x and to the elements accessible from x. Sometimes we use also the more direct notation $\uparrow x$ instead of $X_{\uparrow x}$.[2]

LEMMA 5.3 *Let* $v : (X, R) \to L$, $w : (Y, S) \to L$ *be* L*-evaluations,* $x \in X$, $y \in Y$, $n \geq 1$. *If* $(v, x) \sim_n (w, y)$ *then* $v_{\uparrow x} \approx_n w_{\uparrow y}$.

Proof. Assume $v : (X, R) \to L$, $w : (Y, S) \to L$, $x \in X$, $y \in Y$ and $(v, x) \sim_n (w, y)$. We shall show that $v_{\uparrow x} \approx_n w_{\uparrow y}$. Pick $x' \in X_{\uparrow x}$. Then either $x = x'$ or xRx'. In the former case $(v_{\uparrow x}, x') \sim_n (w_{\uparrow y}, y)$. In the latter case, there is $y' \in Y$ such that yRy' and $(v_{\uparrow x}, x') \sim_{n-1} (w_{\uparrow y}, y')$. Thus by Proposition 5.2(iii), $v_{\uparrow x} \leq_n w_{\uparrow y}$. Similarly, we can show that $w_{\uparrow y} \leq_n v_{\uparrow x}$. Hence by Proposition 5.2(iv) $v_{\uparrow x} \approx_n w_{\uparrow y}$, as required. □

The converse of Lemma 5.3 is not true even if we assume that $v(x) = w(y)$. To see this, let $X = \{0\}$, $Y = \{0, 1\}$ $R = \emptyset$, $S = \{< 0, 1 >\}$, $x = y = 0$,

[2] In the case of modal logic, we prefer to use the customary notation for Kripke semantics, that's why $\uparrow x$ plays now the same role played by $\downarrow x$ in the previous Chapter. In fact, for intuitionistic logic, we use $\leq$ instead of the customary $\geq$ and consequently we take downward closed subsets instead of upward closed ones; in the modal case, howvever, we turn to usual upward closed subsets.

$L = \{a\}$, $v : (X, R) \to L$, $w : (Y, S) \to L$. Then $v_{\uparrow x} \approx_1 w_{\uparrow y}$, $v(x) = w(y)$ but not $(v, x) \sim_1 (w, y)$. A partial converse of Lemma 5.3 is the following statement: if $v_{\uparrow x} \approx_{n+1} w_{\uparrow y}$ and $v(x) = w(y)$, then $(v, x) \sim_n (w, y)$.

The relations $\sim_\infty$ and $\approx_\infty$ may have infinitely many equivalence classes. However, we have

PROPOSITION 5.4 *Let $n \in \omega$. Then the equivalence relations $\approx_{n+1}$ and $\sim_n$ on L-evaluations and L-evaluations with distinguished nodes, respectively, have finite index.*

Proof. Let I_n and J_n be the numbers of equivalence classes of $\sim_n$ and $\approx_n$, respectively. By Proposition 5.2, the equivalence class $[v]_{\approx_{n+1}}$ of an L-evaluation $v : (X, R) \to L$ is uniquely determined by the set $\{[(v, x)]_n : x \in X\}$. Thus, if I_n is finite then so is J_{n+1}. It remains to show that I_n is finite, for $n \in \omega$. By Proposition 5.2, we have that I_0 is the number of elements of L, finite by assumption, and $I_{n+1} \leq I_0 \times 2^{I_n}$. Hence I_n is finite for all $n \in \omega$, as required. □

If $v_i : (X_i, R_i) \to L$ are L-evaluations, for $i \in m$, then

$$[v_0, \ldots, v_{m-1}] = [v_i]_{i \in m} : \coprod_{i \in m} (X_i, R_i) \longrightarrow L$$

denotes the L-evaluation on the disjoint sum $\coprod_{i \in m}(X_i, R_i)$, such that $[v_i]_{i \in m}(x) = v_i(x)$, for $x \in X_i$.

PROPOSITION 5.5 *Let $h : (X, R) \to (Y, S)$ be a morphism of frames, $u : (Y, S) \to L$ an L-evaluation, then*

(i) $(u, h(x)) \sim_\infty (u \circ h, x)$, *for $x \in X$;*

(ii) $u \circ h \leq_\infty u$;

(iii) *if h is surjective, then $u \circ h \approx_\infty u$;*

(iv) *for $m \geq 1$, let $\{h_i : (X_i, R_i) \to (Y, S)\}_{i \in m}$ be a jointly surjective family of morphisms in $\mathbf{F_S}$, then $u \approx_\infty [u \circ h_i]_{i \in m}$;*

(v) $u \approx_\infty [u_{\uparrow y}]_{y \in Y}$.

Proof. (ii) and (iii) follow from (i) and Proposition 5.2. (iv) is reduced to (iii) (because if $\{h_i\}_i$ is jointly surjective, then $[h_i]_i$ is surjective and $[u \circ h_i]_i = u \circ [h_i]_i$); (v) is a special case of (iv). To prove (i), we need to define a strategy for Player II in the ∞-game on $u \circ h$ and u starting in nodes x and $h(x)$, for $x \in X$. Let the last chosen pair in the game be $x' \in X$ and $y' \in Y$ with $h(x') = y'$. If Player I chooses a node $x'' \in X$, then Player II plays $h(x'') \in Y$. If Player I chooses a node $y'' \in Y$ then, since $y' \, S \, y''$ and h is a frame morphism, there is $x'' \in X$ such that $x' \, R \, x''$ and $h(x'') = y''$. This gives the required strategy in the ∞-game. □

Corollary 5.6 *Let h_0, h_1 be morphisms in $\mathbf{F_S}$, v_0, v_1 be a pair of L-evaluations, such that $v_0 \circ h_0 = v_1 \circ h_1$, i.e. the diagram*

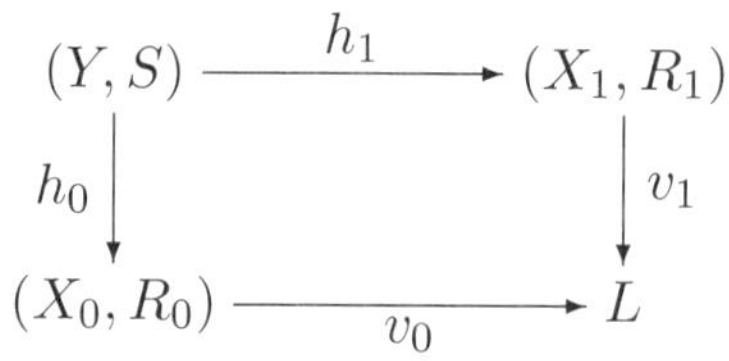

commutes. Then, we have:

(i) $(v_0, h_0(y)) \sim_\infty (v_1, h_1(y))$, *for $y \in Y$;*

(ii) *if h_0 is onto then $v_0 \leq_\infty v_1$;*

(iii) *if both h_0 and h_1 are onto then $v_0 \approx_\infty v_1$.* □

The converse to the statements of Corollary 5.6 are not true in general. In the next Section we shall discuss this problem. However we have the following Proposition.

An evaluation $v : (X, R) \to L$ is said to be *irreducible* iff for any $x, x' \in X$, if $(v, x) \sim_\infty (v, x')$ then $x = x'$.

Proposition 5.7 *Let $v_i : (X_i, R_i) \to L$, for $i = 0, 1$ be L-evaluations. Then*

(i) *there is an irreducible evaluation $v : (X, R) \to L$ and a surjective morphism of frames $h_1 : (X_1, R_1) \to (X, R)$ such that $v_1 = v \circ h_1$;*

(ii) *if $v_0 \leq_\infty v_1$ then there are an L-evaluation $v : (X, R) \to L$ and morphisms $h_i : (X_i, R_i) \to (X, R)$ in $\mathbf{F_S}$, for $i = 0, 1$, such that h_1 is surjective and $v_i = v \circ h_i$, for $i = 0, 1$, i.e. the triangles*

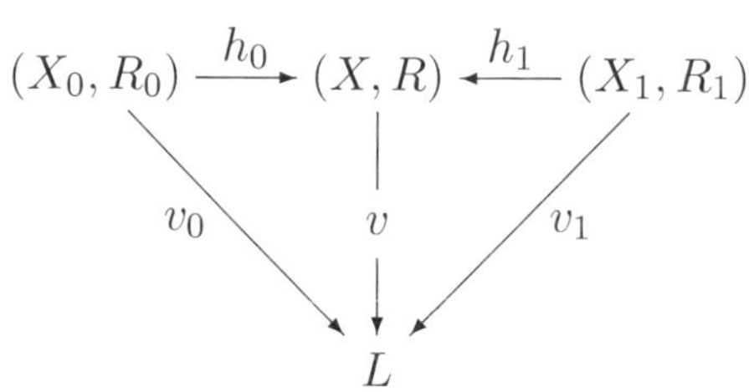

commute;

(iii) *if $v_0 \approx_\infty v_1$ then there are an L-evaluation $v : (X, R) \to L$ and surjections $h_i : (X_i, R_i) \to (X, R)$ in $\mathbf{F_S}$, for $i = 0, 1$, such that $v_i = v \circ h_i$, for $i = 0, 1$.*

Proof. Ad (i). Define on X_1 the equivalence relation

$$x \sim y \text{ iff } (u, x) \sim_\infty (u, y).$$

Let X be the quotient set $X_1/_\sim$ and define the relation R on it by

$$[x]_\sim R[y]_\sim \text{ iff } \exists z\,(xR_1z \,\&\, (u, z) \sim_\infty (u, y)).$$

Define also $v([x]_\sim) = v_1(x)$ and $h_1(x) = [x]_\sim$, for $x \in X_1$. It is immediate to check that the definitions are sound and that the canonical map h_1 is a surjective frame morphism such that $v \circ h_1 = v_1$. Finally, v is irreducible, because if $(v, h_1(x)) \sim_\infty (v, h_1(y))$ then, by Proposition 5.5(i) $(v_1, x) \sim_\infty (v_1, y)$, hence $h_1(x) = h_1(y)$.

Ad (ii) and (iii). Let h_1 and v be as in (i). We put $h_0(y) = [x]_\sim$, where $x \in X_1$ and $(v_1, x) \sim_\infty (v_0, y)$. Such x always exists since $v_0 \leq_\infty v_1$. Again the verification that $v_0 = v \circ h_0$ and that h is a frame morphism are easy. Finally, if $v_0 \approx_\infty v_1$, then in fact h_0 is surjective, as required. □

COROLLARY 5.8 *Let (X, R), and (Y, S) be frames in $\mathbf{F_S}$, $id_X : (X, R) \to X$, and $u : (Y, S) \to X$ be X-evaluations. Then*

(i) *if $u \leq_\infty id_X$ iff $u : (Y, S) \to (X, R)$ is a morphism of frames;*

(ii) *if $u \approx_\infty id_X$ iff $u : (Y, S) \to (X, R)$ is a surjective morphism of frames.* □

Let (X, R) be an **S**-frame. Let R^+ denote the least reflexive relation on X containing R, $\asymp_R$ denotes the equivalence relation $R^+ \cap (R^+)^{op}$ on X. A *cluster* in (X, R) is an equivalence class of $\asymp_R$. Let $\leq_R$ be the partial order relation on the set of clusters $X/\asymp_R$, induced by the relation R. The cluster containing x is denoted by $[x]_R$. Let $n \in \omega$. We say that $[x]_R$ is an *n-cluster* iff the set $\{y \in [x]_R : xRy\}$ has n elements. In particular 0-clusters are equivalence classes of irreflexive points. The *height of frame* (X, R) is the height of $(X/\asymp_R, \leq_R)$. For $x \in X$, the *height of a point* x is the height of the cluster $[x]_R$ in $(X/\asymp_R, \leq_R)$. We denote $ht(X, R)$ and $ht(x, X, R)$ (or simply $ht(x)$ if the frame (X, R) is understood) the height of a frame (X, R) and the height of a point x, respectively.

PROPOSITION 5.9 *Let* $u : (X, R) \to L$ *be an L-evaluation,* $x \in X$, $ht(x) = n$. *Then for any L-evaluation* $v : (Y, S) \to L$ *and any* $y \in Y$, *if* $(u, x) \sim_{2n} (v, y)$ *then* $(u, x) \sim_\infty (v, y)$.

Proof. We argue by induction on n. $n = 1$ means that x is in a $\leq_R$-minimal cluster. If $(u, x) \sim_2 (v, y)$ then $u(x) = v(y)$ and either both x and y are irreflexive or for every $y' \in Y$ such that yR^+y' elements accessible from y' have the same labels as the elements in the cluster $[x]_R$. From that description it easily follows that $(u, x) \sim_\infty (v, y)$.

Now we assume that $n > 1$, and that the Lemma holds for m smaller than n. We have $ht(x) = n$ and $(u, x) \sim_{2n} (v, y)$. It is enough, if we show, by induction on $k \geq 2n - 1$, that for every $x' \in X$, $y' \in Y$,

$$\text{if } x' \in [x]_R,\ yR^+y' \text{ and } (u, x') \sim_{2n-1} (v, y') \text{ then } (u, x') \sim_k (v, y').$$

It follows, in particular, that $(u, x) \sim_k (v, y)$ for all $k \in \omega$, hence $(u, x) \sim_\infty (v, y)$.

For $k = 2n - 1$, there is nothing to prove. Assume that $k > 2n - 1$.

- *Case 1:* if Player I takes x'' such that $x'Rx''$, then there is y'' such that $(u, x'') \sim_{2n-2} (v, y'')$. If $x'' \notin [x]_R$, we can apply induction on n because $ht(x'') < n$, so that $(u, x'') \sim_\infty (v, y'')$. Suppose that $x'' \in [x]_R$; as $(u, x) \sim_{2n} (v, y)$ there is x''' such that xRx''', hence also $x''Rx'''$, and $(u, x''') \sim_{2n-1} (v, y'')$.

- *Subcase 1.1:* If $ht(x''') < n$, then $(u, x''') \sim_{2n-1} (v, y'') \sim_{2n-2} (u, x'')$, so $(u, x''') \sim_\infty (v, y'')$ and $(u, x''') \sim_\infty (u, x'')$, hence $(u, x'') \sim_\infty (v, y'')$, so y'' is a good answer for Player II.

- *Subcase 1.2:* If $ht(x''') = n$, then $x''' \in [x]_R$ so in particular $x''Rx'''$ and $x'''Rx''$; as $u(x''') = v(y'') = u(x'')$, then $(u, x''') \sim_\infty (u, x'')$. In conclusion we have $(u, x'') \sim_\infty (u, x''') \sim_{2n-1} (v, y'')$. By induction on k, we get $(u, x'') \sim_{k-1} (v, y'')$, so again y'' is a good answer for Player II.

- *Case 2:* If Player I takes y'' such that $y'Ry''$, then as $x' \in [x]_R$ and $(u, x) \sim_{2n} (v, y)$, there is x'' such that $(u, x'') \sim_{2n-1} (v, y'')$ and xRx'' and also $x'Rx''$. If $x'' \in [x]_R$ then, by induction on k, we get $(u, x'') \sim_{k-1} (v, y'')$. If $x'' \notin [x]_R$ then $ht(x'') < n$ and, by induction on n, $(u, x'') \sim_\infty (v, y'')$. □

COROLLARY 5.10 *Let* $v : (X, R) \to L$, $u : (Y, S) \to L$ *be L-evaluations,* $ht(X, R) = n$. *Then*

(i) $v \leq_{2n+1} u$ iff $v \leq_\infty u$;

(ii) $u \leq_{2n+1} v$ iff $u \leq_\infty v$;

(iii) $u \approx_{2n+1} v$ iff $u \approx_\infty v$.

Proof. (iii) follows from (i), (ii) and Proposition 5.2. The proofs of (i) and (ii) are similar. We shall show (i).

Let v and u be as in the Corollary. Pick $x \in X$. Then there is a $y \in Y$ such that $(v,x) \sim_{2n} (u,y)$. Since $ht(x) \leq ht(X,R) \leq n$, by Proposition 5.9, $(v,x) \sim_\infty (u,y)$. Thus by Proposition 5.2, $v \leq_\infty u$. □

LEMMA 5.11 *Let $v : (X,R) \to L$, $u : (Y,S) \to L$ be L-evaluations, $n \geq 1$. Then*

(i) *if $v \leq_{n+1} u$ then there is in $\mathbf{F_S}$ an inclusion morphism $h : (Y',S') \hookrightarrow (Y,S)$, such that $v \approx_n u \circ h$;*

(ii) *if $v \leq_\infty u$ then there is in $\mathbf{F_S}$ an inclusion morphism $h : (Y',S') \hookrightarrow (Y,S)$, such that $v \approx_\infty u \circ h$.*

Proof. We shall show (i). (ii) can be proved similarly.

Let v and u be as in the Lemma. We put

$$Y' = \{y' \in Y \mid \exists x \in X\ \exists y \in Y\ \ (v,x) \sim_n (u,y) \text{ and } yS^+y'\}$$

and $S' = S \cap (Y' \times Y')$. Clearly (Y',S') is a generated subframe of (Y,S), i.e. the inclusion $h : (Y',S') \hookrightarrow (Y,S)$ is a morphism in $\mathbf{F_S}$. We need to verify that $v \approx_n u'$, where $u' = u \circ h$. We will use Proposition 5.2(iii,iv).

For any $x \in X$ there is a $y \in Y$ such that $(v,x) \sim_n (u,y)$. But then $y \in Y'$ and $(v,x) \sim_n (u',y)$.

Now, let $y' \in Y'$. By definition of Y', there are $x \in X$ and $y \in Y$ such $(v,x) \sim_n (u,y)$ and yS^+y'. If $y = y'$ we are done. If ySy', there is $x' \in X$ such that xRx' and $(v,x') \sim_{n-1} (u,y') \sim_\infty (u',y')$. Hence by Proposition 5.2, $v \approx_n u \circ h$. □

2. The category of finite S-frames

We fix an arbitrary equational theory **S** containing **K4** for the whole section.

In this section we shall study the category $Alg(\mathbf{S})_{fin}$ and its dual $\mathbf{F_S}$. We prove (Theorem 5.14) that jointly surjective families form a basis for the canonical topology on $\mathbf{F_S}$ iff the amalgamation property (AP)[3] holds in $Alg(\mathbf{S})_{fin}$.

In the following Lemma we collect some facts concerning $\mathbf{F_S}$.

LEMMA 5.12 (i) *A morphism in $\mathbf{F_S}$ is epi iff it is surjective;*

[3]See Section 2.6 for the definition of (AP) and the related notions.

(ii) *The forgetful functor* $U : \mathbf{F_S} \to Set_{fin}$ *preserves finite colimits;*

(iii) *Pullback of an injective morphism along any morphism exists in* $\mathbf{F_S}$*, is injective and is preserved by the forgetful functor* U*. In particular, pullback of a surjective morphism in* $\mathbf{F_S}$ *by an injective morphism is surjective;*

(iv) *Every morphism in* $\mathbf{F_S}$ *factors as a surjective morphism followed by an injective morphism;*

(v) *A morphism in* $\mathbf{F_S}$ *is mono iff it is injective.*

Proof. To see (i) and (ii) consider the following commutative diagram of categories and functors

$$\begin{array}{ccc} Alg(\mathbf{S})_{fin} & \xrightarrow{\cong} & \mathbf{F_S}^{op} \\ {\scriptstyle U'}\downarrow & & \downarrow{\scriptstyle U^{op}} \\ BA_{fin} & \xrightarrow[\cong]{} & Set_{fin}^{op} \end{array}$$

where BA_{fin} is the category of finite Boolean algebras, the horizontal functors are equivalences of categories, establishing dualities between respective categories and U' is the forgetful functor. Since U' preserves finite limits, by duality, U preserves finite colimits. This shows (ii). Since a morphism is a monomorphism in a category iff it is an epimorphism in its dual, the duality functor sends monos in $Alg(\mathbf{S})_{fin}$ to epis in $\mathbf{F_S}$. Moreover, from the way the duality functor $Alg(\mathbf{S})_{fin} \to \mathbf{F_S}$ is constructed it is easy to see that it sends injective morphisms in $Alg(\mathbf{S})_{fin}$ to surjective morphisms in $\mathbf{F_S}$. Thus, since monos are injective in $Alg(\mathbf{S})_{fin}$ (this is true for any equational theory $\mathbf{S}$), epis are surjective in $\mathbf{F_S}$. This shows (i). (iv) is easy. We shall comment on (iii).

We need to show that for any pair of morphisms $m : (X_1, R_1) \to (X, R)$, $f : (X_2, R_2) \to (X, R)$ in $\mathbf{F_S}$ with m injective, there is a pullback of m along f. We assume that m is an inclusion. Then an easy verification shows that the square

$$\begin{array}{ccc} (Y, S) & \xrightarrow{m'} & (X_2, R_2) \\ {\scriptstyle f'}\downarrow & & \downarrow{\scriptstyle f} \\ (X_1, R_1) & \xrightarrow[m]{} & (X, R) \end{array}$$

is a pullback in $\mathbf{F_S}$, where $Y = f^{-1}(X_1)$ and $S = R_2 \cap (Y \times Y)$, m' an inclusion and f' a restriction of f.

Finally, we show (v). In any category, a morphism is mono iff the square

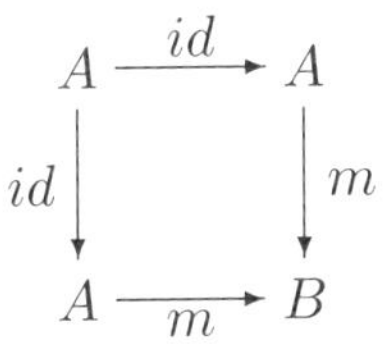

is a pullback. By (iii), such pullbacks are preserved by the forgetful functor U. Thus if m is a mono in $\mathbf{F_S}$ then it is mono in Set_{fin}, as well, i.e. it is injective. □

Let (X, R) be an $\mathbf{S}$-frame, $x, x' \in X$. Let $h : (X, R) \to (Y, S)$ be a morphism in $\mathbf{F_S}$, $x \in X$. We say that x is *h-minimal point* iff for any $x' \in X$, if xRx' and $h(x')S\,h(x)$ then $x'Rx$. In other words, a point x is h-minimal iff any node strictly below it is sent by h strictly below $h(x)$. It is easy to see that, if an element in a cluster is h-minimal, so are the other elements in that cluster. Note that h sends clusters into clusters (i.e. $h([x]_R) \subseteq [h(x)]_S$ for $x \in X$) and clusters containing h-minimal points onto clusters.

PROPOSITION 5.13 *The following are equivalent:*

(i) (AP^*) *holds in* $\mathbf{F_S}$;

(ii) (IT^*) *holds in* $\mathbf{F_S}$;

(iii) (AP^*_{st}) *holds in* $\mathbf{F_S}$.

Proof. (iii) ⇒ (i) is trivial.

(i) ⇒ (ii). Let f be a morphism in $\mathbf{F_S}$, s and i a surjective-injective factorization of f. Consider the following commutative diagram in $\mathbf{F_S}$:

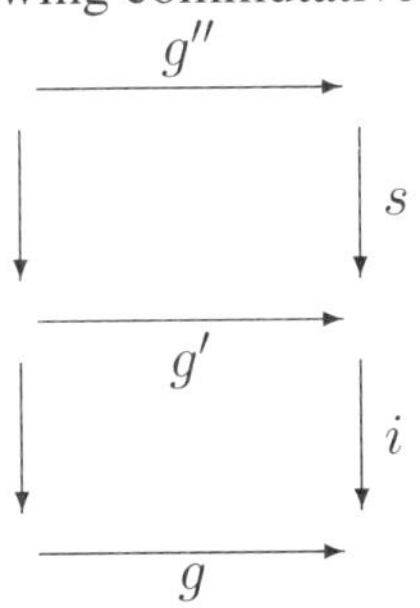

where g is a surjective morphism in $\mathbf{F_S}$, the bottom square is a pullback in $\mathbf{F_S}$ with g' surjective by Lemma 5.12(iii), and the top square given by (AP^*), with g'' surjective. Therefore, the outer square commutes, with g'' surjective, i.e. (IT^*) holds in $\mathbf{F_S}$.

(ii) ⇒ (iii). Assume that (IT^*) holds in $\mathbf{F_S}$. First we prove

Claim. For any pair of morphisms $h_i : (X_i, R_i) \to (X, R)$, $i = 0, 1$ in $\mathbf{F_S}$ and any pair of nodes $x_0 \in X_0$, $x_1 \in X_1$ with x_0 h_0-minimal and

$h(x_0) = h(x_1)$ there is a pair of morphisms $g_i : (Y, S) \to (X_i, R_i)$ in $\mathbf{F_S}$ $i = 0, 1$, and $y \in Y$ such that $h_0 \circ g_0 = h_1 \circ g_1$ and $g_0(y) = x_0$, $g_1(y) = x_1$.

Let h_0, h_1, x_0, x_1 be as in the Claim. We put, for $i = 0, 1$

$$(X', R') = (X_{\uparrow h_0(x_0)}, R_{\uparrow h_0(x_0)}),$$

$$(X_i', R_i') = (X_{\uparrow x_i}, R_{\uparrow x_i}), \qquad h_i' = h_i \lceil X_i' : (X_i', R_i') \longrightarrow (X', R').$$

Then h_0', h_1' are epimorphisms in $\mathbf{F_S}$. By (IT^*), there are morphisms g_0' and g_1' in $\mathbf{F_S}$ with g_1' epi such that the diagram

$$\begin{array}{ccc} (Y,S) & \xrightarrow{\;g_0'\;} & (X_0', R_0') \\ {\scriptstyle g_1'}\downarrow & & \downarrow{\scriptstyle h_0'} \\ (X_1', R_1') & \xrightarrow[\;h_1'\;]{} & (X', R') \end{array}$$

commutes. Fix $y \in Y$ such that $g_1'(y) = x_1$. Let $x_0' = g_0'(y)$. Since x_0 is h_0-minimal, it is h_0'-minimal as well. We have that $x_0(R_1')^+x_0'$ and

$$h_0'(x_0') = h_0'(g_0'(y)) = h_1'(g_1'(y)) = h_1'(x_1) = h_0'(x_0)$$

Therefore $x_0' \in [x_0]_{R_1'}$.[4]

Let $\tau : (X_0', R_0') \to (X_0, R_0)$ be a frame morphism that interchanges x_0 and x_0' leaving the other points in X_0' fixed.

Put $g_0 = \tau \circ g_0'$. Then the diagram (g_1 is g_1' followed by an inclusion)

$$\begin{array}{ccc} (Y,S) & \xrightarrow{\;g_0\;} & (X_0, R_0) \\ {\scriptstyle g_1}\downarrow & & \downarrow{\scriptstyle h_0} \\ (X_1, R_1) & \xrightarrow[\;h_1\;]{} & (X, R) \end{array}$$

commutes and $g_0(y) = x_0$, $g_1(y) = x_1$. This shows the Claim.

Now let $h_0 : (X_0, R_0) \to (X, R)$, $h_1 : (X_1, R_1) \to (X, R)$ be a pair of surjections in $\mathbf{F_S}$. Put

$$I = \{(x_0, x_1) \in X_0 \times X_1 \;:\; h_0(x_0) = h_1(x_1) \text{ and}$$

[4] If $x_0 = x_0'$ this is obvious; if $x_0 R_1' x_0'$, then $h_0'(x_0) = h_0'(x_0')$ is a reflexive point (because h_0' preserves the frame relation) and $x_0' R_1' x_0$ follows from $h_0'(x_0') R' h_0'(x_0)$ and h_0-minimality.

either x_0 is h_0−minimal or x_1 is h_1−minimal}

By the above Claim, for $(x_0, x_1) \in I$ and $i = 0,1$ there is $g_i^{(x_0,x_1)} : (Y^{(x_0,x_1)}, S^{(x_0,x_1)}) \to (X, R)$ in $\mathbf{F_S}$ and $y^{(x_0,x_1)} \in Y^{(x_0,x_1)}$ such that $h_0 \circ g_0^{(x_0,x_1)} = h_1 \circ g_1^{(x_0,x_1)}$ and $x_i = g_i^{(x_0,x_1)}(y^{(x_0,x_1)})$, for $i = 0,1$. We put $(Y,S) = \coprod_{(x_0,x_1)\in I}(Y^{(x_0,x_1)}, S^{(x_0,x_1)})$ and $g_i = [g_i^{(x_0,x_1)}]_{(x_0,x_1)\in I}$, for $i = 0,1$. Then the square

$$\begin{array}{ccc} (Y,S) & \xrightarrow{\;g_0\;} & (X_0,R_0) \\ {\scriptstyle g_1}\downarrow & & \downarrow{\scriptstyle h_0} \\ (X_1,R_1) & \xrightarrow[\;h_1\;]{} & (X,R) \end{array}$$

commutes. We shall show that both g_0 and g_1 are epi. Let $x_0 \in X_0$. Since h_1 is epi, there is $x_1 \in X_1$ such that $h_0(x_0) = h_1(x_1)$. Let $\check{x_1} \in X_1$ be h_1-minimal such that $h(x_1) = h(\check{x_1})$ and either $x_1 = \check{x_1}$ or $x_1 R \check{x_1}$. Such $\check{x_1}$ exists since X_1 is finite and h_1 is a frame morphism. Thus $g_0(y^{(x_0,\check{x_1})}) = x_0$, i.e. g_0 is epi. The argument for g_1 is similar.

We need to show that the diagram above is a pushout in $\mathbf{F_S}$. Let $q_0 : (X_0, R_0) \to (X', R')$, $q_1 : (X_1, R_1) \to (X', R')$ be a pair of morphisms in $\mathbf{F_S}$ such that $q_0 \circ g_0 = q_1 \circ g_1$. Let $x_0, x_0' \in X_0$ and $h_0(x_0) = h_0(x_0')$. There is $x_1 \in X_1$ such that x_1 is h_1-minimal and $h_1(x_1) = h_0(x_0)$. Then

$$q_0(x_0) = q_0 \circ g_0(y^{(x_0,x_1)}) = q_1 \circ g_1(y^{(x_0,x_1)}) =$$

$$= q_1(x_1) = q_1 \circ g_1(y^{(x_0',x_1)}) = q_0 \circ g_0(y^{(x_0',x_1)}) = q_0(x_0')$$

Thus, we can define $q : (X,R) \to (X',R')$ so that $q(x) = q_0(x_0)$ if $h_0(x_0) = x$ for some $x_0 \in X_0$. By the above considerations $q(x)$ does not depend on the choice of x_0. Since h_0 is epi, q is a total function and unique such that $q \circ h_0 = q_0$. Now, since h_0 is epi and both h_0 and q_0 are frame morphisms it is easy to verify that q is a frame morphism as well.

It remains to show that $q \circ h_1 = q_1$. We have

$$q \circ h_1 \circ g_1 = q \circ h_0 \circ g_0 = q_0 \circ g_0 = q_1 \circ g_1$$

But g_1 is epi, so $q \circ h_1 = q_1$ holds. □

THEOREM 5.14 *The following are equivalent*

(i) (AP) *holds in* $Alg(\mathbf{S})_{fin}$*;*

(ii) *finite jointly surjective families form a basis for the canonical topology* $\mathbf{J_S}$ *on* $\mathbf{F_S}$*;*

(iii) *finite jointly surjective families form a basis for a topology* $\mathbf{J_S}$ *on* $\mathbf{F_S}$.

Proof. (ii) $\Rightarrow$ (iii) is trivial.

(i) $\Rightarrow$ (ii). For the family of finite jointly surjective families to be a basis for a topology we only need to check stability condition, because the remaining ones are obviously true. Suppose that $\{h_i : (X_i, R_i) \to (X, R)\}_{i \in I}$ is a finite jointly surjective family in $\mathbf{F_S}$, $f : (Y, S) \to (X, R)$ is a morphism in $\mathbf{F_S}$. Then the morphism $[h_i]_{i\in I} : \coprod_{i\in I}(X_i, R_i) \longrightarrow (X, R)$ is surjective. By Proposition 5.13, (IT^*) holds in $\mathbf{F_S}$. Therefore we can construct a commutative diagram in $\mathbf{F_S}$, for $i \in I$,

$$\begin{array}{ccccc}
(Y_i, S_i) & \xrightarrow{\beta_i} & (Y', S') & \xrightarrow{h} & (Y, S) \\
\downarrow & & \downarrow & & \downarrow f \\
(X_i, R_i) & \xrightarrow[\alpha_i]{} & \coprod_{i\in I}(X_i, R_i) & \xrightarrow[{[h_i]_{i\in I}}]{} & (X, R)
\end{array}$$

where h is epi (by (IT^*)), α_i is the i-th injection into coproduct and the left hand square is a pullback with β_i injective (it exists by Lemma 5.12(iii)). Since $\{\alpha_i\}_{i\in I}$ is a jointly surjective family, by Lemma 5.12(iii), $\{\beta_i\}_{i\in I}$ is a jointly surjective family. Thus $\{h \circ \beta_i\}_{i\in I}$ is a jointly surjective family. But $h_i = [h_i]_{i\in I} \circ \alpha_i$ for $i \in I$, so the finite jointly surjective families are stable and hence they form a basis for a topology.

Since frames in $\mathbf{F_S}$ are finite, every jointly surjective family contains a finite jointly surjective subfamily. Thus the topology on $\mathbf{F_S}$ generated by the finite jointly surjective families of morphisms consists of jointly surjective sieves. Since epimorphisms in $\mathbf{F_S}$ are surjective, the sieves of the canonical topology on $\mathbf{F_S}$ must be jointly surjective. Thus, to show that the topology whose basis is formed by finite jointly surjective families is canonical, it is enough to show that the representable functors have the sheaf property with respect to jointly surjective families of morphisms. Thus, since frames in $\mathbf{F_S}$ are finite, it is enough to show that representable functors the have sheaf property with respect to finite jointly surjective families of morphisms in $\mathbf{F_S}$.

Let (Y, S) be an object of $\mathbf{F_S}$. We shall show that the functor represented by (Y, S) has the sheaf property with respect to finite jointly surjective families. Let $\{h_i : (X_i, R_i) \to (X, R)\}_{i\in I}$ be a finite jointly surjective family in $\mathbf{F_S}$, $\{f_i : (X_i, R_i) \to (Y, S)\}_{i\in I}$ a family of morphism in $\mathbf{F_S}$. Then, it is easy to see that $\{f_i\}_{i\in I}$ is a compatible family on $\{h_i\}_{i\in I}$ iff $\{[f_i]_{i\in I}\}$[5] is a compatible (one-element) family on (one-element) cover $\{[h_i]_{i\in I}\}$. Moreover the morphism

[5]If (X, R) is the empty frame $(\emptyset, \emptyset)$, I maybe empty. In that case $[f_i]_{i\in I}$ is the unique morphism from $(\emptyset, \emptyset)$ to (Y, S) and $[f_i]_{i\in I}$ is the identity on $(\emptyset, \emptyset)$.

$g : (X, R) \to (Y, S)$ is the extension of $\{f_i\}_{i \in I}$ iff it is the extension of $\{[f_i]_{i \in I}\}$. Thus it is enough to show that if $h : (X', R') \to (X, R)$ is a surjective morphism in $\mathbf{F_S}$ and if $\{f : (X', R') \to (Y, S)\}$ is a compatible family on the cover $\{h\}$ then there is a morphism $g : (X, R) \to (Y, S)$ such that $g \circ h = f$. By the dual of Lemma 2.17, we have a pair of morphisms h_0, h_1 such that

$$(X'', R'') \overset{h_0}{\underset{h_1}{\rightrightarrows}} (X', R') \xrightarrow{\ h\ } (X, R)$$

is a coequalizer in $\mathbf{F_S}$. Since $\{f\}$ is compatible on $\{h\}$, we have $f \circ h_0 = f \circ h_1$. So, by the property of coequalizers, there is a morphism $g : (X, R) \to (Y, S)$ such that $g \circ h = f$, as required.

(iii) $\Rightarrow$ (i). By Proposition 5.13, it is enough to show that (IT^*) holds in $\mathbf{F_S}$. Let $h : (X', R') \to (X, R)$, $f : (Y, S) \to (X, R)$ be morphisms in $\mathbf{F_S}$ with h surjective. Then $\{h\}$ is a jointly surjective family. By (iii), jointly surjective families are stable, i.e. there is a jointly surjective family $\{k_i\}_{i \in I}$ such that the diagram

$$\begin{array}{ccc}
(Y_i, S_i) & \xrightarrow{\ k_i\ } & (Y, S) \\
\downarrow & & \downarrow f \\
(X', R') & \xrightarrow[\ h\ ]{} & (X, R)
\end{array}$$

commutes for $i \in I$. Hence the diagram

$$\begin{array}{ccc}
\coprod_{\in I}(Y_i, S_i) & \xrightarrow{\ [k_i]_{i \in I}\ } & (Y, S) \\
\downarrow & & \downarrow f \\
(X', R') & \xrightarrow[\ h\ ]{} & (X, R)
\end{array}$$

commutes as well, and $[h_i]_{i \in I}$ is surjective. Thus (IT^*) holds in $\mathbf{F_S}$. $\Box$

The converses of the statements of Corollary 5.6 are not true in general. We have

PROPOSITION 5.15 *The following are equivalent:*

(i) $Alg(\mathbf{S})_{fin}$ *satisfies* (AP);

(ii) *for any* L*-evaluations* $v_i : (X_i, R_i) \to L$, *with* (X_i, R_i) *in* $\mathbf{F_S}$ *i=0,1, if* $v_0 \leq_\infty v_1$ *then there are a surjection* $h_0 : (Y, S) \to (X_0, R_0)$ *and a morphism* $h_1 : (Y, S) \to (X_1, R_1)$ *in* $\mathbf{F_S}$ *such that* $v_0 \circ h_0 = v_1 \circ h_1$, *i.e. the square*

$$\begin{array}{ccc} (Y,S) & \xrightarrow{h_1} & (X_1,R_1) \\ {\scriptstyle h_0}\downarrow & & \downarrow{\scriptstyle v_1} \\ (X_0,R_0) & \xrightarrow[v_0]{} & L \end{array}$$

commutes;

(iii) *for any L-evaluations* $v_i : (X_i, R_i) \to L$, *with* (X_i, R_i) *in* $\mathbf{F_S}$ *for* $i = 0, 1$, *if* $v_0 \approx_\infty v_1$ *then there are surjections* $h_i : (Y, S) \to (X_i, R_i)$ *for* $i = 0, 1$ *in* $\mathbf{F_S}$ *such that* $v_0 \circ h_0 = v_1 \circ h_1$.

Proof. (i) $\Rightarrow$ (ii). Assume (i). Let $v_i : (X_i, R_i) \to L$ be L-evaluations, for $i = 0, 1$, with $v_0 \leq_\infty v_1$. By Proposition 5.7(ii), there are morphisms $k_i : (X_i, R_i) \to (X, R)$ in $\mathbf{F_S}$ and an evaluation $v : (X, R) \to L$ such that $v \circ k_i = v_i$ for $i = 0, 1$ and k_1 surjective. By Proposition 5.13, (IT^*) holds in $\mathbf{F_S}$. Thus, there are a surjection $h_0 : (Y, S) \to (X_0, R_0)$ and a morphism $h_1 : (Y, S) \to (X_1, R_1)$ in $\mathbf{F_S}$ such that $k_0 \circ h_0 = k_1 \circ h_1$. Then

$$v_0 \circ h_0 = v \circ k_0 \circ h_0 = v \circ k_1 \circ h_1 = v_1 \circ h_1$$

as required.

(ii) $\Rightarrow$ (iii). Assume (ii). Let $v_i : (X_i, R_i) \to L$ be L-evaluations (X_i, R_i) with in $\mathbf{F_S}$ for $i = 0, 1$, with $v_0 \approx_\infty v_1$. By (ii), there are morphisms $g_i : (Y_i, S_i) \to (X_0, R_0)$ and $k_i : (Y_i, S_i) \to (X_1, R_1)$ such that $v_1 \circ k_i = v_0 \circ g_i$ for $i = 0, 1$ with g_0 and k_1 surjective. Thus, we have a commuting diagram

$$\begin{array}{ccc} (Y_0,S_0) \sqcup (Y_1,S_1) & \xrightarrow{[k_0,k_1]} & (X_1,R_1) \\ {\scriptstyle [g_0,g_1]}\downarrow & & \downarrow{\scriptstyle v_1} \\ (X_0,R_0) & \xrightarrow[v_0]{} & L \end{array}$$

with both $[k_0, k_1]$ and $[g_0, g_1]$ surjective, as required.

(iii) $\Rightarrow$ (i). Note that if morphisms $h_i : (X_i, R_i) \to (X, R)$ in $\mathbf{F_S}$ are surjective, for $i = 0, 1$, then, by Corollary 5.8(ii), $h_0 \approx_\infty h_1$ (as X-evaluations). The remaining details are easy. □

The above Proposition says that the converses to Corollary 5.6 (ii) and (iii) holds iff the amalgamation property holds in $Alg(\mathbf{S})_{fin}$. We want to point out that the converse of Corollary 5.6 (i) is equivalent (in the above sense) to an even stronger property of $Alg(\mathbf{S})_{fin}$, called superamalgamation, cf. [Ma2].

3. The category $\mathbf{M_S}$

In this section we assume again that $\mathbf{S}$ is an equational theory containing **K4**, moreover it has the finite model property, and $Alg(\mathbf{S})_{fin}$ has the amalgamation property. Thus, by Theorem 5.14, finite jointly surjective families form a basis for the canonical topology $\mathbf{J_S}$ on $\mathbf{F_S}$. By a sheaf we will always mean a $\mathbf{J_S}$-sheaf. We define the category $\mathbf{M_S}$ which in the following Sections will be proved to be equivalent to the dual of the category $Alg(\mathbf{S})_{fp}$.

For any finite set L, we define a presheaf (contravariant functor)

$$h_L : \mathbf{F_S}^{op} \longrightarrow Set$$

such that $h_L(X, R) = L^X$ and $h_L(k)(v) = v \circ k$ for any $\mathbf{S}$-frame (X, R), any morphism $k : (Y, S) \to (X, R)$ in $\mathbf{F_S}$ and any L-evaluation $v : (X, R) \to L$.

LEMMA 5.16 *h_L is a sheaf.*

Proof. Let $\{f_i : (X_i, R_i) \to L\}_{i\in I}$ be the compatible family on a finite jointly surjective cover $\{h_i : (X_i, R_i) \to (X, R)\}_{i\in I}$. We need to show that there is a unique L-evaluation $g : (X, R) \to L$ such that $g \circ h_i = f_i$, for $i \in I$. The uniqueness follows from the fact that $\{h_i\}_{i\in I}$ is jointly surjective. We shall prove the existence of g.

Let $f : (X', R') \to L$ denote $[f_i]_{i\in I}$ and $h : (X', R') \to (X, R)$ denote $[h_i]_{i\in I}$. Clearly h is surjective. It is easy to see that $\{f\}$ is a compatible family on $\{h\}$ and that an L-evaluation g is an extension of f on the cover $\{h\}$ iff g is an extension of $\{f_i\}_{i\in I}$ on $\{h_i\}_{i\in I}$. By Proposition 5.13, (AP^*_{st}) holds in $\mathbf{F_S}$, and hence epis are regular. Thus, there are morphisms h_0 and h_1, such that the diagram

$$(X'', R'') \underset{h_1}{\overset{h_0}{\rightrightarrows}} (X', R') \xrightarrow{\ h\ } (X, R)$$

is a coequalizer in $\mathbf{F_S}$. So, by Lemma 5.12(ii), it is a coequalizer in Set_{fin}. Since f is compatible $f \circ h_0 = f \circ h_1$. Thus, there is a function $g : X \to L$ such that $g \circ h = f$. Hence $g : (X, R) \to L$ is the required extension. □

Let C^L be a subsheaf of h_L. C^L is *$\leq_n$-closed* (*$\approx_n$-closed*) iff $\forall\, v \in C^L\ \forall w \in h_L\ w \leq_n v$ ($w \approx_n v$) implies $w \in C^L$. If C^L is $\leq_n$-closed then n is said to be an *index* of C^L. C^L is *closed* iff it is $\leq_n$-closed for some $n \in \omega$.

An object of $\mathbf{M_S}$ is a pair $< L, C >$ (denoted C^L for short), where L is a finite set and C is a closed subsheaf of h_L. A morphism $f : C^L \to D^M$ in $\mathbf{M_S}$ is a natural transformation with an index, i.e. f is a natural transformation between sheaves C^L and D^M and there is $n \in \omega$ (called *index*), such that, for all $m \in \omega$, all L-evaluations v, u in C^L and all $x \in dom(v)$, $y \in dom(u)$

$$\text{if } (v, x) \sim_{n+m} (u, y) \text{ then } (f(v), x) \sim_m (f(u), y).$$

Note that if an object C^L or a morphism f has an index n then it has index n', for any $n' \geq n$. In the following we will be interested in computing indexes of objects and morphisms in terms of indexes of some other objects and morphisms. Even if it makes sense to ask about the least of such indexes, we will never insist on it. Just pure existence will be always enough for us.

The following three Lemmas give some useful characterizations and connections between the notions introduced above.

LEMMA 5.17 *Let C^L be a subsheaf of h_L, $n \in \omega$. Then C^L is $\leq_n$-closed iff C^L is $\approx_n$-closed.*

Proof. It is clear that, if C^L is $\leq_n$-closed then it is $\approx_n$-closed.

Assume that C^L is $\approx_n$-closed. Let $v : (X, R) \to L \in C^L$, $w : (Y, S) \to L \in h_L$ and $w \leq_n v$. Then, the inclusion map $i : (Y, S) \to (X \sqcup Y, R \sqcup S)$ is a morphism in $\mathbf{F_S}$ and $[v, w] \approx_n v$. Since C^L is $\approx_n$-closed, $[v, w] \in C^L$. Since C^L is a presheaf, $w = [v, w] \circ i \in C^L$, as required. □

As a consequence of Lemma 5.17, we say that sheaf C^L is *n-closed* or has index n iff it is either $\leq_n$-closed or $\approx_n$-closed.

LEMMA 5.18 *Let C^L be an object of $\mathbf{M_S}$ with index $n+1$. If $v : (X, R) \to L$ is an evaluation such that, for each $x \in X$ there are an evaluation $v_x : (Y_x, S_x) \to L \in C^L$ and a point $y_x \in Y_x$ such that $(v, x) \sim_n (v_x, y_x)$ for $x \in X$, then $v \in C^L$.*

Proof. Let v and v_x for $x \in X$ be as in the Lemma. Since C^L is a sheaf, $[v_x]_{x \in X}$ is in C^L and by, Proposition 5.2(iii), $v \leq_{n+1} [v_x]_{x \in X}$. Since $n+1$ is an index of C^L, $v \in C^L$. □

LEMMA 5.19 *Let $n \in \omega$ and let $f : C^L \to D^M$ be a transformation*[6] *between objects of $\mathbf{M_S}$ such that for $v, u \in C^L$ and $x \in dom(v)$, $y \in dom(u)$*

$$(*) \qquad (v, x) \sim_n (u, y) \quad \textit{implies} \quad (f(v), x) \sim_0 (f(u), y).$$

Then f is a natural transformation and has index n, i.e. it is a morphism in $\mathbf{M_S}$.

Proof. Assume $(*)$. We shall show by induction on $m \in \omega$ that

$$(*)_m \qquad (v, x) \sim_{n+m} (u, y) \quad \textit{implies} \quad (f(v), x) \sim_m (f(u), y).$$

[6] By a *transformation* f between sheaves C and D on $\mathbf{F_S}$ we mean any family of functions $\{f_{(X,R)} : C(X, R) \to D(X, R)\}_{(X,R) \in \mathbf{F_S}}$.

holds. For $m = 0$, $(*)_m$ is just $(*)$. Assume $(*)_m$. We shall show $(*)_{m+1}$. Let $v : (X, R) \to L$, $u : (Y, S) \to L \in C^L$, $x \in X$, $y \in Y$, and $(v, x) \sim_{n+m+1} (u, y)$. Clearly, $(f(v), x) \sim_0 (f(u), y)$. Pick $x' \in X$ such that $x\, R\, x'$. By assumption, there is a $y' \in Y$ such that $y\, S\, y'$ and $(v, x') \sim_{n+m} (u, y')$. Hence by $(*)_m$, $(f(v), x') \sim_m (f(u), y')$. If we pick $y' \in Y$ first then we can argue similarly. So, by Proposition 5.2(ii), $(f(v), x) \sim_{m+1} (f(u), y)$, and $(*)_{m+1}$ holds.

It remains to show that f is natural. Let $k : (X, R) \to (Y, S)$ be a morphism in $\mathbf{F_S}$ and let $u : (Y, S) \to L$ be an evaluation in C^L. We need to show that $f(u) \circ k = f(u \circ k)$. Let $x \in X$. We shall show that

$$f(u) \circ k(x) = f(u \circ k)(x).$$

By Proposition 5.5(i), $(u, k(x)) \sim_\infty (u \circ k, x)$; by $(*)$ we get $(f(u), k(x)) \sim_0 (f(u \circ k), x)$. Hence

$$f(u \circ k)(x) \; = \; f(u)(k(x)) \; = \; f(u) \circ k(x)$$

as required. □

We have an embedding functor

$$\mathbf{\Psi_S} : \mathbf{M_S} \longrightarrow Sh(\mathbf{F_S}, \mathbf{J_S})$$

where $Sh(\mathbf{F_S}, \mathbf{J_S})$ is a Grothendieck topos on the site $(\mathbf{F_S}, \mathbf{J_S})$.

The following Proposition describes finite limits in $\mathbf{M_S}$. They are computed pointwise, as in sheaves.

PROPOSITION 5.20 *$\mathbf{M_S}$ has finite limits and $\mathbf{\Psi_S}$ preserves them. In particular, $h_L \times h_M \cong h_{L\times M}$ for any finite sets L and M; h_1 is the terminal object (here $1 = \{0\}$).*

Proof. Let C^L and D^M be objects of $\mathbf{M_S}$. Then we can define their product on the object (X, R) of $\mathbf{F_S}$ as

$$C^L \times D^M(X, R) =$$

$$\{< v, u >: (X, R) \to L \times M : v \in C^L(X, R),\ u \in D^M(X, R)\}$$

and on morphisms in the obvious way by composition. $C^L \times D^M$ is a subfunctor of $h_{L\times M}$ with index being the maximum of indexes of C^L and D^M. It is also clear that $C^L \times D^M$ is a sheaf, since both C^L and D^M are. The projections are defined in the obvious way and they have index 0. Finally, the index of a universal morphism into the product $< f, g >$, for given two morphisms f, g with a common domain into C^L and D^M, respectively, is equal to the maximum of the indexes of f and g.

An equalizer of a pair of morphisms $f, g : C^L \to D^M$ in $\mathbf{M_S}$ is defined by

$$E^L = \{v \in C^L \ : \ f(v) = g(v)\}$$

The index of E^L is the maximum of the index of C^L plus 1 and of the indices of f and g. The remaining details are left as an exercise. □

We shall describe now an $\mathbf{S}$-algebra structure on h_2 in $\mathbf{M_S}$. Note that for $(X, R) \in \mathbf{F_S}$, $h_2(X, R) = 2^X$, i.e. it is the universe of the $\mathbf{S}$-algebra $< 2^X, \wedge_X, \vee_x, \rightarrow_X, \neg_X, 0_X, 1_X, \Box_R >$. Moreover, for a morphism $k : (X, R) \to (Y, S)$ in $\mathbf{F_S}$, $h_2(k) : 2^Y \to 2^X$ given by $h_2(k)(v) = v \circ k$ for $v \in 2^Y$, is an $\mathbf{S}$-algebra morphism, i.e. it commutes with all boolean operations and □. This is equivalent to saying that the transformations

$$\wedge \vee, \rightarrow : h_2 \times h_2 \longrightarrow h_2$$

$$\neg, \Box : h_2 \longrightarrow h_2$$

$$\mathbf{0}, \mathbf{1} : h_1 \longrightarrow h_2$$

given for (X, R) in $\mathbf{F_S}$ by

$$\wedge_{(X,R)} = \wedge_X, \qquad \vee_{(X,R)} = \vee_X, \qquad \rightarrow_{(X,R)} = \rightarrow_X,$$

$$\neg_{(X,R)} = \neg_X, \qquad \Box_{(X,R)} = \Box_R, \qquad \mathbf{0}_{(X,R)} = 0_X, \qquad \mathbf{1}_{(X,R)} = 1_X$$

are natural. We have

LEMMA 5.21 *The natural transformations $\wedge$, $\vee$, $\rightarrow$, $\neg$, $\mathbf{0}$, $\mathbf{1}$ have index 0, and □ has index 1. Hence they are morphisms in $\mathbf{M_S}$.*

Proof. It is easy to see that all these transformations but □ have index 0. For example, for $v, u \in h_2(X, R)$ and $x \in X$ we have that $(v \wedge_{(X,R)} u)(x) = 1$ iff $v(x) = 1$ and $u(x) = 1$. So, the value $(v \wedge_{(X,R)} u)(x)$ depends on values of v and u at node x only.

Let $v \in h_2(X, R)$, $u \in h_2(Y, S)$, $x \in X$, $y \in Y$, so that $(v, x) \sim_1 (u, y)$. Then $\Box_{(X,R)}(v)(x) = 1$ iff for all $x' \in X$, if $x \, R \, x'$ then $v(x') = 1$. From this and $(v, x) \sim_1 (u, y)$ we get that for all $y' \in Y$, if $y \, R \, y'$ then $u(y') = 1$. This means that $\Box_{(X,R)}(u)(y) = 1$, i.e. □ has index 1. □

PROPOSITION 5.22 *The object h_2 together with the morphisms $\wedge$, $\vee$, $\rightarrow$, $\neg$, $\mathbf{0}$, $\mathbf{1}$, □ forms an internal $\mathbf{S}$-algebra in $\mathbf{M_S}$.*

Proof. Let $t(x_1, \ldots, x_n)$ be a term in the language of modal algebras. Let the morphism $\mathbf{t} : h_2^n \to h_2$ in $\mathbf{M_S}$ be the interpretation of the term t in the internal algebra $< h_2, \wedge, \ldots, \Box >$, and $t^{<2^X, \Box_R>} : (2^X)^n \to 2^X$ be the interpretation of t in $< 2^X, \Box_R >$ for any $(X, R) \in \mathbf{F_S}$. Since the interpretations of all connectives in h_2 are defined at any (X, R) in $\mathbf{F_S}$ as the corresponding connectives in the $\mathbf{S}$-algebra $< 2^X, \Box_R >$, by induction on the complexity of term t we can prove that the same thing holds for terms, i.e. $\mathbf{t}_{(X,R)} = t^{<2^X,\Box_R>}$.

In order to prove the Proposition we need to show that if an equation $t_0(x_1, \ldots, x_n) = t_1(x_1, \ldots, x_n)$ of terms holds in all $\mathbf{S}$-algebras (in Set) then it holds in the internal algebra $< h_2, \wedge, \ldots, \Box >$, as well. Clearly, if an equations holds is all $\mathbf{S}$-algebras then it holds, in particular, in all finite $\mathbf{S}$-algebras. So, for (X, R) in $\mathbf{F_S}$ we have

$$\mathbf{t_0}_{(X,R)} = t_0^{<2^X,\Box_R>} = t_1^{<2^X,\Box_R>} = \mathbf{t_1}_{(X,R)}$$

But this means that $\mathbf{t_0} = \mathbf{t_1}$, as required. $\Box$

It is a general fact that if an object Σ of a category C with finite products together with a collection of morphisms is an internal algebra in C of a certain kind then the contravariant functor representable by Σ lifts to the category of algebras of that kind. Spelling this out in our situation, we have

COROLLARY 5.23 *For any C^L in $\mathbf{M_S}$, the set $\mathbf{M_S}(C^L, h_2)$ carries an $\mathbf{S}$-algebra structure induced via compositions by the one on h_2. Moreover, for any morphisms $f : C^L \to D^M$ in $\mathbf{M_S}$, the map of composing with f*

$$f^* : \mathbf{M_S}(D^M, h_2) \longrightarrow \mathbf{M_S}(C^L, h_2)$$

is a morphism of $\mathbf{S}$-algebras. $\Box$

We shall denote the set $\mathbf{M_S}(C^L, h_2)$ as well as the $\mathbf{S}$-algebra on that set by $\mathbf{T_S}(C^L)$. Thus, by the above Corollary, we have a functor

$$\mathbf{T_S} : \mathbf{M_S} \longrightarrow Alg(\mathbf{S})^{op}$$

such that, for $f : C^L \to D^M$ in $\mathbf{M_S}$

$$\mathbf{T_S}(f) = f^* : \mathbf{T_S}(D^M) \longrightarrow \mathbf{T_S}(C^L).$$

We define the object Ω by the following equalizer in $\mathbf{M_S}$:

$$\Omega \xrightarrow{\ \varepsilon\ } h_2 \underset{id_{h_2} \wedge \Box}{\overset{id_{h_2}}{\rightrightarrows}} h_2$$

Using our description of equalizers in $\mathbf{M_S}$ we have for $(X, R) \in \mathbf{F_S}$

$$\Omega(X, R) =$$

$$\{v \in h_2(X,R) : \forall x, y \in X \ \text{ if } \ xRy \text{ and } v(x) = 1 \ \text{ then } \ v(y) = 1\}$$

and ε is an inclusion morphism. Thus for any $g : C^L \to h_2$ in $\mathbf{M_S}$, $g = g \wedge \Box g$ iff g factors through ε, i.e. iff there is g' in $\mathbf{M_S}$ making the triangle

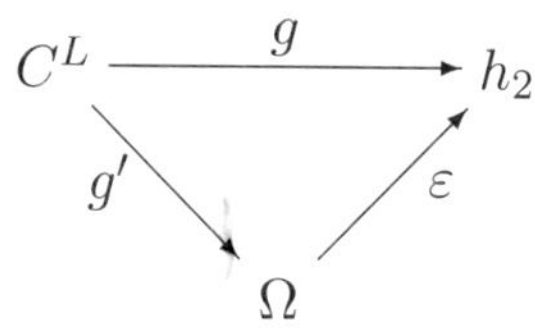

to commute. Note also that, since $\Box \leq \Box\Box$, for any $g : C^L \to h_2$ in $\mathbf{M_S}$, the morphism $g \wedge \Box g$ factors through ε. In particular, since $\Box \mathbf{1} = \mathbf{1}$, we have a morphism $\top : h_1 \to \Omega$ such that $\varepsilon \circ \top = \mathbf{1}$.

PROPOSITION 5.24 *The morphism $\top : h_1 \to \Omega$ is a subobject classifier is $Sh(\mathbf{F_S}, \mathbf{J_S})$. Moreover, it classifies regular monomorphisms in $\mathbf{M_S}$.*[7]

Proof. Let denote by Ω' the subobject classifier in $Sh(\mathbf{F_S}, \mathbf{J_S})$. Thus $\Omega'(X,R)$ contains $\mathbf{J_S}$-closed sieves on (X,R) in $\mathbf{F_S}$. Note, that a sieve S on (X,R) defines a generated subframe (X_S, R_S) of (X,R) such that $X_S = \bigcup_{f \in S} im(f)$. Let $i_S : (X_S, R_S) \hookrightarrow (X,R)$ denote the inclusion. Then, the sieve $i_S^*(S)$ on (X_S, R_S) is jointly surjective. Thus, if S is $\mathbf{J_S}$-closed $i_S \in S$. But this means that a morphism $f : (X', R') \to (X, Y)$ belongs to S iff it factors through i_S. Now, it is easy to verify, that the transformation

$$\tau : \Omega' \longrightarrow \Omega$$

such that $\tau_{(X,R)}(S) : (X,R) \to 2$ is given by

$$\tau_{(X,R)}(S)(x) = \begin{cases} 1 & \text{if } \ x \in X_S \\ 0 & \text{otherwise} \end{cases}$$

is natural and that $\tau_{(X,R)}$ is a bijection for (X,R) in $\mathbf{F_S}$. Thus Ω' is isomorphic to Ω in $Sh(\mathbf{F_S}, \mathbf{J_S})$. Moreover, the triangle

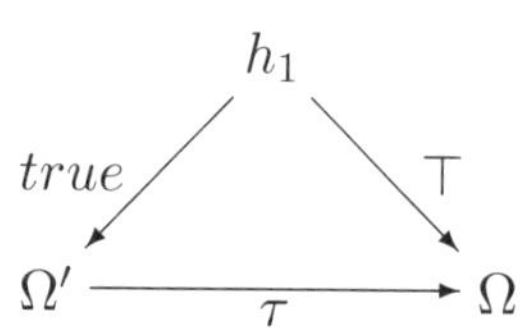

commutes. This shows that $\top : h_1 \to \Omega$ is a subobject classifier in $Sh(\mathbf{F_S}, \mathbf{J_S})$.

[7] We shall see in Section 5.6 that all monomorphisms are regular in $\mathbf{M_S}$, thus Ω turns out to be a subobject classifier in $\mathbf{M_S}$ as well.

We shall show that $\top : h_1 \to \Omega$ classifies regular subobjects in $\mathbf{M_S}$. Recall that a morphism is a regular monomorphism iff it is an equalizer. By our description of equalizers in Proposition 5.20, any equalizer of a pair of morphisms is isomorphic to an inclusion. We shall prove that for any inclusion morphism $m : C^L \hookrightarrow D^L$ in $\mathbf{M_S}$, there is a unique morphism $\chi_m : D^L \to \Omega$ making the following square

$$\begin{array}{ccc} C^L & \xrightarrow{\;!\;} & h_1 \\ {\scriptstyle m}\downarrow & & \downarrow{\scriptstyle \top} \\ D^L & \xrightarrow[\chi_m]{} & \Omega \end{array}$$

a pullback in $\mathbf{M_S}$. Moreover, we shall show that any inclusion morphism is a regular monomorphism in $\mathbf{M_S}$.

Let $m : C^L \hookrightarrow D^L$ be an inclusion in $\mathbf{M_S}$. Since $\top : h_1 \to \Omega$ is a subobject classifier in $Sh(\mathbf{F_S}, \mathbf{J_S})$, we already know that there is a unique natural transformation χ_m making the above square a pullback in $Sh(\mathbf{F_S}, \mathbf{J_S})$. The morphism χ_m is given, for $v : (X, R) \to L$ in D^L and $x \in X$, by

$$\chi_m(v)(x) = 1 \quad \text{iff} \quad v_{\uparrow x} \in C^L. \tag{5.3}$$

Note that, if $x, y \in X$, $v_{\uparrow x} \in C^L$ and xRy then $v_{\uparrow y} \in C^L$. Hence, $\chi_m(v) \in \Omega$. We need to verify, that χ_m is in fact a morphism in $\mathbf{M_S}$, i.e. that χ_m has an index.

Let n be an index of C^L, $v, v' \in D^L$, $x \in dom(v)$, $x' \in dom(v')$, and $(v, x) \sim_n (v', x')$. By Lemma 5.3, $v_{\uparrow x} \approx_n v'_{\uparrow x'}$. Therefore, either both $v_{\uparrow x}$, $v'_{\uparrow x'}$ belongs to C^L or none. Hence $\chi_m(v)(x) = \chi_m(v')(x')$. This, together with Lemma 5.19, shows that χ_m is a morphism in $\mathbf{M_S}$ with index n.

Finally, note that the diagram in $\mathbf{M_S}$

$$C^L \xrightarrow{\;m\;} D^L \underset{\top \circ !}{\overset{\chi_m}{\rightrightarrows}} \Omega$$

is an equalizer in $Sh(\mathbf{F_S}, \mathbf{J_S})$, hence also in $\mathbf{M_S}$. Thus any inclusion in $\mathbf{M_S}$ is a regular mono. □

Let

$$\mathcal{S} : \mathbf{M_S}^{op} \longrightarrow Set$$

be the pullback functor operating on regular subobjects, i.e. $\mathcal{S}(C^L)$ is the set of the regular subobjects of the object C^L in $\mathbf{M_S}$, and the function $\mathcal{S}(f) : \mathcal{S}(D^M) \longrightarrow \mathcal{S}(C^L)$ (denoted as f^*) pulls subobjects of D^M back along $f : C^L \longrightarrow D^M$.

COROLLARY 5.25 *The functor $\mathcal{S} : \mathbf{M_S}^{op} \longrightarrow Set$ is representable by Ω.*

Proof. We shall show that there is a natural isomorphism

$$\nu : \mathcal{S} \longrightarrow \mathbf{M_S}(-, \Omega)$$

For an inclusion $m : C^L \hookrightarrow D^L$ in $\mathbf{M_S}$ we put $\nu_{D^L}(m) = \chi_m$.

By Proposition 5.24, ν_{D^L} is bijective for $D^L \in \mathbf{M_S}$. Thus, it remains to show that ν is natural, i.e. that for any morphism $f : E^M \to D^L$ in $\mathbf{M_S}$ the diagram

$$\begin{array}{ccc} \mathcal{S}(D^L) & \xrightarrow{\nu_{D^L}} & \mathbf{M_S}(D^L, \Omega) \\ {\scriptstyle f^*}\downarrow & & \downarrow{\scriptstyle \mathbf{M_S}(f,\Omega)} \\ \mathcal{S}(E^M) & \xrightarrow[\nu_{E^M}]{} & \mathbf{M_S}(E^M, \Omega) \end{array}$$

commutes. Consider the diagram

$$\begin{array}{ccccc} f^*(C^L) & \longrightarrow & C^L & \xrightarrow{!} & h_1 \\ {\scriptstyle m'}\downarrow & & \downarrow{\scriptstyle m} & & \downarrow{\scriptstyle \top} \\ E^M & \xrightarrow[f]{} & D^L & \xrightarrow[\chi_m]{} & \Omega \end{array}$$

where the right hand square is a pullback with χ_m the characteristic morphism of m, and the left hand square is a pullback of m along f. Thus, $m' = f^*(m)$. Since the outer square is a pullback as well, by Proposition 5.24, $\chi_{m'} = \chi_m \circ f$. We have

$$\mathbf{M_S}(f, \Omega) \circ \nu_{D^L}(m) = \mathbf{M_S}(f, \Omega)(\chi_m) =$$
$$= \chi_m \circ f = \chi_{m'} = \nu_{E^M}(m') = \nu_{E^M} \circ f^*(m)$$

i.e. ν is natural. □

There is an internal Heyting algebra structure on Ω in $\mathbf{M_S}$. There are two (independent) reasons for that. As we have seen, $\top : h_1 \to \Omega$ is a subobject classifier in the topos $Sh(\mathbf{F_S}, \mathbf{J_S})$. Thus, there are operations in $Sh(\mathbf{F_S}, \mathbf{J_S})$ on Ω making Ω into an internal Heyting algebra in $Sh(\mathbf{F_S}, \mathbf{J_S})$. It can be shown that these operations have an index, so that Ω is in fact an internal algebra in $\mathbf{M_S}$. On the other hand, the elements of any **K4**-algebra satisfying the equation $x = x \wedge \Box x$ form a Heyting algebra. Thus the fact that Ω is an internal Heyting algebra in $\mathbf{M_S}$ can be deduced from the presentation of Ω as an equalizer of id_{h_2} and $id_{h_2} \wedge \Box$. We shall describe the operations on Ω explicitly. Since $\Box\mathbf{1} = \mathbf{1}$ and $\Box\mathbf{0} = \mathbf{0}$, both constants $\mathbf{1}$ and $\mathbf{0}$ factorizes through Ω. Moreover, operations $\wedge, \vee : h_2 \times h_2 \longrightarrow h_2$ when applied to the elements of $\Omega \times \Omega$ takes values in Ω. We shall denote the restriction of these operations by the same symbols. Ω together with $\wedge$, $\vee$, $\mathbf{1}$, $\mathbf{0}$ is an internal distributive lattice in $\mathbf{M_S}$. We define an operation

$$\Rightarrow : \Omega \times \Omega \longrightarrow \Omega$$

by $v{\Rightarrow}w = \Box^{+}(v{\rightarrow}w)$, for $v, w \in \Omega(X, R)$ and $(X, R) \in \mathbf{F_S}$, (where $\Box^{+}x = x \wedge \Box x$). It is easy to verify that $\Rightarrow$ is the operation of relative pseudo-complement. Thus we have

PROPOSITION 5.26 *Ω together with the operations $\wedge$, $\vee$, $\Rightarrow$, $\mathbf{1}$, $\mathbf{0}$ form an internal Heyting algebra in $\mathbf{M_S}$.* □

The following fact is a consequence of Proposition 2.16 and of the remarks at the end of Section 2.5. Here we present another proof.

PROPOSITION 5.27 *$\mathbf{M_S}$ has stable joins and stable implications among regular subobjects.*

Proof. We need to show that the contravariant functor $\mathcal{S}$ from $\mathbf{M_S}$ takes values in the category of Heyting algebras. But, by Corollary 5.25, $\mathcal{S}$ is isomorphic to the functor $\mathbf{M_S}(-, \Omega)$ represented by Ω. By Proposition 5.26, Ω carries a structure of an internal Heyting algebra in $\mathbf{M_S}$. So, the functor $\mathbf{M_S}(-, \Omega)$, and any functor isomorphic to it, lifts to a functor from $\mathbf{M_S}$ to the category of Heyting algebras (see the remarks before Corollary 5.23). □

PROPOSITION 5.28 *The object h_2 is a regular injective in $\mathbf{M_S}$.*

Proof. Let m be a regular monomorphism in $\mathbf{T_S}(C^L)$. We can assume that it is an inclusion $m : C^L \rightarrow D^L$. We need to show that any morphism $g : C^L \rightarrow h_2$ lifts to a morphism $\hat{g} : D^L \rightarrow h_2$ so that the triangle

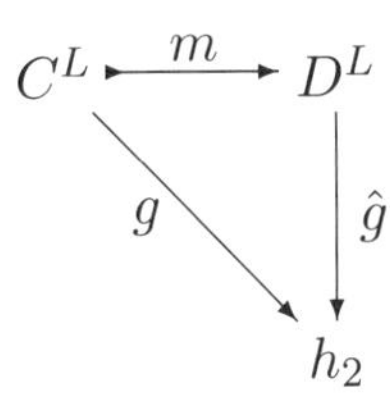

commutes. We put, for $v : (X, R) \rightarrow L \in D^L$ and $x \in X$

$$\hat{g}(v)(x) = \begin{cases} g(v_{\uparrow x})(x) & \text{if} \quad v_{\uparrow x} \in C^L \\ 0 & \text{otherwise} \end{cases}$$

The verification that $\hat{g}$ is the required morphism is left for the reader. □

The next observation will be used in the following Section. It relates the $\mathbf{S}$-algebras $\mathbf{T_S}(h_L)$ and $\mathbf{T_S}(C^L)$ for any inclusion (regular monomorphism) $m : C^L \to h_L$ in $\mathbf{M_S}$. Such m induces a morphism of $\mathbf{S}$-algebras $m^* : \mathbf{T_S}(h_L) \to \mathbf{T_S}(C^L)$ and we have a commuting diagram

$$\begin{array}{ccccc} C^L & \xrightarrow{\;!\;} & h_1 & & \\ {\scriptstyle m}\downarrow & & \downarrow{\scriptstyle \top} & \searrow{\scriptstyle \mathbf{1}} & \\ h_L & \xrightarrow[\chi_m]{} & \Omega & \xrightarrow[\varepsilon]{} & h_2 \end{array}$$

with the square being a pullback. The top element in $\mathbf{T_S}(C^L)$ is $\mathbf{1}\circ!$. Let $\bar{\chi}_m = \varepsilon \circ \chi_m$. Thus $m^*(\bar{\chi}_m) = 1_{\mathbf{T_S}(C^L)}$ and m^* induces a homomorphism

$$\alpha_m : \mathbf{T_S}(h_L)/_{\bar{\chi}_m} \longrightarrow \mathbf{T_S}(C^L)$$

from the quotient algebra. We have

LEMMA 5.29 *The homomorphism α_m defined above is an isomorphism.*

Proof. α_m is a surjection as a consequence of Proposition 5.28.

As $\bar{\chi}_m = \bar{\chi}_m \wedge \Box\bar{\chi}_m$, a morphism $g : h_L \to h_2$ in $\mathbf{M_S}$ represents the top element in $\mathbf{T_S}(h_L)/_{\bar{\chi}_m}$ iff $\bar{\chi}_m \leq g$. Assume that $m^*(g) = \mathbf{1}\circ!$. Then for $v \in C^L$ and $x \in dom(v)$, we have

$$g \circ m(v)(x) = \mathbf{1}\circ!(v)(x) = 1 \qquad (5.4)$$

Comparing (5.4) with the condition (5.3) defining χ_m, we get that $\bar{\chi}_m(v)(x) \leq g(v)(x)$, for all $v \in h_L$ and $x \in dom(v)$. Therefore $\bar{\chi}_m \leq g$ whenever $m^*(g) = \mathbf{1}\circ!$, i.e. α_m is injective as well. □

For the later reference we note the following

LEMMA 5.30 *For any C^L in $\mathbf{M_S}$, the inclusion $\mathcal{S}(C^L) \hookrightarrow Sub(C^L)$ is a Heyting algebra morphism (where $Sub(C^L)$ is the Heyting algebra of subsheaves of C^L).*

Proof. The Lemma follows from the fact that all operations in both Heyting algebras $\mathcal{S}(C^L)$ and $Sub(C^L)$ are induced by those on Ω. For example, if $a : A \hookrightarrow C^L$, $b : B \hookrightarrow C^L$ are elements either of $\mathcal{S}(C^L)$ or of $Sub(C^L)$, then their conjunction $A \cap B \hookrightarrow C^L$ in any of these algebras is defined by the following pullback square

$$\begin{array}{ccc} A \cap B & \longrightarrow & 1 \\ \downarrow & & \downarrow{\scriptstyle \top} \\ C^L & \xrightarrow[\wedge\circ<\chi_a,\chi_b>]{} & \Omega \end{array}$$

□

4. Finitely presented S-algebras

In this section, as before, **S** is an equational theory containing **K4**, has the finite model property, and $Alg(\mathbf{S})_{fin}$ has the amalgamation property. We shall prove below that any finitely presented **S**-algebra is isomorphic to one of form $\mathbf{T_S}(C^L)$, for some C^L in $\mathbf{M_S}$.

First we introduce some auxiliary morphisms.

Let C^L be in $\mathbf{M_S}$, (X, R) in $\mathbf{F_S}$, $v \in h_L(X, R)$, $x \in X$, $n \in \omega$. We define a morphism in $\mathbf{M_S}$

$$\lceil v \rceil_x^n : C^L \longrightarrow h_2$$

putting for $w \in C^L$, $\lceil v \rceil_x^n(w)(y) = 1$ iff $(v, x) \sim_n (w, y)$.

Clearly $\lceil v \rceil_x^n$ is a morphism in $\mathbf{M_S}$ with index n. Moreover

LEMMA 5.31 *For any v, x, n as above, we have*

$$\lceil v \rceil_x^{n+1} = \lceil v \rceil_x^0 \wedge \Box(\textstyle\bigvee_{x' \in R(x)} \lceil v \rceil_{x'}^n) \wedge (\bigwedge_{x' \in R(x)} \Diamond \lceil v \rceil_{x'}^n)$$

where $R(x) = \{x' \in X \mid x \, R \, x'\}$.

Proof. Recall that for $f : C^L \to h_2$ in $\mathbf{M_S}$, $w : (Y, S) \to L$ in C^L and $y \in Y$ we have

$$\Box f(w)(y) = 1 \quad \text{iff} \quad \forall y' \in Y \quad y \, S \, y' \ \text{ implies } \ f(w)(y') = 1$$

$$\Diamond f(w)(y) = 1 \quad \text{iff} \quad \exists y' \in Y \quad y \, S \, y' \ \text{ and } \ f(w)(y') = 1$$

To show the claim, pick arbitrary w and y as above. We have

$$\lceil v \rceil_x^0(w)(y) = 1 \ \text{ iff } \ v(x) = w(y)$$

Moreover, we have

$$\Box(\textstyle\bigvee_{x' \in R(x)} \lceil v \rceil_{x'}^n)(w)(y) = 1$$

iff

$$\forall y' \in Y \quad y \, S \, y' \ \text{ implies } \ (\textstyle\bigvee_{x' \in R(x)} \lceil v \rceil_{x'}^n)(w)(y') = 1$$

iff

$$\forall y' \in Y \quad y \, S \, y' \ \text{ implies } \ \exists x' \in R(x) \ \lceil v \rceil_{x'}^n(w)(y') = 1$$

iff

$$\forall y' \in Y \, \exists x' \in X \quad y \, S \, y' \ \text{ implies } \ x \, R \, x' \ \text{ and } \ \lceil v \rceil_{x'}^n(w)(y') = 1$$

iff

$$\forall y' \in Y \, \exists x' \in X \quad y \, S \, y' \ \text{ implies } \ x \, R \, x' \ \text{ and } \ (v, x') \sim_n (w, y')$$

In a similar way we can prove

$$(\textstyle\bigwedge_{x' \in R(x)} \Diamond \lceil v \rceil_{x'}^n)(w)(y) = 1$$

iff

$$\forall x' \in X \, \exists y' \in Y \quad x \, R \, x' \text{ implies } y \, S \, y' \text{ and } (v, x') \sim_n (w, y')$$

Using the above three equivalences, Proposition 5.2(ii), and the definition of $\lceil v \rceil_x^{n+1}$ we obtain

$$(\lceil v \rceil_x^0 \wedge \Box(\bigvee_{x' \in R(x)} \lceil v \rceil_{x'}^n) \wedge (\bigwedge_{x' \in R(x)} \Diamond \lceil v \rceil_{x'}^n))(w)(y) = 1 \text{ iff}$$

$$(v, x) \sim_{n+1} (w, y) \text{ iff } \lceil v \rceil_x^{n+1}(w)(y) = 1$$

Since w and y were arbitrary we obtain the claim. □

For any $C^L \in \mathbf{M_S}$, we have a Boolean homomorphism

$$\iota_L : 2^L \longrightarrow \mathbf{T_S}(C^L)$$

where $\iota_L(k) : C^L \to h_2$ is the natural transformation such that for v in C^L, $\iota_L(k)(v) = k \circ v$.

PROPOSITION 5.32 *The image of the Boolean homomorphism $\iota_L : 2^L \to \mathbf{T_S}(C^L)$ generates $\mathbf{T_S}(C^L)$ as an $\mathbf{S}$-algebra.*

Proof. Note that, if $k \in 2^L$ then $\iota_L(k)$ has index 0. Let us fix $g : C^L \to h_2$ in $\mathbf{M_S}$ with index n. If $n = 0$ we can define $k_g \in 2^L$ so that, for $l \in L$

$$k_g(l) = \begin{cases} g(v)(x) & \text{if there are } v \in C^L, x \in X, \text{ s.t. } v(x) = l; \\ 0 & \text{otherwise.} \end{cases}$$

As g has index 0, k_g is well defined. Then, for any evaluation $v : (X, R) \to L$ in C^L and $x \in X$, we have $g(v) = k_g \circ v$, i.e. $g = \iota_L(k_g)$. Thus, g has index 0 iff g is in the image of ι_L.

Now assume that $n > 0$. Let $\{(v_i, x_i)\}_{i \in I}$ be a set of representants of all equivalence classes of the relation $\sim_n$ defined on L-evaluations in C^L with a distinguished node. By Proposition 5.4, the set I is finite. Let $I_0 = \{i \in I : g(v_i)(x_i) = 1\}$. Then, for any L-evaluation w and $y \in dom(w)$ there is an $i \in I$, such that $(w, y) \sim_n (v_i, x_i)$. Since index of g is n, $g(w)(y) = g(v_i)(x_i)$. Thus, $g(w)(y) = 1$ iff $i \in I_0$. Hence $g = \bigvee_{i \in I_0} \lceil v_i \rceil_{x_i}^n$. Using Lemma 5.31, we get by induction that any morphism of form $\lceil v \rceil_x^n$ is expressible in terms of elements of index 0. Therefore, g is expressible in terms of elements of index 0. But elements of index 0 are in the image of ι_L, hence the image of ι_L generates $\mathbf{T_S}(C^L)$. □

Let $C^L \in \mathbf{M_S}$, $v : (X, R) \to L$ be an L-evaluation in C^L and let $\alpha : \mathbf{T_S}(C^L) \longrightarrow (2^X, \Box_R)$ be an $\mathbf{S}$-algebra morphism. We define an $\mathbf{S}$-algebra morphism

$$ev_v : \mathbf{T_S}(C^L) \longrightarrow (2^X, \Box_R)$$

such that, for $g \in \mathbf{T_S}(C^L)$, $ev_v(g) = g(v)$ and an L-evaluation

$$\overline{\alpha} : (X, R) \to L$$

as the dual of the Boolean algebra morphism

$$2^L \xrightarrow{\iota_L} \mathbf{T_S}(C^L) \xrightarrow{\alpha} (2^X, \Box_R)$$

Thus we have (writing $(-)^\circ$ for duals) $(\overline{\alpha})^\circ = \alpha \circ \iota_L$.

The following proposition shows that C^L is the sheaf of **S**-algebra morphisms from $\mathbf{T_S}(C^L)$ to all finite **S**-algebras.

PROPOSITION 5.33 *For C^L in $\mathbf{M_S}$, and (X, R) in $\mathbf{F_S}$ as above, there is a bijective correspondence between the* **S***-algebra morphisms $\alpha : \mathbf{T_S}(C^L) \to (2^X, \Box_R)$ and the L-evaluations $v : (X, R) \to L$ in C^L, given by the mutually inverse maps*

$$\alpha \mapsto \overline{\alpha} \qquad v \mapsto ev_v$$

so that the triangles

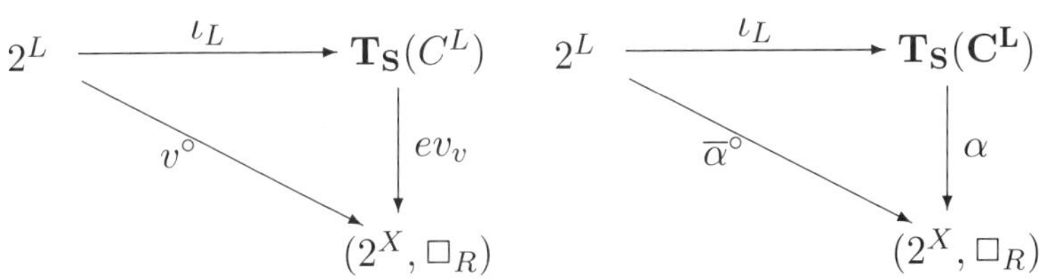

commute. The correspondence is natural in (X, R).

Proof. We verify first that, the function $ev_v : \mathbf{T_S}(C^L) \to (2^X, \Box_R)$ is indeed a homomorphism of **S**-algebras, for $v \in C^L$. We check only that ev_v preserves $\Box$ leaving the rest for the reader. For g in $\mathbf{T_S}(C^L)$ and $v \in C^L$ we have

$$ev_v(\Box g) = (\Box g)(v) = \Box_R(g(v)) = \Box_R(ev_v(g)).$$

For $k \in 2^L$ we have

$$(ev_v \circ \iota_L)(k) = \iota_L(k)(v) = k \circ v = v^\circ(k)$$

i.e. the left hand triangle commutes and $v = \overline{ev_v}$.

The right hand triangle commutes by definition of $\overline{\alpha}$. By Proposition 5.32, the homomorphism α is uniquely determined by its composition with ι_L. Hence $\alpha = ev_{\overline{\alpha}}$, as well. $\Box$

PROPOSITION 5.34 *Let L be a finite set. Then $\mathbf{T_S}(h_L)$ is the free* **S***-algebra over the finite Boolean algebra 2^L.*

Proof. Let us write F (for the time being) for this free $\mathbf{S}$-algebra and let $\eta : 2^L \to F$ be the related universal map.

Proposition 5.33 together with the duality for finite boolean algebras is saying that $\iota_L : 2^L \to \mathbf{T_S}(h_L)$ satisfy the universal property of $\eta : 2^L \to F$ with respect to finite $\mathbf{S}$-algebras. We shall show that this together with Proposition 5.32 and the finite model property implies that that ι_L satisfy the universal property of η with respect to all $\mathbf{S}$-algebras, i.e. $\mathbf{T_S}$ isomorphic to F.

As F is a finitely presented $\mathbf{S}$-algebra, by Lemma 2.19, there is an embedding $\langle \beta_i \rangle_i : F \to \prod_{i \in I}(2^{X_i}, \Box_{R_i})$ where the (X_i, R_i) are finite $\mathbf{S}$-frames. Let $v_i : (X_i, R_i) \to L$ be the L-evaluation whose dual is $\beta_i \circ \eta : 2^L \to 2^{X_i}$, for $i \in I$. Thus we can form a diagram

$$\begin{array}{ccc} 2^L & \xrightarrow{\quad\eta\quad} & F \\ {\scriptstyle\iota}\downarrow & \overset{\iota}{\swarrow} & \downarrow{\scriptstyle\langle\beta_i\rangle_i} \\ \mathbf{T_S}(h_L) & \xrightarrow[\langle ev_{v_i}\rangle_i]{} & \prod_i(2^{X_i}, \Box_{R_i}) \end{array}$$

in which the diagonal $\mathbf{S}$-algebra homomorphism ι is the unique morphism existing by the universal property of $\eta : 2^L \to F$. Thus the left hand triangle commutes and ι is surjective as a consequence of 5.32. The outer square commutes by Proposition 5.33. Thus again by universal property of η the right hand triangle commutes. Hence ι is injective as the first factor of an injective morphism $\langle \beta_i \rangle_i$. This means that ι is an isomorphism, as required. □

COROLLARY 5.35 *An $\mathbf{S}$-algebra is finitely presented iff it is isomorphic to one of form $\mathbf{T_S}(C^L)$, for some $C^L \in \mathbf{M_S}$.*

Proof. If L is a finite set, then 2^{2^L} is the free Boolean algebra on L. Thus, by Proposition 5.34, $\mathbf{T_S}(h_{2^L})$ is a free $\mathbf{S}$-algebra on L. Therefore any finitely presented $\mathbf{S}$-algebra is isomorphic to one of form $\mathbf{T_S}(h_{2^L})/g$, for a finite set L and $g \in \mathbf{T_S}(h_{2^L})$ such that $g = g \wedge \Box g$. Taking a pullback in $\mathbf{M_S}$

$$\begin{array}{ccc} C^{2^L} & \xrightarrow{\;!\;} & h_1 \\ {\scriptstyle m}\downarrow & & \downarrow{\scriptstyle \mathbf{1}} \\ h_{2^L} & \xrightarrow[g]{} & h_2 \end{array}$$

we get a regular mono m such that $g = \varepsilon \circ \chi_m$ (where $\varepsilon : \Omega \to h_2$ is the inclusion map defined in the previous Section). By Lemma 5.29, $\mathbf{T_S}(h_{2^L})/g$ is isomorphic to $\mathbf{T_S}(C^{2^L})$. Hence any finitely presented $\mathbf{S}$-algebra is isomorphic to one of form $\mathbf{T_S}(C^{2^L})$ for some C^{2^L} in $\mathbf{M_S}$.

On the other hand, again by Lemma 5.29, for any inclusion $m : C^L \to h_L$ in $\mathbf{M_S}$, the $\mathbf{S}$-algebra $\mathbf{T_S}(C^L)$ is isomorphic to $\mathbf{T_S}(h_L)/(\varepsilon \circ \chi_m)$. By Proposition 5.34 it is a finitely presented $\mathbf{S}$-algebra, and hence so is $\mathbf{T_S}(C^L)$. □

5. Duality

In this Section we assume that **S** is a theory containing **K4**, having the finite model property and such that $Alg(\mathbf{S})_{fin}$ has the amalgamation property. For such a theory **S** we establish a duality between the category $Alg(\mathbf{S})_{fp}$ of finitely presented **S**-algebras and the category $\mathbf{M_S}$.

Let C^L, D^M be objects of $\mathbf{M_S}$, $\mu : \mathbf{T_S}(C^L) \to \mathbf{T_S}(D^M)$ be a morphism of **S**-algebras. We define, for $(X, R) \in \mathbf{F_S}$, a function

$$\mu^*_{(X,R)} : D^M(X, R) \longrightarrow C^L(X, R)$$

such that, for $u \in D^M(X, R)$

$$\mu^*_{(X,R)}(u) = \overline{ev_u \circ \mu}.$$

By the Proposition 5.33, we have that $\mu^*_P(u) \in C^L(X, R)$ and that $\mu^* : D^M \to C^L$ is a natural transformation. Moreover

PROPOSITION 5.36 *With the notation as above, we have*

(i) $ev_{\mu^*(u)} = ev_u \circ \mu$, *for* $u \in D^M$.

(ii) $\mu^* : D^M \to C^L$ *is a morphism in* $\mathbf{M_S}$.

(iii) $\mu = \mu^{**}$.

(iv) *For any morphism* $f : D^M \to C^L$ *in* $\mathbf{M_S}$, *we have* $f = f^{**}$.

Proof. Ad (i).Using the bijective correspondence of Proposition 5.33 and definition of $\mu^*(u)$ we have $ev_{\mu^*(u)} = ev_{\overline{ev_u \circ \mu}} = ev_u \circ \mu$.

Ad (ii). We need to verify that μ^* has an index. Let fix $n \in \omega$ so that it is an index of all morphisms $\mu(\iota_L(k))$ for $k \in 2^L$. We shall show that n is an index of μ^*.

Let $u, v \in D^M$, $x \in dom(u)$, $y \in dom(v)$, and $(u, x) \sim_n (v, y)$. The for any $k \in 2^L$ we have

$$k(\mu^*(u)(x)) = (k \circ \mu^*(u))(x) =$$

$$= (\mu^*(u)^\circ(k))(x) = (ev_u \circ \mu \circ \iota_L(k))(x) =$$

$$= \mu(\iota_L(k))(u)(x) = \mu(\iota_L(k))(v)(y) =$$

$$= (ev_v \circ \mu \circ \iota_L(k))(y) = (\mu^*(v)^\circ(k))(y) =$$

$$= (k \circ \mu^*(v))(y) = k(\mu^*(v)(y))$$

Since $k \in 2^L$ was arbitrary we get that $\mu^*(u)(x) = \mu^*(v)(y)$. Thus, by Lemma 5.19, μ^* has index n.

Ad (iii). Let $g \in \mathbf{T_S}(C^L)$ and $v \in D^M$. Using (i) we obtain

$$\mu^{**}(g)(v) = (g \circ \mu^*)(v) = g(\mu^*(v)) =$$

$$= ev_{\mu^*(v)}(g) = ev_v \circ \mu(g) = \mu(g)(v)$$

Since g and v were arbitrary, we have $\mu = \mu^{**}$.

Ad (iv). Let $g \in \mathbf{T_S}(C^L)$ and $v \in D^M$. Then, again using (i), we have

$$ev_{f^{**}(v)}(g) = ev_v \circ f^*(g) =$$

$$= ev_v(g \circ f) = g(f(v)) = ev_{f(v)}(g).$$

As g was arbitrary $ev_{f^{**}(v)} = ev_{f(v)}$. But both $f^{**}(v)$, $f(v)$ belong to C^L, so by Proposition 5.33, we have that $f^{**}(v) = f(v)$. Since v was arbitrary, as well, we obtain that $f^{**} = f$, as required. □

In this way we get

THEOREM 5.37 (DUALITY THEOREM) *Assume that* **S** *is a theory containing* **K4** *in the language of* **K4**, *having the finite model property and such that* $Alg(\mathbf{S})_{fin}$ *has the amalgamation property. Then, the functor*

$$\mathbf{T_S} : \mathbf{M_S} \longrightarrow Alg(\mathbf{S})^{op}$$

is full, faithful and its essential image consists of finitely presented **S**-*algebras. In particular, the categories* $\mathbf{M_S}$ *and* $Alg(\mathbf{S})^{op}_{fp}$ *are equivalent, via functor* $\mathbf{T_S}$.

Proof. By Corollary 5.35, $\mathbf{T_S}$ is essentially surjective, and by Proposition 5.36 (iii, iv) $\mathbf{T_S}$ is full and faithful. □

6. Regularity of monomorphisms

This section contains a digression for a side result: we prove here that monos are all regular in the category $\mathbf{M_S}$, so that it turns out that $\mathbf{M_S}$ is a r-Heyting category iff it is a Heyting category. Although the knowledge of this fact is not directly relevant to our main goal (to be r-Heyting is the necessary and sufficient condition for model-completability, independently on what happens with all monos), we recall that regularity of monos corresponds to Beth theorem (see Proposition 2.14), so it is an important condition by itself.

Beth theorem in fact holds for all varieties of $K4$-modal algebras, this is a deep result shown in [Ma5]. Although Maksimova's result is more general (because the finite model property is not assumed), the proof below has some independent interest, being of essentially combinatorial nature. We assume in this Section that **S** is a theory containing **K4** in the language of **K4**, having the finite model property and such that $Alg(\mathbf{S})_{fin}$ has the amalgamation property (this last condition is not really needed for the proof below, just use the 'set-based dualities' mentioned in the exercises).

We make the following additional definitions and conventions:

- for $n \in \omega$, evaluations v, w and points $p \in dom(v), q \in dom(w)$, $(v,p) \approx_{n+1} (w,q)$ means that for every $\bar{p}$ such that $pR\bar{p}$ there is $\bar{q}$ such that $qR\bar{q}$ and $(v,\bar{p}) \sim_n (w,\bar{q})$ and vice versa (notice that we do not require $v(p) = w(q)$); [8] $[(v,q)]_{n+1}$ denotes the equivalence class of (v,q) with respect to the relation $\sim_{n+1}$;
- for an evaluation v, a point $p \in dom(v)$ and $m > 0$, $rk_m(v,p)$ is the cardinality of the set of the equivalence classes $\{[(v,q)]_{m-1} : pRq\}$ ($rk_m(v,p)$ is called the *m-rank of the point* p in v);
- in what follows, we fix a monic $f : C^L \to D^M$ in $\mathbf{M_S}$, and $n > 0$ which is an index for both f and C.

LEMMA 5.38 *Let* $w : (X,R) \to L$ *be an evaluation in* C^L, $p, q \in X$ *such that* pRq, $(w,p) \approx_{n+1} (w,q)$ *and* $f(w)(p) = f(w)(q)$. *We have that* $w(p) = w(q)$ *(hence also* $(w,p) \sim_{n+1} (w,q)$*).*

Proof. We suppose on the contrary that $w(p) \neq w(q)$. We distinguish two cases: (a) $q \in [p]_R$; (b) $q \notin [p]_R$ ($[p]_R$ is the cluster of p in (X,R)). In each case we shall construct two different evaluations in C on which f has the same value. This contradicts the fact that f is mono.

Case (a). Let $\tau : (X,R) \to (X,R)$ be the frame morphism interchanging p and q and leaving other points fixed. Since $w(p) \neq w(q)$ and $f(w)(p) = f(w)(q)$, we have that $w \neq w \circ \tau$ and $f(w \circ \tau) = f(w) \circ \tau = f(w)$, as required.

Case (b). We define an evaluation $v : \uparrow p \to L$ as follows: $v(x) = w(x)$ for $x \neq p$, $v(p) = w(q)$. We shall show that $w_{\uparrow p}$ and v are the needed evaluations. Clearly $w_{\uparrow p} \neq v$.

We show by induction on $k = 0, \dots, n$ that:

(i) $(v,p) \sim_k (w,q)$;

(ii) for all $s \in [p]_R$, $s \neq p$, we have $(v,s) \sim_k (w,s)$.

[8]Recall that, for two evaluations w, v the notation $v \approx_{n+1} w$ means that for every $p \in dom(v)$ there is $q \in dom(w)$ such that $(v,p) \sim_n (w,q)$ and vice versa.

Facts (i) and (ii) will be proved simultaneously (that is, the induction hypothesis of each of them is used within the proof of the other fact). For $k = 0$ both (i) and (ii) are obvious.

Ad (i). If Player I chooses $\bar{q}$ in w such that $qR\bar{q}$, Player II answers $\bar{q}$ and wins as there is no way to go back to p (the only point within $\uparrow p$ in which the two evaluations differ), because $q \notin [p]_R$. If Player I chooses $\bar{p}$ in v such that $pR\bar{p}$, then there is $\bar{q}$ such that $qR\bar{q}$ and $(w, \bar{p}) \sim_n (w, \bar{q})$. Again there are no problems if $\bar{p} \notin [p]_R$ (Player II answers $\bar{q}$ and wins); if $\bar{p} \in [p]_R$, we have two subcases, namely $p \neq \bar{p}$ and $p = \bar{p}$. In the former case, by the induction hypothesis of (ii), we have $(v, \bar{p}) \sim_{k-1} (w, \bar{p}) \sim_n (w, \bar{q})$, so again $\bar{q}$ is a winning answer. In case $p = \bar{p}$, from $(w, p) \sim_n (w, \bar{q})$ and pRq, it follows that there is q' such that $\bar{q}Rq'$ and $(w, q) \sim_{n-1} (w, q')$. By induction hypothesis of (i), we get $(v, p) \sim_{k-1} (w, q)$, hence q' is a good answer for Player II.

Ad (ii). Here the situation is as follows. Player II always answer the same point played by Player I, except in two cases: 1) Player I chooses p within v, where answer is q (Player II wins by the induction hypothesis of (i)); 2) Player I chooses p within w, where answer is any $\bar{q}$ such that $qR\bar{q}$ and $(w, p) \sim_{n-1} (w, \bar{q})$ (such point exists as $(w, p) \approx_{n+1} (w, q)$, hence we can find the desired $\bar{q}$ in two moves, from p to s and back to p).

From (i) it follows that $v \in C$: in fact we have $(v, p) \sim_n (w, q)$, hence also $v \approx_n w_{\uparrow q}$ and finally $v \in C$, because n is an index for C.[9] Moreover, from (i) and (ii) we get that $f(w_{\uparrow p}) = f(v)$: in fact n is an index for f, so by (ii) $(f(w), s) \sim_0 (f(v), s)$ for all $s \in cl(p), s \neq p$, moreover, according to our assumptions, $(f(w), p) \sim_0 (f(w), q) \sim_0 (f(v), p)$ by (i), as required. □

Let (X, R) be a frame. $x \in X$ is a *root* of (X, R) iff xRy, for all $y \in X$. (X, R) is a *rooted frame* iff it has a root. An L-evaluation $w : (X, R) \to L$ is *rooted* iff (X, R) is rooted.

Let $w : (X, R) \to L$ be a rooted evaluation with specified root, say, ρ; a point $p \in X$ is said to be a *full rank point* iff $rk_{n+2}(w, \rho) = rk_{n+2}(w, p)$.

LEMMA 5.39 *Let $w : (X, R) \to L$ be a rooted evaluation in C and let $p, q \in X$ be full rank points. We have:*

(i) $(w, p) \approx_{n+2} (w, q)$;

(ii) *if $f(w)(p) = f(w)(q)$, then $w(p) = w(q)$ (hence also $(w, p) \sim_{n+2} (w, q)$).*

[9] This is the only point in the whole proof of the regularity of f needing the fact that n is an index for C (in other passages it would be sufficient to ask only for $n + 2$ to be an index for C.).

Proof. Ad (i): let ρ be a root of (X, R). We always have $\{[(w, \bar{q})]_{n+1} : qR\bar{q}\} \subseteq \{[(w, \bar{q})]_{n+1} : \rho R\bar{q}\}$; as such finite sets have the same cardinality (because q is full rank), they must coincide, which means that $(w, \rho) \approx_{n+2} (w, q)$. Analogously, we have $(w, \rho) \approx_{n+2} (w, p)$ and finally $(w, p) \approx_{n+2} (w, q)$ by transitivity.

Ad (ii): if p is a root, then either there is nothing to prove (because $p = q$) or we are in the situation of Lemma 5.38 (because pRq and $(w, p) \approx_{n+2} (w, q)$ according to (i) just proved). If p is not a root, then we have that, say, ρ is a root and ρRp; as q is a full rank point, there must be q' such that qRq' and $(w, p) \sim_{n+1} (w, q')$. We can now apply Lemma 5.38 to the evaluation w relatively to the points q, q': in fact, we have $(w, q') \approx_{n+1} (w, p) \approx_{n+2} (w, q)$ (by (i)) and $f(w)(q) = f(w)(p) = f(w)(q')$, as n is an index for f. So $w(q) = w(q') = w(p)$, where the first equality holds by Lemma 5.38. □

PROPOSITION 5.40 *Let $u : P \to L$ be a rooted evaluation in C with selected root ρ. Let r be $rk_{n+2}(u, \rho)$. We have that*

(i) *for every rooted evaluation $v : Q \to M$ (with selected root σ) such that $(f(u), \rho) \sim_{2r+1} (v, \sigma)$, there is $v^* : Q \to L$ such that $(v^*, \sigma) \sim_{n+2} (u, \rho)$ (consequently $v^* \in C$) and $f(v^*) = v$. Such a v^* is called a lifting of v (and is unique as f is mono).*

(ii) *for every rooted evaluation $u' : P' \to L$ (with selected root ρ') in C such that $(f(u), \rho) \sim_{2r+1} (f(u'), \rho')$, we have that $(u, \rho) \sim_{n+2} (u', \rho')$.*

Proof. First notice that (ii) easily follows from (i): take (v, σ) to be $(f(u'), \rho')$, then we have $(v^*, \sigma) \sim_{n+2} (u, \rho)$ and $f(v^*) = f(u')$, hence $(v^*, \sigma) = (u', \rho') \sim_{n+2} (u, \rho)$ because f is mono. Nevertheless, we shall prove (i) by induction on r and we shall be free to use (ii) too when calling for the inductive hypothesis.

Assume that (i) is true for all (u', ρ') having rank strictly less than r. Recall that $p \in P$ has full rank iff $rk_{n+2}(u, p) = r$.

We say $q \in Q$ is *low* iff there is a $p \in P$ such that $(v, q) \sim_{2r} (f(u), p)$ and p does not have full rank.

We say $q \in Q$ is *high* iff there is a $p \in P$ such that $(v, q) \sim_{2r} (f(u), p)$ and p has full rank.

Notice that a point q cannot be simultaneously low and high: otherwise there would be points $p, p' \in P$ such that p has full rank, p' does not have full rank and $(f(u), p) \sim_{2r} (f(u), p')$; applying induction hypothesis of (ii) to $(u_{\uparrow p'}, p')$ would give $(u_{\uparrow p'}, p') \sim_{n+2} (u_{\uparrow p}, p)$ which cannot be as rank is different. Thus we have a partition of Q into two disjoint sets formed by high point and by low points.

If $q \in Q$ is low, then there is $p \in P$ such that $rk_{n+2}(u,p) < r$ and $(v,q) \sim_{2r} (f(u),p)$. Moreover, by inductive hypothesis, there is $(v_{\uparrow q})^* : \uparrow q \longrightarrow L$ such that $(u_{\uparrow p}, p) \sim_{n+2} ((v_{\uparrow q})^*, q)$ and $f((v_{\uparrow q})^*) = v_{\uparrow q}$. By uniqueness of liftings, for any low $q, q' \in Q$ the liftings $(v_{\uparrow q})^*$ and $(v_{\uparrow q'})^*$ must agree on the intersection of their domains.

We define v^* as follows

$$v^*(q) \quad = \quad \begin{cases} (v_{\uparrow q})^*(q) & \text{if } q \text{ is low} \\ u(p) & \text{if } q \text{ is high and } (f(u),p) \sim_{2r} (v,q) \end{cases}$$

Notice that the above definition is correct: if, for high q, we have $(f(u),p) \sim_{2r} (v,q) \sim_{2r} (f(u),p')$, then p and p' must both have the full rank of the root, so that $u(p) = u(p')$ according to Lemma 5.39. In order to complete the proof of the Proposition we only need to show the following

Claim. For any high $q \in Q$ and $p \in P$ such that $(f(u),p) \sim_{2r} (v,q)$, we have that $(u,p) \sim_{n+2} (v^*,q)$.

We prove the Claim by showing that $(v^*,q) \sim_k (u,p)$ holds by a secondary induction on $k = 0, \ldots, n+2$. For $k = 0$ the statement is just the definition of v^*. For $k > 0$, notice that if Player I picks $\bar{p}$ such that $pR\bar{p}$, then Player II realizes that there is $\bar{q}$ such that $qR\bar{q}$ and $(f(u),\bar{p}) \sim_{2r-1} (v,\bar{q})$ and answers this $\bar{q}$. Similarly, if Player I picks $\bar{q}$ such that $qR\bar{q}$, then Player II realizes that there is $\bar{p}$ such that $pR\bar{p}$ and $(f(u),\bar{p}) \sim_{2r-1} (v,\bar{q})$ and answers this $\bar{p}$.[10] We need to justify his choice. Clearly the answer is winning if $\bar{p}$ has rank less than r (we just apply induction hypothesis on the rank parameter); so let us suppose that $\bar{p}$ has full rank. Notice that, as we supposed that $(f(u),\rho) \sim_{2r+1} (v,\sigma)$, there must be some p' such that $(f(u),p') \sim_{2r} (v,\bar{q})$. If p' does not have full rank, then from $(f(u_{\uparrow p'}),p') \sim_{2r} (v_{\uparrow \bar{q}},\bar{q}) \sim_{2r-1} (f(u_{\uparrow \bar{p}}),\bar{p})$ we can apply twice induction hypothesis on rank, thus getting $(u,p') \sim_{n+2} (v^*,\bar{q})$ (by (i)) and $(u,p') \sim_{n+2} (u,\bar{p})$ (by (ii)) and $(v^*,\bar{q}) \sim_{n+2} (u,\bar{p})$, justifying Player II's answer. If p' has full rank too, then we have $f(u)(\bar{p}) = v(\bar{q}) = f(u)(p')$, hence by Lemma 5.39 $(u,\bar{p}) \sim_{n+2} (u,p')$. By induction on the second parameter, we have $(u,p') \sim_{k-1} (v^*,\bar{q})$, yielding $(u,\bar{p}) \sim_{k-1} (v^*,\bar{q})$ which justifies Player II's answer once again. □

THEOREM 5.41 *All monomorphisms are regular in* $\mathbf{M_S}$.

Proof. We show that our f is regular. Consider $\exists_f(C) \hookrightarrow D$; let R be the maximum possible $(n+2)$-rank; we first show that $\exists_f(C)$ has index $2R+2$.

---

[10]Notice that $2r-1>0$, unless $r=0$. Thus a special care should be taken to the case in which the rank is 0. However, as we have $(f(u),\rho) \sim_{2r+1} (v,\sigma)$, this can happen only in case both P and Q consists of a single irreflexive point: in such a trivial case, u itself is a lifting of v.

In fact, let $v \approx_{2R+2} f(u)$ for some $u \in C$; we show that $v_{\uparrow q} \in \exists_f(C)$ for all $q \in dom(v)$. For such q, there is $p \in dom(u)$ such that $(v_{\uparrow q}, q) \sim_{2R+1} (f(u_{\uparrow p}), p)$. By the previous Proposition there is $v_q^* : \uparrow q \to L$ in C such that $f(v_q^*) = v_{\uparrow q}$, hence $v_{\uparrow q} \in \exists_f(C)$.

Next, we show that C is isomorphic to the regular mono $\exists_f(C) \hookrightarrow D$; clearly the restriction of f in the codomain realizes such an isomorphism, because by (ii) in Proposition 5.40, $f^{-1} : \exists_f(C) \to C$ has index $2R + 1$. □

As a consequence of the above Theorem, we have that the subobject classifier of $Sh(\mathbf{F_S}, \mathbf{J_S})$ is also a subobject classifier in $\mathbf{M_S}$ (see Proposition 5.24).

7. Combinatorial Conditions

In this section $\mathbf{S}$, unless otherwise stated, denotes an equational theory containing **K4**, satisfying the finite model property and amalgamation for finite algebras.

In Chapter 3 (Theorem 3.11), we have shown that for an equational theory T, if the category $Alg(T)_{fp}^{op}$ (i.e. the opposite of the category of finitely presented T-algebras) has stable joins and stable implications (an assumption which is always true for equational extensions of **K4**), then T admits a model completion iff $Alg(T)_{fp}^{op}$ is an r-Heyting category. In Theorem 5.37 we have shown that the category $Alg(\mathbf{S})_{fp}^{op}$ is equivalent to the category $\mathbf{M_S}$. Thus we know that the theory $\mathbf{S}$ admits a model completion iff $\mathbf{M_S}$ is an r-Heyting category (or equivalently iff it is a Heyting category by the results of the last Section).

In this Section we establish some combinatorial conditions (Theorem 5.49) necessary and sufficient for $\mathbf{M_S}$ (hence also for $Alg(\mathbf{S})_{fp}^{op}$) to be a Heyting category. This will reduce the problem of the existence of a model completion of the theory $\mathbf{S}$ to that of deciding the validity of these combinatorial conditions.

Let $f : D^L \to E^M$ be a morphism in $\mathbf{M_S}$. The strategy we adopt is explained by the following picture and by few words below it.

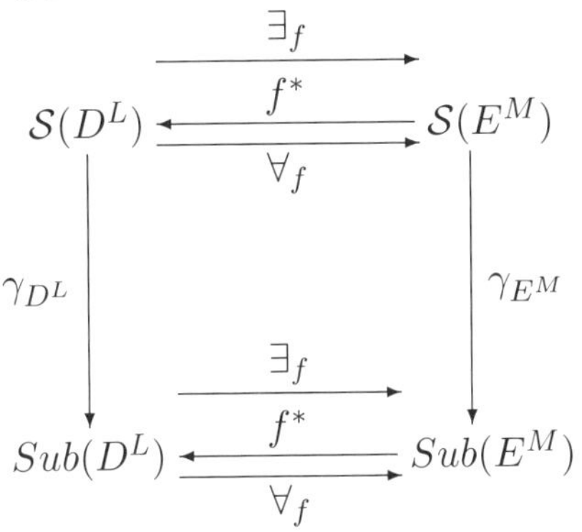

The inclusion $\gamma_{E^M} : \mathcal{S}(E^M) \hookrightarrow Sub(E^M)$ is a lattice morphism. Moreover, the pullback functor f^* operates on $\mathcal{S}(E^M)$ as well taking values into $\mathcal{S}(D^L)$, with $f^* \circ \gamma_{EM} = \gamma_{DL} \circ f^*$. The adjoints to f^*, $\exists_f \dashv f^* \dashv \forall_f$ always exist in Grothendieck toposes. Then we will show that images and dual images in $\mathbf{M_S}$, if they exist, must coincide with $\exists_f$ and $\forall_f$ in sheaves, respectively. This, together with the standard explicit description of $\exists_f$ and $\forall_f$ in sheaves, gives us eventually the required conditions equivalent to the existence of images and dual images in $\mathbf{M_S}$. Finally, it will be immediate to prove that images satisfy Beck-Chevalley condition in $\mathbf{M_S}$ in case they exist.

LEMMA 5.42 *Consider a commutative square*

$$\begin{array}{ccc} P_1 & \xrightarrow{\ f\ } & P_2 \\ {\scriptstyle\gamma_1}\downarrow & & \downarrow{\scriptstyle\gamma_2} \\ A_1 & \xrightarrow[\ g\ ]{} & A_2 \end{array}$$

in the category of lattices, with γ_1, γ_2 inclusions, A_1, A_2 complete lattices and g complete lattice homomorphism. Denote by $\exists_g, \forall_g$ the left and right adjoints to g, respectively. Then

(i) *if every element in A_1 is a meet of elements in P_1 and if the left adjoint $\exists_f$ to f exists, then $\gamma_1 \circ \exists_f = \exists_g \circ \gamma_2$;*

(i) *if every element in A_1 is a join of elements in P_1 and if the right adjoint $\forall_f$ to f exists, then $\gamma_1 \circ \forall_f = \forall_g \circ \gamma_2$.*

Proof. Both cases are similar. We shall prove (i). Fix $a \in P_2$. Let $b \in A_1$ and $\{a_i\}_{i \in I} \subseteq P_1$ be such that $b = \bigwedge_{i \in I} \gamma_1(a_i)$. We have the following sequence of equivalent statements:

$$\begin{array}{c}
\gamma_1 \circ \exists_f(a) \ \leq \ b \\ \hline
\gamma_1 \circ \exists_f(a) \ \leq \ \bigwedge_{i \in I} \gamma_1(a_i) \\ \hline
\forall i \in I \quad \gamma_1 \circ \exists_f(a) \ \leq \ \gamma_1(a_i) \\ \hline
\forall i \in I \quad \exists_f(a) \ \leq \ a_i \\ \hline
\forall i \in I \quad a \ \leq \ f(a_i) \\ \hline
\forall i \in I \quad \gamma_2(a) \ \leq \ \gamma_2 \circ f(a_i) \\ \hline
\forall i \in I \quad \gamma_2(a) \ \leq \ g \circ \gamma_1(a_i) \\ \hline
\forall i \in I \quad \exists_g \circ \gamma_2(a) \ \leq \ \gamma_1(a_i) \\ \hline
\exists_g \circ \gamma_2(a) \ \leq \ \bigwedge_{i \in I} \gamma_1(a_i) \\ \hline
\exists_g \circ \gamma_2(a) \ \leq \ b
\end{array}$$

Since b was arbitrary, we get $\exists_g \circ \gamma_2(a) = \gamma_1 \circ \exists_f(a)$, i.e. (i) holds. □

Let D^L be an object in $\mathbf{M_S}$, $v \in D^L$, $x \in dom(v)$, $n \in \omega \cup \{\infty\}$, $n \geq 1$. We define subpresheaves of D^L:

$$\|v\|_n^D = \{v' \in D^L \ : \ v' \leq_n v\}$$

$$\rangle\!\rangle v, x \langle\!\langle_n^D = \{v' \in D^L \ : \ \forall x' \in dom(v') \ \ (v, x) \not\prec_n (v', x')\}$$

We have

LEMMA 5.43 *Let D^L be an object in $\mathbf{M_S}$, $v : (X, R) \to L \in D^L$, $x \in X$. Then $\|v\|_n^D$ and $\rangle\!\rangle v, x \langle\!\langle_n^D$ are objects of $\mathbf{M_S}$.*

Proof. Note that the morphism $\lceil v \rceil_x^n : C^L \longrightarrow h_2$ defined on page 143 make sense even for $n = \infty$ and that, by Proposition 5.9, $\lceil v \rceil_x^\infty$ has index $2k$, where k is the height of x in (X, R). For $1 \leq n \leq \infty$, both $\|v\|_n^D$ and $\rangle\!\rangle v, x \langle\!\langle_n^D$ arise as the following pullbacks in $\mathbf{M_S}$

$$\begin{array}{ccc} \|v\|_n^D & \xrightarrow{\ !\ } & h_1 \\ \downarrow & & \downarrow \mathbf{1} \\ D^L & \xrightarrow[\bigvee_{x \in dom(v)} \lceil v \rceil_x^{n-1}]{} & h_2 \end{array} \qquad \begin{array}{ccc} \rangle\!\rangle v, x \langle\!\langle_n^D & \xrightarrow{\ !\ } & h_1 \\ \downarrow & & \downarrow \mathbf{1} \\ D^L & \xrightarrow[\neg \circ \lceil v \rceil_x^n]{} & h_2 \end{array}$$

(where $\infty - 1$ is ∞). Thus they are indeed objects of $\mathbf{M_S}$. □

LEMMA 5.44 *Let D^L be an object in $\mathbf{M_S}$, $1 \leq n \leq \infty$ and let C be $n+1$-closed subsheaf of D^L*[11]*. Then, for $u \in D^L$*

$$u \notin C \text{ iff } \exists x \in dom(v) \ \ C \subseteq \rangle\!\rangle u, x \langle\!\langle_n^D$$

Proof. Using Lemma 5.18, we have

$$u \notin C$$

iff

$$\exists x \in dom(u) \ \forall v \in C \forall x \in dom(v) \ (u, x) \not\prec_n (v, y)$$

iff

$$\exists x \in dom(u) \ \forall v \in C \ v \in \rangle\!\rangle v, x \langle\!\langle_n^D$$

iff

$$\exists x \in dom(u) \ C \subseteq \rangle\!\rangle v, x \langle\!\langle_n^D$$

as required. □

[11] In case $n = \infty$ it just mean that C is any subsheaf of D^L.

LEMMA 5.45 *Let D^L be an object of* $\mathbf{M_S}$.

(i) *The inclusion morphism $\gamma_{D^L} : \mathcal{S}(D^L) \hookrightarrow Sub(D^L)$ is a lattice morphism.*

(ii) *Every object of $Sub(D^L)$ is a meet as well as a join of objects from $\mathcal{S}(D^L)$.*

Proof. (i) is known from Lemma 5.30. Let us prove (ii).

Let $C \in Sub(D^L)$. For $v \in C$, we have $v \in \|v\|^D_\infty \subseteq C$ and, by Lemma 5.43, $\|v\|^D_\infty \in \mathcal{S}(D^L)$. Then $C = \bigvee_{v \in C} \|v\|^D_\infty$.

To show that C is a meet of elements of $\mathcal{S}(D^L)$, let us fix, for each $u \in D^L$ such that $u \notin C$ a point $x_u \in dom(u)$ so that $C \subseteq \rangle\!\rangle u, x_u \langle\!\langle^D_\infty$. Such x_u exist by previous Lemma. Again, by Lemma 5.43, any such $\rangle\!\rangle u, x_u \langle\!\langle^D_\infty$ is in $\mathcal{S}(D^L)$. Since $u \notin \rangle\!\rangle u, x \langle\!\langle^D_\infty$, for any $x \in dom(u)$, we have

$$C = \bigcap \{ \rangle\!\rangle u, x_u \langle\!\langle^D_\infty : u \in D^L \text{ and } u \notin C\}$$

as required. □

LEMMA 5.46 *Let $n > 0$, $v : (X, R) \to L$, $w : (Y, S) \to L$ be evaluations, $w \approx_{n+1} v$ and let $\{h_i : (X_i, R_i) \to (X, R)\}_{i \in I}$ be a cover. Then there is a cover $\{g_i : (Y_i, S_i) \to (Y, S)\}_{i \in I}$ such that $v \circ h_i \approx_n w \circ g_i$ for every $i \in I$.*

Proof. Let $\{g_i : (Y_i, S_i) \hookrightarrow (Y, S)\}_{i \in I}$ be a family of inclusions of generated subframes such that Y_i is the following set

$$\{y' \in Y \ : \ \exists_{y \in Y} \exists_{x \in X_i} ((w, y) \sim_n (v \circ h_i, x) \ \& \ (ySy' \text{ or } y = y'))\}$$

for $i \in I$. Since $w \approx_{n+1} v$ and $\{h_i\}_{i \in I}$ is a cover, so is $\{g_i\}_{i \in I}$. Moreover, from the definition of Y_i, we easily get that $v \circ h_i \approx_n w \circ g_i$, for $i \in I$. □

PROPOSITION 5.47 *Let $f : D^L \to E^M$ be a morphism in $\mathbf{M_S}$. Then the following conditions are equivalent*

(LA) *the pullback functor $f^* : \mathcal{S}(E^M) \longrightarrow \mathcal{S}(D^L)$ has a left adjoint $\exists_f : \mathcal{S}(D^L) \longrightarrow \mathcal{S}(E^M)$;*

$(\exists)$ $\quad \forall n \in \omega \, \exists N \in \omega \, \forall v \in E^M \, \forall u \in D^L$

$$v \approx_N f(u) \quad \text{implies} \quad \exists u' \in D^L \ \ u' \leq_n u \text{ and } v \approx_\infty f(u')$$

Proof. $(LA) \Rightarrow (\exists)$. By Lemmas 5.42, and 5.45, $\exists_f(B)$ must be defined as in the category of sheaves and have index for any $B \in \mathcal{S}(D^L)$. Fix $n \in \omega$. Let N be an index of all sheaves $\exists_f(\|u\|^D_n)$ for $u \in D^L$ (recall that there are in general only finitely many distinct n-closed subsheaves, so as each of the sheaves $\exists_f(\|u\|^D_n)$ has an index there is one common to all of them). Suppose

$u \in D^L$ and $v : (X,R) \to M \in E^M$ and $v \approx_N f(u)$. As $f(u) \in \exists_f(\|u\|_n^D)$ we have that $v \in \exists_f(\|u\|_n^D)$, and, by definition of images in $Sh(\mathbf{F_S}, \mathbf{J_S})$, there are a cover $\{h_i : (X_i, R_i) \to (X,R)\}_{i \in I}$ and L-evaluations $u_i : (X_i, R_i) \to L$ such that $u_i \leq_n u$ and $v \circ h_i = f(u_i)$ for all $i \in I$. Then putting $u' = [u_i]_{i \in I}$, we get $u' \leq_n u$ and $v \approx_\infty f(u')$, as required.

$(\exists) \Rightarrow (LA)$. Let $C^L \hookrightarrow D^L$ be an inclusion in $\mathbf{M_S}$, with C^L n-closed. We shall show that the sheaf-image $\exists_f(C^L)$ has index $N+1$, where N is as in the condition $(\exists)$ for n. Suppose that $v : (X,R) \to L \in E^M$, $w : (Y,S) \to M \in \exists_f(C^L)$ and $v \approx_{N+1} w$. Thus, there are a cover $\{h_i : (Y_i, S_i) \to (Y,S)\}_{i \in I}$ and L-evaluations $u_i : (Y_i, S_i) \to L$ in C^L such that $f(u_i) = w \circ h_i$, for $i \in I$. By Lemma 5.46, there is a cover $\{g_i : (X_i, R_i) \to (X,R)\}_{i \in I}$ such that $v \circ g_i \approx_N w \circ h_i$, for $i \in I$. So we have $f(u_i) \approx_N v \circ g_i$. Thus by $(\exists)$ there are $u_i' \in C^L$ such that $u_i' \leq_n u$ and $v \circ g_i \approx_\infty f(u_i')$. Then $v \circ g_i \in \exists_f(C^L)$, as $\exists_f(C^L)$ is closed for $\approx_\infty$ being a subsheaf. As $\{g_i\}_i$ is a cover, $v \in \exists_f(C^L)$ as well by the sheaf property. □

PROPOSITION 5.48 *Let $f : D^L \to E^M$ be a morphism in $\mathbf{M_S}$. Then the following conditions are equivalent*

(RA) *the pullback functor $f^* : \mathcal{S}(E^M) \longrightarrow \mathcal{S}(D^L)$ has a right adjoint $\forall_f : \mathcal{S}(D^L) \longrightarrow \mathcal{S}(E^M)$;*

$(\forall)$ $\quad \forall n \in \omega \, \exists N \in \omega \, \forall v \in E^M \, \forall u \in D^L$

$$f(u) \approx_N v \quad \text{implies} \quad \exists u' \in D^L \;\; u \approx_n u' \text{ and } f(u') \leq_\infty v$$

Proof. $(RA) \Rightarrow (\forall)$. By Lemmas 5.42, and 5.45, $\forall_f(B)$ must be defined as in the category of sheaves and have index for any $B \in \mathcal{S}(D^L)$. Fix $n \in \omega$. Let N being an index of all $\forall_f(\rangle\!\rangle u, x \langle\!\langle_n^D)$ for $u \in D^L$ and $x \in dom(u)$. Let $u : (X,R) \to L \in D^L$, $v : (Y,S) \to M \in E^M$ and $v \approx_N f(u)$. For $x \in X$, $f(u) \notin \forall_f(\rangle\!\rangle u, x \langle\!\langle_n^D)$, and hence $v \notin \forall_f(\rangle\!\rangle u, x \langle\!\langle_n^D)$. Thus, for every $x \in X$, there are a frame morphism $h_x : (Y_x, S_x) \to (Y,S)$, an L-evaluation $u_x : (Y_x, S_x) \to L$ and $y_x \in Y_x$ such that $f(u_x) = v \circ h_x$ and $(u_x, y_x) \sim_n (u, x)$. Putting $u'' = [u_x]_{x \in X} : \coprod_{x \in X}(Y_x, S_x) \to L$, we get $u \leq_{n+1} u''$. By Lemma 5.11, there is a frame morphism $h : (X', R') \to \coprod_{x \in X}(Y_x, S_x)$ such that, for $u' = u'' \circ h$, we have $u \approx_n u'$. Then we have

$$f(u') = f(u'') \circ h \leq_\infty f(u'') = [v \circ h_x]_{x \in X} \leq_\infty v,$$

as required.

$(\forall) \Rightarrow (RA)$. Let $n \geq 1$, $C^L \hookrightarrow D^L$ be an inclusion in $\mathbf{M_S}$, with n being an index of C^L. We need to show that the dual image $\forall_f(C^L)$ in sheaves has index $N+1$, where N is as in the condition $(\forall)$ for n. Suppose that $w : (X,R) \to M \in \forall_f(C^L)$, $v : (Y,S) \to M \in E^M$ and $v \approx_{N+1} w$. We shall show that $v \in \forall_f(C^L)$. Let $h : (Y', S') \to (Y, S)$ be a frame morphism and suppose that $v \circ h = f(u)$, for some $u \in D^L$. It remains to show that $u \in C^L$. By Lemma 5.11, there is a frame morphism $g : (X', R') \to (X, R)$ such that $v \circ h \approx_N w \circ g$. So $f(u) \approx_N w \circ g$. By $(\forall)$ there is $u' \in D^L$ such that $u \approx_n u'$ and $f(u') \leq_\infty w \circ g$. Since $\forall_f(C^L)$ is a sheaf, $f(u') \in \forall_f(C^L)$ and then $u' \in C^L$. Hence $u \in C^L$, as well. □

Combining together the results of this and of the former Sections we obtain necessary and sufficient conditions for the existence of model completions.

THEOREM 5.49 *Let* **S** *be an equational theory containing* **K4** *and having the finite model property. Then,* **S** *admits model completion iff the following two conditions are satisfied:*

(AP) *the category* $Alg(\mathbf{S})_{fin}$ *has the amalgamation property;*

$(\forall\exists)$ *for every arrow* $f : D^L \to E^M$ *in the category* $\mathbf{M_S}$ *we have that:*
$\forall n \in \omega\ \exists N \in \omega\ \forall v \in E^M\ \forall u \in D^L$

$$v \approx_N f(u) \quad \text{implies} \quad \exists u' \in D^L \;\; u' \approx_n u \text{ and } v \approx_\infty f(u')$$

Proof. We already observed in previous section that $Alg(\mathbf{S})^{op}_{fp}$ has stable joins and stable implications, thus Theorem 3.11 applies.

Assume that (AP) and $(\forall\exists)$ hold. Now as (AP) holds, by Theorem 5.37, $\mathbf{M_S}$ is equivalent to $Alg(\mathbf{S})^{op}_{fp}$. The condition $(\forall\exists)$ is stronger than both $(\exists)$ and $(\forall)$, so that in $\mathbf{M_S}$ left and right adjoints to inverse images of (regular) monos both exist, by Propositions 5.47, and 5.48. Finally, Beck-Chevalley condition holds because direct and inverse images of (regular) monos are computed in $\mathbf{M_S}$ as in sheaves. Thus $\mathbf{M_S}$ (and also $Alg(\mathbf{S})^{op}_{fp}$ which is equivalent to it) is r-Heyting and a model completion exists.

Conversely, if a model completion exists, then finite algebras satisfy the amalgamation property, by Proposition 2.20. So again, we can form the category $\mathbf{M_S}$, which must be r-Heyting by the duality Theorem 5.37 and by Theorem 3.11. This means in particular that images and dual images must both exist in $\mathbf{M_S}$, so that $(\exists)$ and $(\forall)$ both hold by Propositions 5.47, and 5.48. We shall show that if the conditions $(\exists)$ and $(\forall)$ hold then $(\forall\exists)$ holds as well. Fix $n \in \omega$. Pick $N \in \omega$ so that it is good for n in both $(\exists)$ and $(\forall)$. Let $v \in E^M$, $u \in D^L$ be such that $v \approx_N f(u)$. Then, by $(\exists)$ and $(\forall)$, there are $u_e, u_a \in D^L$ such that $u_e \leq_n u$, $v \approx_\infty f(u_e)$, $u_a \approx_n u$ and $f(u_a) \leq_\infty v$. Put $u' = [u_e, u_a]$. Then $u' \approx_n u$ and $f(u') = [f(u_a), f(u_e)] \approx_\infty v$. So u' is as required for the condition $(\forall\exists)$. □

We conclude this Section with a sheaf representation theorem for finitely presented **S**-algebras, similar to the one given in Chapter 4, for finitely presented Heyting algebras.

The category $Sh(\mathbf{F_S}, \mathbf{J_S})$ is a Heyting category, cf. the Appendix. Lemmas 5.42, and 5.45 show that both images and dual images in $\mathbf{M_S}$, if they exist, are computed as in the category of sheaves $Sh(\mathbf{F_S}, \mathbf{J_S})$. In view of Proposition 5.20, we can rephrase this as follows: the embedding functor

$$\mathbf{\Psi_S} : \mathbf{M_S} \longrightarrow Sh(\mathbf{F_S}, \mathbf{J_S})$$

preserves all the Heyting category structure that exists is $\mathbf{M_S}$. We have a functor

$$\mathbf{\Phi_S} : Alg(\mathbf{S})^{op}_{fp} \longrightarrow Sh(\mathbf{F_S}, \mathbf{J_S})$$

such that for any finitely presented $\mathbf{S}$-algebra A, $\mathbf{\Phi_S}(A)$ is the $\mathbf{J_S}$-sheaf such that for an $\mathbf{S}$-frame (X, R) we have

$$\mathbf{\Phi_S}(A)(X, R) \;=\; Alg(\mathbf{S})(A, (2^X, \Box_R))$$

Functors $\mathbf{\Phi_S}(A)$ and $\mathbf{\Phi_S}$ acts on morphisms in the obvious way, by composition. Similarly as in Section 4.3, we can show that the triangle of categories and functors

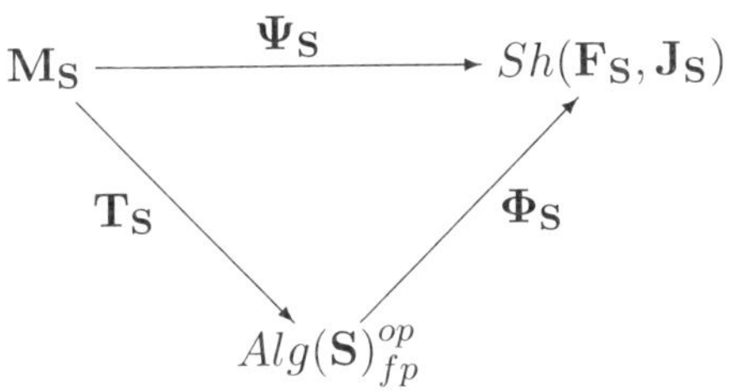

commutes up to an isomorphism (i.e. there is a natural isomorphism between the functors $\mathbf{\Phi_S} \circ \mathbf{T_S}$ and $\mathbf{\Psi_S}$). By Theorem 5.37, $\mathbf{T_S}$ is an equivalence of categories, so we obtain the following Sheaf Representation for finitely presented $\mathbf{S}$-algebras.

THEOREM 5.50 *Assume that* $\mathbf{S}$ *is a theory containing* **K4**, *having the finite model property and that* $Alg(\mathbf{S})_{fin}$ *has the amalgamation property. Then, the functor*

$$\mathbf{\Phi_S} : Alg(\mathbf{S})^{op}_{fp} \longrightarrow Sh(\mathbf{F_S}, \mathbf{J_S})$$

preserves finite limits, subobject classifier and moreover all images and dual images that exist in $Alg(\mathbf{S})^{op}$. *In particular, if* $Alg(\mathbf{S})^{op}$ *is a Heyting category,* $\mathbf{\Phi_S}$ *is a Heyting category morphism.* $\Box$

8. Exercises

1 Show that the functors (5.1) and (5.2) defined in section 5.1, indeed establish equivalence of categories $Alg(\mathbf{K4})_{fin}$ and $\mathbf{F}^{op}$.

2 The *degree* $d_\Box(\varphi)$ of a modal formula φ is defined as follows.

For atomic sentences p, $d_\Box(p) = 0$, moreover

$$d_\Box(\alpha \vee \beta) = d_\Box(\alpha \wedge \beta) = d_\Box(\alpha \to \beta) = max(d_\Box(\alpha), d_\Box(\beta)),$$

$d_\Box(\Box\alpha) = d_\Box(\alpha) + 1.$

Let $L = \mathcal{P}(\vec{p})$ and let $v : (X, R) \to L$ be an evaluation; for $x \in X$ and for a modal formula $\alpha(\vec{p})$, define the *forcing relation* $v, x \models \alpha(\vec{p})$ by the following clauses:

(i) $v, x \models p_i$ iff $p_i \in v(x)$ for $i = 1, \ldots, n$;

(ii) $v, x \models \alpha(\vec{p}) \wedge \beta(\vec{p})$ iff $v, x \models \alpha(\vec{p})$ and $v, x \models \beta(\vec{p})$;

(iii) $v, x \models \alpha(\vec{p}) \vee \beta(\vec{p})$ iff $v, x \models \alpha(\vec{p})$ or $v, x \models \beta(\vec{p})$;

(iv) $v, x \models \alpha(\vec{p}) \to \beta(\vec{p})$ iff $v, x \models \alpha(\vec{p})$ implies $v, x \models \beta(\vec{p})$;

(v) $v, x \models \neg\alpha(\vec{p})$ iff not $v, x \models \alpha(\vec{p})$;

(vi) not $v, x \models 0$;

(vii) $v, x \models 1$;

(viii) $v, x \models \Box\alpha(\vec{p})$ iff $\forall y \in X \;\; (xRy \;\text{ implies }\; v, y \models \alpha(\vec{p}))$;

(ix) $v, x \models \Diamond\alpha(\vec{p})$ iff $\exists y \in X \;\; (xRy \;\text{ and }\; v, y \models \alpha(\vec{p}))$.

Put also $v \models \alpha(\vec{p})$ iff $\forall x \in X \;\; v, x \models \alpha(\vec{p})$. Now let $d_\Box(\varphi(\vec{x})) = n$ and let v, u be L-evaluations. Show that

(a) if $x \in dom(u)$, $y \in dom(v)$ and $(u, x) \sim_n (v, y)$ then $u, x \models \varphi(\vec{x})$ iff $v, y \models \varphi(\vec{x})$;

(b) if $u \approx_{n+1} v$ then $u \models \varphi(\vec{x})$ iff $v \models \varphi(\vec{x})$;

Conclude that, if $d_\Box(\varphi(\vec{x})) = n$, then the subsheaf of h_L of those L-evaluations u that $u \models \varphi(\vec{x})$ has index $n + 1$.

3 Show that any class **C** of finite **K4**-algebras closed under finite products, quotients and subalgebras is the class of finite algebras in a subvariety of $Alg(\mathbf{K4})$. Below we give some hints for this problem.

(a) Let A be a finite, subdirectly irreducible **K4**-algebra, m the element of A, such that $m = 1$ generates the least non-trivial congruence on A. Define modal formulas

$$D_A = \bigwedge_{a,b \in A} \Box^+(p_a \wedge p_b \leftrightarrow p_{a \wedge b}) \wedge$$

$$\wedge \bigwedge_{a \in A} ((\Box^+(\neg p_a \leftrightarrow p_{\neg a}) \wedge (\Box^+(\Box p_a \leftrightarrow p_{\Box a}))$$

and

$$J_A = D_A \to p_m$$

Show that for any **K4**-algebra B, $B \not\models J_A = 1$ iff there are homomorphisms e and m, such that

$$\begin{array}{ccc} B & \xrightarrow{\;e\;} & B' \\ & & \uparrow m \\ & & A \end{array}$$

with e onto and m one-to-one.

(b) Show that if A is finite, subdirectly irreducible **K4**-algebra not in **C**, then $B \models J_A = 1$, for any B in **C**.

(c) Conclude that any finite **K4**-algebra A not in **C**, is not in the variety generated by **C**.

4 Prove that if **S** is a system enjoying the finite model property, then **S** is locally finite iff there exists a number k such that $ht(X, R) \leq k$ holds for all $(X, R) \in \mathbf{F_S}$. (*Hint:* Use Proposition 5.9).

5 *Set-based dualities.* The following sequence of exercises develops a duality, similar to those given in this Chapter, for the category of finitely presented algebras of an equational theory **S**, containing the modal system **K4**, having the finite model property, but without the assumption that the category of finite **S**-algebras has the amalgamation property. This duality does not give rise to a sheaf representation like cf. Theorem 5.50.

Assume that **S** is an equational theory containing **K4**. By a frame we mean here an **S**-frame. For a finite set L, by h_L we denote the set of all L-evaluations. Let C be a subset of h_L. C is *n-closed* iff $\forall v \in C \forall w \in h_L\ w \leq_n v$ implies $w \in C$. C is *closed* iff it is n-closed for some $n \in \omega$. C has the *extension property* iff for any finite jointly surjective family $\{h_i : (X_i, R_i) \longrightarrow (X, R)\}_{i \in I}$ of injections in $\mathbf{F_S}$ and any L-evaluation $v : (X, R) \longrightarrow L$, if $v \circ h_i \in C$ for $i \in I$, then $v \in C$.

The category $\mathbf{N_S}$ has as objects pairs $< L, C >$ (denoted C^L for short), where L is a finite set and C is a closed subset of h_L with the extension property. A morphism $f : C^L \longrightarrow D^M$ in $\mathbf{N_S}$ is a function preserving domains of evaluations and having an index, i.e. there is $n \in \omega$ (called *index* of f) such that for all $m \in \omega$, all L-evaluations v, u in C^L and all $x \in dom(v)$, $y \in dom(u)$ if $(v, x) \sim_{n+m} (u, y)$ then $(f(v), x) \sim_m (f(u), y)$.

(a) Show that $\mathbf{N_S}$ so defined is indeed a category with finite limits.

(b) Show that there is a structure of an internal **S**-algebra on h_2.

(c) Show the fullness of the functor $\mathbf{T_S}$ obtained by homming into the $\mathbf{S}$-algebra h_2 (in other words

$$\mathbf{T_S} : \mathbf{N_S} \longrightarrow Alg(\mathbf{S})^{op}$$

is defined, for $f : C^L \longrightarrow D^M$ in $\mathbf{N_S}$, as

$$\mathbf{T_S}(f) : \mathbf{N_S}(D^M, h_2) \longrightarrow \mathbf{N_S}(C^L, h_2)$$

where $\mathbf{T_S}(f)(\alpha) = \alpha \circ f$ for $\alpha \in \mathbf{N_S}(D^M, h_2)$).

(d) Show that, if **S** has the finite model property then $\mathbf{T_S}$ is full and faithful and its essential image consists of finitely presented **S**-algebras, i.e. the category $Alg(\mathbf{S})^{op}_{fp}$ is equivalent to $\mathbf{N_S}$.

6 Let **S** be an equational theory, containing the modal system **K4**, and having the finite model property. Using the previous exercise, prove that finitely presented **S**-algebras are atomic.

7 Show that the functor $\mathbf{\Phi_S} : Alg(\mathbf{S})^{op}_{fp} \longrightarrow Sh(\mathbf{F_S}, \mathbf{J_S})$, defined on page 159, is conservative.

8 Show that $Alg(\mathbf{S})_{fp}$ has disjoint and universal finite coproducts and that $\mathbf{\Phi_S}$ preserves them.

Chapter 6

MODEL COMPLETIONS IN MODAL LOGIC

1. Negative results

In this section, we shall use Theorem 5.49 in order to obtain negative results, i.e. in order to show that a model completion does not exist for many equational theories containing $\mathbf{K4}$. Let $\mathbf{n}$ be the n-elements linearly ordered reflexive chain, that is $\mathbf{n}$ is the frame consisting of the set $\{1, 2, \ldots, n\}$ ordered by the natural order $\leq$. $\mathbf{n}^*$ is the frame obtained by $\mathbf{n}$ by replacing 1 by a two-elements cluster, that is $\mathbf{n}^*$ has $\{1^+, 1^-, 2, \ldots, n\}$ as underlying set and the order is defined by cases by $p \leq q$ iff i) either $p = 1^+$ or ii) $p = 1^-$ or iii) $(p, q \geq 2$ and $p \leq q)$.

LEMMA 6.1 *Let* $\mathbf{F_S}$ *satisfy* (AP^*)*; if* $\mathbf{n} \in \mathbf{F_S}$ *for every n and if* $\mathbf{2}^* \in \mathbf{F_S}$*, then* $\mathbf{n}^* \in \mathbf{F_S}$ *for every n.*

Proof. This Lemma is a special case of a general fact explained in Exercise 1; we nevertheless prove it directly. Fix $n > 2$; let $k : \mathbf{n} \to \mathbf{2}$ be the frame surjective morphism associating 1 with 1 and 2 with any $m > 1$. Let $h : \mathbf{2}^* \to \mathbf{2}$ be the frame surjective morphism associating 2 with 2 and 1 with both $1^+, 1^-$. Consider any frame (X, R) in $\mathbf{F_S}$ for which there are surjective frame morphisms π_1, π_2 making the square

$$\begin{array}{ccc} (X,R) & \xrightarrow{\pi_1} & \mathbf{n} \\ {\scriptstyle \pi_2}\downarrow & & \downarrow{\scriptstyle k} \\ \mathbf{2}^* & \xrightarrow[h]{} & \mathbf{2} \end{array}$$

to commute. Now it is clear that $\mathbf{n}^*$ must be a quotient of (X, R): define the related surjective frame morphism q for every $x \in X$ by i) $q(x) = 1^+$

iff $\pi_1(x) = 1^+, \pi_2(x) = 1$, ii) $q(x) = 1^-$ iff $\pi_1(x) = 1^-, \pi_2(x) = 1$, iii) $q(x) = m$ (for $m > 1$) iff $\pi_1(x) = 2, \pi_2(x) = m$. $\square$

THEOREM 6.2 *Let* $\mathbf{S}$ *be a variety of* $\mathbf{K4}$ *algebras having the finite model property. If* $\mathbf{n} \in \mathbf{F_S}$ *for every n and if* $\mathbf{2}^* \in \mathbf{F_S}$*, then* $\mathbf{S}$ *does not admit a model completion.*

Proof. Lemma 6.1 tells us that also $\mathbf{n}^*$ is in $\mathbf{F_S}$ for every $n \in N$. We show that condition $(\forall\exists)$ of Theorem 5.49 fails.

Let M be the two element-set $\{a, b\}$ and let L be $\{a, b, a^+, b^+\}$. Let $f : h_L \to h_M$ be the composition with the function mapping a, a^+ into a and b, b^+ into b (this is clearly an arrow in $\mathbf{M_S}$ having index 0). We introduce for any $k \geq 1$, an M-evaluation $v^k : (Y^k, S^k) \to M$ and an L-evaluation $u^k : (X^k, R^k) \to L$. (Y^k, S^k) is $\mathbf{2k}$ and $v^k(i) = a$ if i is odd and $v^k(i) = b$ if i is even. (X^k, R^k) is $\mathbf{2k}^*$; $u^k(x)$ is equal to $v^k(x)$ (no plus!) if $x \geq 2$ and $u^k(1^+) = a^+$, $u^k(1^-) = b^+$.

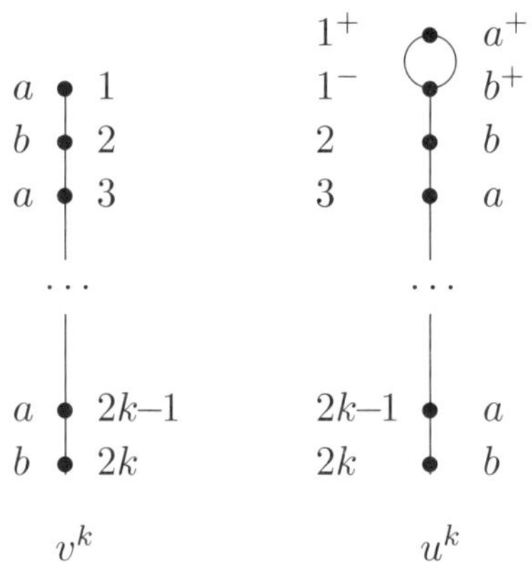

We collect some easy preliminary facts.

(i) *If* $i \geq 2$*, then* $(v^k, i) \sim_\infty (f(u^k), i)$: this is trivial.

(ii) *If* $i \geq j \geq 2$ *and* i, j *are both odd or even, then* $(v^k, i) \sim_{2k-i} (f(u^k), j)$: this is shown by induction on i. For $i = 2k$, the claim is trivial; for $i < 2k$, the only relevant case is when Player I chooses $s \in X^k$ such that $s < i$ (otherwise the obvious answer is suggested by (i)). In this case, Player II answers i or $i + 1$, depending on the fact that s is even or odd, and wins by induction hypothesis.

(iii) *For every* $i \in Y^k$ *and for* $x = 1^+, 1^-$, $(v^k, i) \sim_0 (f(u^k), x)$ *implies* $(v^k, i) \sim_{2k-i} (f(u^k), x)$: this is clear for $i = 2k, 2k - 1$. Suppose that

$i < 2k - 1$. Again, by (i), we can examine only the case in which Player I plays $s \in X^k$ for $s < i$ or $s = 1^+, 1^-$. In the latter case, the answer is i or $i + 1$: Player II wins by induction hypothesis (or we played an identical move). In the former case, the answer is again i or $i + 1$, depending if s is even or odd. Player II wins by (ii).

Using (iii), we easily get that $f(u^k) \approx_{2k-1} v^k$.

We show that $(\forall\exists)$ fails for $n = 3$. For reductio, suppose that N exists so that $(\forall\exists)$ is satisfied. Take k such that $2k - 1 \geq N$, $u = u^k$, $v = v^k$. Now $f(u) \approx_N v$ holds by the above analysis. We show that there cannot be any L-evaluation $u' : (X', R') \to L$ such that $u' \approx_3 u$ and $f(u') \approx_\infty v$. We prove that there must be $x_1, x_2 \in X'$ such that $x_1 R' x_2$, $x_2 R' x_1$, $u'(x_1) = a^+$, $u'(x_2) = b^+$: this is in contradiction to $f(u') \approx_\infty v^k$, because in v^k Player II cannot play points labelled $a, b, a, b, \ldots$ for ever. To prove our claim, we use the fact that $u' \approx_3 u$. In particular, there must be some point x_1 in X' whose u'-value is a^+. As our frames are finite, we can pick such an x_1 of minimal height, i.e. x_1 must be such that $u'(x_1) = a^+$ and for every x', (if $xR'x'$ and $u(x') = a^+$, then $x'R'x_1$). Player II has a 3-strategy, so he must be able to answer successfully to the following moves: first, Player I plays $x_1 \in X'$, then (as the forced answer is 1^+), he plays 1^- in X^k and finally 1^+ in X^k. If x_2, x' are the answers of Player II, we have $x_1 R' x_2 R' x'$, $u'(x_2) = b^+$, $u'(x') = a^+$ and also $x' R' x_1$ by the minimality of x_1. In conclusion, we have $x_1 R' x_2$ and $x_2 R' x_1$ with $u'(x_1) = a^+$ and $u'(x_2) = b^+$, as claimed. □

Theorem 6.2 shows in particular that **S4** and **K4** algebras do not have a model completion.[1] We give another example of the use of Theorem 5.49 for negative results:

PROPOSITION 6.3 *The system* **S4.3.Grz** *does not admit a model completion.*[2]

Proof. It is well-known that the variety in question has the finite model property, like any variety extending **S4.3** (see [Bu]). Let M be the three-elements set $\{a, b, c\}$ and let L be $\{a, b, c, a^+, b^+\}$. Let $f : h_L \to h_M$ be the composition with the function mapping a, a^+ into a and b, b^+ into b and c into c. Consider, for any $k \geq 1$, the evaluations $u^k : \mathbf{2k+1} \to L$ and $v^k : \mathbf{2k+2} \to M$ so specified: $u^k(1) = c, u^k(2) = a^+, u^k(3) = b^+, u^k(m) = a$ (for $m > 3$, m even), $u^k(m) = b$ (for $m > 3$, m odd), $v^k(1) = c, v^k(m) = b$ (for $m \geq 2$, m even), $v^k(m) = a$ (for $m \geq 2$, m odd).

[1] The fact that the first order theory of interior algebras does not admit a model completion was first proved in [Li].

[2] This result follows also from the fact that the amalgamation fails in this variety (see [Ma4]). It is quite interesting to observe that the amalgamation does hold in the restricted subcategory of finite algebras.

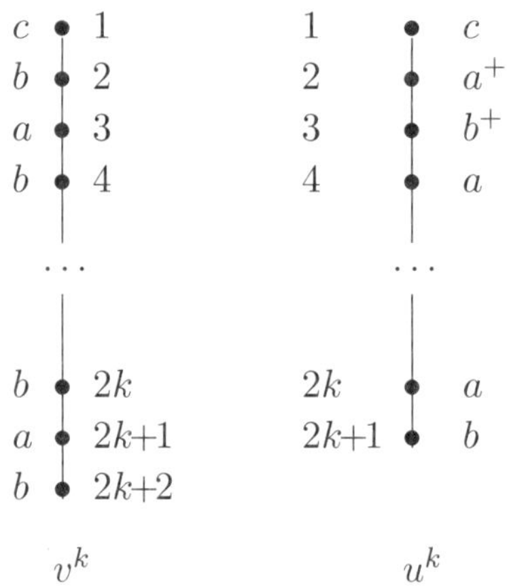

It is not difficult to see that $f(u^k) \approx_{2k-1} v^k$. Again, condition $(\forall\exists)$ fails for $n = 3$. Suppose it holds for a certain N and take $u = u^k$, $v = v^k$ such that $N \leq 2k - 1$. Let u' be such that $f(u') \approx_\infty v$ and $u' \approx_3 u$. We shall find a contradiction, using the fact that the domain (X', R') of u' must be a finite partially ordered locally linear set (notice that in such frames we can speak of *the successor* of a non terminal point). Let $x' \in X'$ be a point of minimal height such that $(f(u'), x') \sim_\infty (v, 1)$. This point cannot be terminal; we claim that for its successor y' we must have

$$u'(y') = b \text{ or } u'(y') = b^+. \tag{6.1}$$

In fact, y' cannot be labelled c by minimality; suppose for reductio that $u'(y') = a$ or that $u'(y') = a^+$. Consider $y > 2$ such that $(f(u'), y') \sim_\infty (v, y)$ and $z' \in X'$ such that $x'Rz'$ and $(f(u'), z') \sim_\infty (v, 2)$ (they both exist because $(f(u'), x') \sim_\infty (v, 1)$). We must have $y'Rz'$ by the fact that y' is the successor of x', hence from $(f(u'), y') \sim_\infty (v, y)$ we get $z \geq y$ which is infinitely equivalent, by transitivity, to 2. However two distinct points in the domain of v cannot be infinitely equivalent. This ends the proof of the claim (6.1).

Now from $u' \approx_3 u$ we get that $(u', x') \sim_2 (u, 1)$: this is a contradiction as in u Player II has no chance of playing a point labelled b or b^+ followed by a point labelled a^+ (the fact that there is a point in u' labelled a^+ above y' comes from the fact that $(u', x') \sim_2 (u, 1)$ and the fact that y' is the successor of x'). □

2. Diagonalizable algebras

Positive results in the application of Theorem 5.49 are obtained through the crucial notion of rank of a point in an evaluation. Recall from section 5.1 that $[(v, x)]_m$ means the equivalence class of (v, x) with respect to $\sim_m$. Given a point $x \in X$ and an evaluation $v : (X, R) \to L$, the *m-rank* of (v, x), written $rk_m(x)$ (for $m \geq 1$), is the cardinality of the set $\{[(v, y)]_{m-1} \mid xRy\}$. By transitivity of R, we always have that if xRy then $rk_m(x) \geq rk_m(y)$. A point x is said to be a *frontier point* [Sh] iff for every y, xRy and $rk_m(x) = rk_m(y)$ imply yRx (i.e. it implies that x and y lie in the same cluster). As our frames are finite, for every x either x is a frontier point or there is a frontier point y such that xRy and $rk_m(x) = rk_m(y)$. An easy property of the notion of rank is provided by the following:

LEMMA 6.4 *Suppose that we are given an evaluation $v : (X, R) \to L$ and two points $x, y \in X$ such that xRy and $rk_m(x) = rk_m(y)$. Then for every x' such that xRx', there is y' such that yRy' and $(v, x') \sim_{m-1} (v, y')$.*

Proof. This is due to the fact that two finite sets of the same cardinality one of which is included in the other must coincide. □

We deal now with diagonalizable algebras, which are the variety **G** mentioned in section 5.1. System **G** is known for its relationship with arithmetic; for this and for the finite model property with respect to finite irreflexive, transitive frames see e.g. [Bo]. For the remaining part of this section we shall deal only with *irreflexive, transitive finite frames*, whose accessibility relation will always be indicated by $<$. The relation $\leq$ will be the reflexive closure or $<$. In irreflexive, transitive frames the notion of a frontier point x in an evaluation $v : (X, <) \to L$ can be so simplified: x is frontier iff for every $y > x$ we have that $rk_m(y) < rk_m(x)$. As the first condition of Theorem 5.49 is easily seen to be satisfied for the class of all irreflexive, transitive frames (for the proof, take the set-theoretical pullback and define the relation componentwise), the existence of the model completion for the first order theory of diagonalizable algebras relies entirely on condition $(\forall\exists)$.[3]

THEOREM 6.5 *The theory of diagonalizable algebras admits a model completion.*

Proof. Let us fix our data $f : C^L \to D^M$ for the condition $(\forall\exists)$. Notice that, by the statement itself, it is sufficient to establish condition $(\forall\exists)$ for infinitely many $n \in \omega$. We shall prove it for every $n \geq 2$ big enough so that n is an index for C and $n-1$ is an index for f. We take as N any number such that $N \geq 2K+2$, where K is the maximum $n-1$-rank that a point can have in an L-evaluation (we recall that $\sim_{n-2}$ is an equivalence relation having finitely many equivalence classes). Suppose now that for $u : (X, <) \to L$ in C and for $v : (Y, <) \to M$ in D, we have that $f(u) \approx_N v$. We wish to build $u' : (X', <) \to L$ and a

[3] A condition similar to $\forall\exists$ (for evaluations in rooted, irreflexive, transitive finite frames) was introduced and proved in [Sh] in connection to questions related to strong interpolation. The proof we give here is a modification of the proof from Chapter 4 for the case of Heyting algebras.

surjective morphism of frames $p : (X', <) \to (Y, <)$ such that $u' \approx_n u$ (hence $u' \in C$) and $f(u') = v \circ p$ (this guarantees that $f(u') \approx_\infty v$).

The difficulty of the proof consists in finding an appropriate definition for X'. We define as X' the set of pairs $(x, y) \in X \times Y$ such that:

(1) $(f(u), x) \sim_{2rk_{n-1}(x)} (v, y)$;

(2) there exist $x_0 \leq x$, $y_0 \leq y$ such that $rk_{n-1}(x_0) = rk_{n-1}(x)$ and $(f(u), x_0) \sim_{2rk_{n-1}(x)+1} (v, y_0)$.

Any pair (x_0, y_0) as in (2) is said to be a *justification* for the pair $(x, y) \in X'$. The irreflexive, transitive relation $<$ in X' is given by

$$(x, y) < (x', y') \text{ iff } (y < y' \ \& \ \forall z' \geq x' \ \exists z > x \ (u, z) \sim_{n-2} (u, z')).$$

We define $u'(x, y)$ to be equal to $u(x)$ and $p(x, y)$ to be equal to y. Notice that, because of the hypothesis $f(u) \approx_N v$, for every $x \in X$ there is $y \in Y$ such that $(x, y) \in X'$ and conversely for every $y \in Y$ there is $x \in X$ such that $(x, y) \in X'$ (a justification is the pair (x, y) itself in both cases). So in order to check that $u' \approx_n u$, that p is a surjective frame morphism and that $f(u') = v \circ p$ we need the following three facts:

(i) for every $(x, y) \in X'$, we have that $(u', (x, y)) \sim_{n-1} (u, x)$;

(ii) for every $(x, y) \in X'$ and for every $y' > y$, there exists $(x', y') \in X'$ such that $(x, y) < (x', y')$;

(iii) for every $(x, y) \in X'$, we have that $(f(u'), (x, y)) \sim_0 (v, y)$.

Proof of (i). The claim is shown by induction on $k = 0, 1, \ldots, n-1$. It holds by the definition of u' in the case $k = 0$. Suppose $n - 1 \geq k > 0$. If Player I picks $(x', y') > (x, y)$, then by induction hypothesis we have that $(u', (x', y')) \sim_{k-1} (u, x')$; by the definition of $<$ in X', taking $z' = x'$, we have that there exists $z > x$ such that $(u, z) \sim_{n-2} (u, x')$. Recalling that $k - 1 \leq n - 2$, we have also $(u', (x', y')) \sim_{k-1} (u, z)$, so z is a good answer for Player II. Suppose now that Player I picks $x' > x$ in X. By Lemma 6.4 and the fact that any point x either is a frontier point or a frontier point of the same rank is accessible from it, there is $x'' \in X$ such that $x < x''$, $(u, x') \sim_{n-2} (u, x'')$ and $rk_{n-1}(x'') < rk_{n-1}(x)$. Consequently, there will be $y' > y$ such that $(x'', y') \in X'$ with justification (x'', y') (notice that $2rk_{n-1}(x'') + 1 \leq 2(rk_{n-1}(x) - 1) + 1 = 2rk_{n-1}(x) - 1$). As $(x, y) < (x'', y')$ trivially holds in X' because $x < x''$ and $y < y'$, Player II answers (x'', y') and wins by induction hypothesis and by the fact that $(u, x') \sim_{n-2} (u, x'')$.

Proof of (ii). Suppose we are given $(x, y) \in X'$ and $y' > y$. Let (x_0, y_0) be a justification for (x, y); by the fact that $(f(u), x_0) \sim_{2rk_{n-1}(x)+1} (v, y_0)$, we get $x' > x_0$ such that $(f(u), x') \sim_{2rk_{n-1}(x)} (v, y')$, so we can take (x', y') in order to show the claim. In fact, in case $rk_{n-1}(x') = rk_{n-1}(x_0)$, (x_0, y_0) is a justification for (x', y') and in case $rk_{n-1}(x') < rk_{n-1}(x_0)$, (x', y') has itself as a justification. Finally, we have $(x, y) < (x', y')$ by Lemma 6.4 and the facts that $x_0 \leq x$, $rk_{n-1}(x_0) = rk_{n-1}(x)$ and $x_0 < x'$.

Proof of (iii). For any given point $(x, y) \in X'$, we have that $(f(u'), (x, y)) \sim_0 (f(u), x)$ by (i) and the fact that $n - 1$ is an index for f and we also have $(f(u), x) \sim_0 (v, y)$ by the definition of X'. □

3. Varieties of interior algebras

In this section we deal only with varieties of interior or **S4**-algebras, consequently we assume that our frames are (finite) *preordered sets*. The preorder relation is always indicated[4]by $\leq$ and $x < y$ stands for $x \leq y$ and $x \neq y$. Given a preordered set $(X, \leq)$, the points $x \in X$ satisfying the condition

$$\forall y \in X \; (x \leq y \text{ implies } y \leq x)$$

are called *external*, the other ones are called *internal*. Similarly a cluster is called external iff all points of its are external and internal iff all points of its are internal (in other words, a cluster is external iff, as an equivalence class in the antisymmetric quotient, it is maximal). X^i will be the set of internal points of $(X, \leq)$ and X^e will be the set of external points of $(X, \leq)$. The *inner dimension* $\delta^i(X, \leq)$ of $(X, \leq)$ is the maximum cardinality of inner clusters of $(X, \leq)$ and the *outer dimension* $\delta^e(X, \leq)$ of $(X, \leq)$ is the maximum cardinality of external clusters of $(X, \leq)$. For instance, for a frame $(X, \leq)$, we have $\delta^i(X, \leq) = 1$ iff the following condition is satisfied for every $x, y, z \in X$:

$$(1) \qquad (x \leq y \;\&\; y \leq x \;\&\; y \leq z) \;\Rightarrow\; (z \leq y \text{ or } x = y);$$

moreover, we have $\delta^e(X, \leq) \leq 2$ iff the following condition is satisfied for every $x, y, z \in X$:

$$(2) \qquad (x \in X^e \;\&\; x < y \;\&\; y < z) \;\Rightarrow\; x = z.$$

Notice also that $(X, \leq)$ belongs to $\mathbf{F}_{\mathbf{Grz}}$ iff $\delta^i(X, \leq) \leq 1$ and $\delta^e(X, \leq) \leq 1$. In addition to the notion of height of a frame given in section 6, we introduce the notion of *width* of a frame $(X, \leq)$ by saying that $wt(X, \leq) \leq k$ iff for every $x \in X$, in $\uparrow x = \{y \,|\, x \leq y\}$ there are at most k mutually incomparable points. Finally we put $e(X, \leq) \leq s$ iff for every $x \in X$, in $\uparrow x$ there are at most s external clusters (notice that $e(X, \leq) \leq 1$ iff $(X, \leq)$ is locally confluent).

[4]Notice that there is a slight change in notation with respect to Chapter 4: what is $\leq$ here, was $\geq$ there. There should be no danger of confusion however, as the modal and the intuitionistic parts of the book are independent.

We know (see Proposition 2.8) that the existence of a model completion implies the amalgamation property; in [Ma2] a finite list including all amalgamable varieties of interior algebras is given. Each variety $\mathbf{S}$ in the list has the finite model property, hence it can be univocally identified once $\mathbf{F_S}$ is specified ($\mathbf{S}$ will just be the variety generated by the finite interior algebras whose dual frames belong to $\mathbf{F_S}$) What is shown in [Ma2] is the following:

THEOREM 6.6 *If $\mathbf{S}$ is an amalgamable variety of interior algebras, then $\mathbf{S}$ has the finite model property; moreover $\mathbf{S} = \mathbf{S4.3.Grz}$ or $\mathbf{F_S}$ must be either*

$$(*) \qquad \{(X, \leq) \mid ht(X, \leq) \leq l \;\&\; wt(X, \leq) \leq k \;\&\; \delta^i(X, \leq) \leq m \;\& \;\&\; \delta^e(X, \leq) \leq n\},$$

for some $l \in \{0, 1, 2\}$ and some $k, m, n \in \{1, 2, \omega\}$ or

$$(**) \qquad \{(X, \leq) \mid \delta^i(X, \leq) \leq m \;\&\; \delta^e(X, \leq) \leq n \;\&\; e(X, \leq) \leq s\},$$

for some $s \in \{1, \omega\}$ and some $m, n \in \{1, 2, \omega\}$.

The presentation of the proof of Theorem 6.6 is beyond the scope of the book; however in the exercises the reader will find some hints on how to prove it if the finite model property is assumed.

Notice that, if the finite model property holds for $\mathbf{S}$ and AP holds in $Alg(\mathbf{S})$ then (AP^*) holds in $\mathbf{F_S}$, as shown in Proposition 2.20, but not conversely. For instance, (AP^*) holds for $\mathbf{F_{S4.3.Grz}}$, but AP fails in $Alg(\mathbf{S4.3.Grz})$, cf. [Ma4]. Taking into account this last fact,[5] Theorem 6.6 gives a list of 49 varieties,[6] among which there are all the amalgamable varieties of interior algebras. The 31 varieties described by the condition (*) are all shown to be amalgamable in [Ma3] and so they admit a model completion because they are locally finite[7] (it is a general fact that for locally finite varieties the existence of a model completion is the same as amalgamability [Wh], see also the Exercises of Chapter 3).[8] We shall in any case turn to these 31 varieties in the exercises. The remaining part of this section is devoted to the proof of the following:

THEOREM 6.7 *There are exactly six non-locally finite varieties $\mathbf{S_1}, \ldots, \mathbf{S_6}$ of interior algebras whose first order theory admits a model completion. They all*

[5] $\mathbf{S4.3.Grz}$ must be eliminated from our search of varieties admitting a model completion also because of Proposition 6.3.

[6] Notice that if $ht(X, \leq) = 1$ iff $wt(X, \leq) = 1$ iff $\delta^i(X, \leq) = 0$, moreover $ht(X, \leq) = 0$ iff $wt(X, \leq) = 0$ iff $\delta^e(X, \leq) = 0$ iff X is empty.

[7] For a system $\mathbf{S}$ enjoying the finite model property, it can be shown (see [CZ] or Exercise 4 in Chapter 5) that $\mathbf{S}$ is locally finite iff there exists a number k such that $ht(X, R) \leq k$ holds for all $(X, R) \in \mathbf{F_S}$.

[8] As a third reason, notice that if height of finite $\mathbf{S}$-frames is bounded, condition $(\forall\exists)$ trivializes by Corollary 5.10.

have the finite model property and the corresponding classes of finite frames are respectively the following ones:

$$\begin{aligned}
\mathbf{F}_{\mathbf{S}_1} &= \{(X, \leq) \,|\, \delta^i(X, R) \leq 1\} \\
\mathbf{F}_{\mathbf{S}_2} &= \{(X, \leq) \,|\, \delta^i(X, R) \leq 1 \;\&\; \delta^e(X, \leq) \leq 2\} \\
\mathbf{F}_{\mathbf{S}_3} &= \{(X, \leq) \,|\, \delta^i(X, R) \leq 1 \;\&\; \delta^e(X, \leq) \leq 1\} \\
\mathbf{F}_{\mathbf{S}_4} &= \{(X, \leq) \,|\, \delta^i(X, R) \leq 1 \;\&\; e(X, R) \leq 1\}\} \\
\mathbf{F}_{\mathbf{S}_5} &= \{(X, \leq) \,|\, \delta^i(X, R) \leq 1 \;\&\; \delta^e(X, \leq) \leq 2 \;\&\; e(X, R) \leq 1\} \\
\mathbf{F}_{\mathbf{S}_6} &= \{(X, \leq) \,|\, \delta^i(X, R) \leq 1 \;\&\; \delta^e(X, \leq) \leq 1 \;\&\; e(X, R) \leq 1\}.
\end{aligned}$$

Proof. Among the 18 varieties listed in (**), the above ones are the only ones which are compatible with Theorem 6.2. On the other hand Lemma 6.8, and Propositions 6.9, 6.12 below show that the conditions of Theorem 5.49 are fulfilled by them. □

LEMMA 6.8 *Condition* (AP^*) *holds in* $\mathbf{F}_{\mathbf{S}_1}, \ldots, \mathbf{F}_{\mathbf{S}_6}$.

Proof. This is a quite elementary but tedious verification. We sketch it. Suppose we are given frames $(X, \leq), (X_1, \leq), (X_2, \leq)$ in $\mathbf{F}_{\mathbf{S}_i}$ $(i = 1, \ldots, 6)$ and surjective morphisms of frames

$$k_1 : (X_1, \leq) \to (X, \leq), \quad k_2 : (X_2, \leq) \to (X, \leq);$$

we need to build a commutative square

$$\begin{array}{ccc}
(Y, \leq) & \xrightarrow{p_1} & (X_1, \leq) \\
{\scriptstyle p_2}\downarrow & & \downarrow{\scriptstyle k_1} \\
(X_2, \leq) & \xrightarrow[k_2]{} & (X, \leq)
\end{array}$$

in $\mathbf{F}_{\mathbf{S}_i}$ with surjective frame morphisms p_1, p_2.

Let us first consider the cases $i \neq 2, 5$. In all these case, we can take Y to be the set-theoretical pullback and define its preorder relation as follows. For $(x, y), (x', y') \in Y$ $(x, y) \leq (x', y')$ holds iff one (or possibly more) of the following cases happen: 1) $x = x'$ and $y = y'$; 2) $x \in X_1^i$ and $x < x'$ and $y \leq y'$; 3) $y \in X_2^i$ and $x \leq x'$ and $y < y'$; 4) $x \in X_1^e$ and $y \in Y_2^e$ and $x \leq x'$ and $y \leq y'$.

Transitivity of $\leq$ so introduced is established by direct inspection of all the possibilities; (1) follows from the fact that if (x, y) and (x', y') are different and lie in the same cluster, then x, x', y, y' must be all external. Moreover, if $(X, \leq), (X_1, \leq), (X_2, \leq)$ are all locally confluent and/or posets, so is $(Y, \leq)$.

Finally we take as p_1, p_2 the two projections; in order to check that they are frame morphisms, it is useful to observe that for an arbitrary frame morphism $f : (Z_1, \leq) \to (Z_2, \leq)$, the fact that $z \in Z_1$ is external implies that so is $f(z)$ and the fact that $f(z)$ is external implies that there is $z' \geq z$ such that $f(z') = f(z)$ which is external too.

Let us finally consider the cases $i = 2, 5$. Here (2) is supposed to hold and we need a slight modification of the above construction: we shall replace Y by a subset Y' of its endowed with the restriction of the preorder relation of Y. In order to introduce Y' we need some preliminary easy definitions. A point x in a frame is said to be terminal iff for all y, $x \leq y$ implies $x = y$. A point in X_1 (resp. X_2) is said to be exceptional iff it is external, not terminal, but $k_1(x)$ (resp. $k_2(x)$) is terminal. Let us color in two different ways exceptional points both in X_1 and in X_2 in such a way that exceptional points related by $<$ get different colors. Y' will be Y minus the pairs (x, y) such that x, y are both exceptional and have different colors.[9] The fact that $(Y', \leq)$ is contained in $(Y, \leq)$ (and that the relation is the restriction) implies that it inherits from $(Y, \leq)$ transitivity of $\leq$ and condition (1). A careful inspection shows that condition (2) holds too. The facts that the restricted projections are still surjective frame morphisms and that $(Y', \leq)$ is locally confluent in case $(X, \leq), (X_1, \leq), (X_2, \leq)$ are, both follow from the following property (recall once again that the preorder relation in Y' is the restriction of the preorder relation in Y): for every $(x, y) \in Y$ there are x', y' such that $(x', y), (x, y') \in Y'$ and such that in Y we have $(x, y) \leq (x', y)$ and $(x, y) \leq (x, y')$. □

As expected, the real problem is condition $(\forall\exists)$. We shall prove it for $\mathbf{S_1}$ and then show that the remaining cases can be reduced to this.

PROPOSITION 6.9 *Condition* $(\forall\exists)$ *holds in* $\mathbf{S_1}$.

Proof. During the proof of this Proposition, we shall deal only with finite preordered sets $(X, \leq)$ satisfying (1). The argument we shall use is of the same kind as the argument we used in the previous section for diagonalizable algebras, however here the situation is more complicated. The reader shuold keep in mind the definitions introduced at the beginning of Section 6.2.

Let us fix our data $f : C^L \to D^M$ for the condition $(\forall\exists)$. Let $n \geq 4$ be any fixed number, big enough to be also an index for C and big enough so that $n-1$ is an index for f. We take as N any number greater or equal to $3K + 3$, where K is the maximum $n-1$-rank that a point can have in an L-evaluation. Suppose now that for $u : (X, \leq) \to L$ in C and for $v : (Y, \leq) \to M$ in D, we have that $f(u) \approx_N v$. We wish to build $u' : (X', \leq) \to L$ and a frame

[9] This trick does not work if we replace condition (2) by the requirement that external dimension is less or equal to some given n for $n > 2$. In fact, for these classes of frames, (AP^*) is easily seen to fail.

surjective morphism $p : (X', \leq) \to (Y, \leq)$ such that $u' \approx_n u$ (hence $u' \in C$) and $f(u') = v \circ p$ (this guarantees that $f(u') \approx_\infty v$).

We take as X' the set of all pairs $(x, y) \in X \times Y$ such that the following two conditions are satisfied:

(i) $(f(u), x) \sim_{3rk_{n-1}(x)+1} (v, y)$ or (x is a frontier point and $(f(u), x) \sim_{3rk_{n-1}(x)} (v, y)$);

(ii) there exist $x_0 \leq x$, $y_0 \leq y$ such that $rk_{n-1}(x_0) = rk_{n-1}(x)$ and $(f(u), x_0) \sim_{3rk_{n-1}(x)+2} (v, y_0)$.

Any pair (x_0, y_0) as in (ii) is said to be a *justification* for the pair $(x, y) \in X'$. We define the relation $\leq$ in X' by saying that $(x, y) \leq (x', y')$ iff one of the following (non mutually exclusive) conditions applies:[10]

(I) $x = x'$ and $y = y'$;

(II) $y \in Y^i$ and $y < y'$ and $u_{\uparrow x} \geq_{n-1} u_{\uparrow x'}$;

(III) $x \in X^i$ and $x < x'$ and $y \leq y'$;

(IV) $x' \in X^e$ and $y' \in Y^e$ and $u_{\uparrow x} \geq_{n-1} u_{\uparrow x'}$ and $y \leq y'$.

Transitivity of $\leq$ just defined is established by direct inspection, i.e. by cases. Suppose in fact that we have $(x, y) \leq (x', y') \leq (x'', y'')$. Now if $(x', y') \leq (x'', y'')$ holds by clause (II) (resp. (IV)), then $(x, y) \leq (x'', y'')$ holds again by clause (II) (resp. (IV)); if $(x', y') \leq (x'', y'')$ holds by clause (III), then $(x, y) \leq (x', y')$ must hold by clause (I), (II) or (III) and so $(x, y) \leq (x'', y'')$ holds by the same clause.

Condition (1) is also true: to show it, suppose that we have $(x, y) \leq (x', y') \leq (x, y)$, $(x, y) \neq (x', y')$ and $(x', y') \leq (x'', y'')$. We must show that $(x'', y'') \leq (x', y')$. We first claim that the relations $(x, y) \leq (x', y') \leq (x, y)$ hold by (IV), so that x, x', y, y' are all external. In fact, if (II) applies e.g. for $(x, y) \leq (x', y')$, we would have $y < y'$ and also $y' \leq y$ (by $(x', y') \leq (x, y)$), so y would not be internal as required by (II). (III) cannot apply also: if it applies e.g. to $(x, y) \leq (x', y')$, then we would have $x < x'$ and x would be internal, so $(x', y') \leq (x, y)$ cannot hold (recall that we just showed that it cannot hold by (II), moreover (I) and (IV) cannot be used and (III) would yield $x < x' < x$ in contrast to the fact that x is internal). So only (IV) applies in order to establish the relations $(x, y)' \leq (x', y') \leq (x, y)$ and consequently x, x', y, y' are all external. This implies that the relation $(x', y') \leq (x'', y'')$ must hold by (I) or (IV). In the former case we immediately get what we want,

[10] Recall from Chapter 5 that $u_{\uparrow x}$ is the restriction of u to the generated subframe $\{z \mid x \leq z\}$.

namely $(x'', y'') \leq (x', y')$. In the latter case, we get $y'' \leq y'$, as y' is external; moreover as x' and x'' are both external and as $u_{\uparrow x'} \geq_{n-1} u_{\uparrow x''}$, we get $u_{\uparrow x'} \approx_\infty u_{\uparrow x''}$, by Corollary 5.10 and easy passages (recall that $n - 1 \geq 3$). This shows that $(x'', y'') \leq (x', y')$ holds by (IV).

We define $u'(x, y)$ to be equal to $u(x)$ and $p(x, y)$ to be equal to y. Notice that, because of the hypothesis $f(u) \approx_N v$, for every $x \in X$ there is $y \in Y$ such that $(x, y) \in X'$ (with justification (x, y)) and conversely for every $y \in Y$ there is $x \in X$ such that $(x, y) \in X'$ (with justification (x, y)). So in order to check that $u' \approx_n u$, that p is a surjective frame morphism and that $f(u') = v \circ p$ the following three facts are sufficient:

(i) for every $(x, y) \in X'$, we have that $(u', (x, y)) \sim_{n-1} (u, x)$;

(ii) for every $(x, y) \in X'$ and for every $y' > y$, there exists $(x', y') \in X'$ such that $(x, y) \leq (x', y')$;

(iii) for every $(x, y) \in X'$, we have that $(f(u'), (x, y)) \sim_0 (v, y)$.

Proof of (i). The claim is shown by induction on $k = 0, 1, \ldots, n - 1$. It holds by the definition of u' in the case $k = 0$. Suppose $n - 1 \geq k > 0$.

Case (i.1) Suppose that Player I picks $(x', y') \geq (x, y)$ in X'. We have $u_{\uparrow x'} \leq_{n-1} u_{\uparrow x}$, so there is $x'' \geq x$ such that $(u, x'') \sim_{n-2} (u, x')$. As $(u', (x', y')) \sim_{k-1} (u, x')$ holds by induction and as $k - 1 \leq n - 2$, x'' is a good answer for Player II.

Case (i.2) Suppose that Player I picks $z > x$ in X (for $z = x$ answer is trivial, because our frames are reflexive). Take a frontier point $x' \geq x$ of the same rank as x and take $y' \geq y$ such that $(f(u), x') \sim_{3rk_{n-1}(x)} (v, y')$ (we take $x = x'$ and $y' = y$ if x is frontier). We can suppose, by Lemma 6.4, that (up to $\sim_{n-2}$-equivalence) $z \geq x'$.

Subcase (i.2.1) x' is external: in this case $(f(u), x') \sim_\infty (v, y')$ by Proposition 5.9 and the fact that rank must be at least 1 in reflexive frames. So from $z \geq x'$ it follows that there is $t \geq y'$ such that $(f(u), z) \sim_\infty (v, t)$. Playing another couple of moves we can assume that t is external too (take an external point above t and go back to z, which is external, in the successive move). Consequently $(z, t) \in X'$ and $(x, y) \leq (z, t)$ holds by clause (IV).

Subcase (i.2.2) x' is internal and $z > x'$. Then the rank of z is strictly less than the rank of x' and consequently there is $t \geq y'$ such that $(z, t) \in X'$ (with justification (z, t) itself). We also have $(x, y) \leq (z, t)$ by clause (III).

Subcase (i.2.3) x' is internal and $z = x'$. In this case, $(z, y') \in X'$ as $x' = z$ is frontier and as $(f(u), z) \sim_{3rk_{n-1}(z)} (v, y')$ (the justification is the same as for (x, y)); moreover, as $z > x$, we have $(z, y') \geq (x, y)$ by clause (III).

Proof of (ii). We have two cases.

Case (ii.1) y' is internal. Let (x_0, y_0) be a justification for (x, y). As $(f(u), x_0) \sim_{3rk_{n-1}(x)+2} (v, y_0)$ and as $y_0 \leq y \leq y'$, there is $x' \geq x_0$ such that $(f(u), x') \sim_{3rk_{n-1}(x)+1} (v, y')$, consequently $(x', y') \in X'$ (if the rank diminishes, a justification is (x', y') itself, if it remains the same, a justification is again (x_0, y_0)). We only have to show that $(x, y) \leq (x', y')$. As y is internal and as $y < y'$, it is sufficient to observe that $u_{\uparrow x'} \leq_{n-1} u_{\uparrow x_0} \approx_{n-1} u_{\uparrow x}$ (the latter is because $rk_{n-1}(x_0) = rk_{n-1}(x)$ and $x_0 \leq x$, see Lemma 6.4), so that (II) applies.

Case (ii.2) y' is external. In this case, there exists $x' \geq x$ such that $(f(u), x') \sim_{3rk_{n-1}(x)-1} (v, y')$ and also $(f(u), x') \sim_\infty (v, y')$, by Proposition 5.9. Playing another couple of moves, we can suppose that x' is external too, so we have $(x', y') \in X'$ and $(x, y) \leq (x', y')$ by (IV).

Proof of (iii). For any given point $(x, y) \in X'$, we have that $(f(u'), (x, y)) \sim_0 (f(u), x)$ by (i) and the fact that $n - 1$ is an index for f; we have also $(f(u), x) \sim_0 (v, y)$ by the definition of X'. □

Recall from section 5.1 that an evaluation $u : (X, \leq) \to L$ is said to be irreducible iff for every $x, y \in X$, if $(u, x) \sim_\infty (u, y)$ then $x = y$. Recall also by Proposition 5.7(i) that for every evaluation $u : (X, \leq) \to L$ there is an irreducible evaluation $u' : (X', \leq) \to L$ such that $u \approx_\infty u'$.

LEMMA 6.10 *Suppose that $u : (X, \leq) \to L$ and $u' : (X', \leq) \to L$ are two evaluations such that $u \approx_3 u'$. If $\delta^e(X, \leq) \leq k$, then there exists another evaluation $u'' : (X'', \leq) \to L$ such that $u' \approx_\infty u''$ and $\delta^e(X'', \leq) \leq k$.*

Proof. Let $u'' : (X'', \leq) \to L$ be an irreducible evaluation such that $u' \approx_\infty u''$. We claim that in $(X'', \leq)$ the external dimension cannot be more than k. We use the facts that u'' is irreducible and that $u \approx_3 u''$. Indeed if the external dimension of $(X'', \leq)$ is more than k, there are external points $x_1, \ldots, x_{k+1} \in X''$ in the same cluster, whose u''-values are pairwise different (in an irreducible evaluation two points in the same cluster must be labelled differently). But in this case Player I wins the 3-game as follows: first he picks x_1, then if x is the answer of Player II, he picks an external point $x' \geq x$. Now the cluster of x' has less than $k + 1$ points, so Player I wins if in the successive move he makes the appropriate choice among $x_1, \ldots, x_{k+1}$. □

LEMMA 6.11 *Suppose that $u : (X, \leq) \to L$ and $u' : (X', \leq) \to L$ are two evaluations such that $u \approx_5 u'$. If $(X, \leq)$ is locally confluent, then there exists another evaluation $u'' : (X'', \leq) \to L$ such that $u' \approx_\infty u''$ and such that $(X'', \leq)$ is locally confluent too.*

Proof. Let again $u'' : (X'', \leq) \to L$ be an irreducible evaluation such that $u' \approx_\infty u''$. We show that $(X'', \leq)$ is locally confluent by using that u'' is irreducible and that $u \approx_5 u''$. Suppose that $x'', y''_1, y''_2 \in X''$ are such that $x'' \leq y''_1$, $x'' \leq y''_2$. In X there are x, y_1, y_2 such that $(u, x) \sim_4 (u'', x'')$,

$x \le y_1, x \le y_2$, $(u, y_1) \sim_3 (u'', y_1'')$, $(u, y_2) \sim_3 (u'', y_2'')$. Take an external point z such that $z \ge y_1, z \ge y_2$. In u'' there will be z_1'', z_2'' such that $y_1'' \le z_1''$, $y_2'' \le z_2''$ and that $(u, z) \sim_2 (u'', z_1'')$, $(u, z) \sim_2 (u'', z_2'')$. By Proposition 5.9 and the fact that z is external, we get $(u, z) \sim_\infty (u'', z_1'')$, $(u, z) \sim_\infty (u'', z_2'')$, that is $z_1'' = z_2''$ by the irreducibility of u''. □

PROPOSITION 6.12 *Condition* $(\forall\exists)$ *holds for* $\mathbf{S_2}, \ldots, \mathbf{S_6}$.

Proof. We know from Proposition 6.9 that $(\forall\exists)$ holds for $\mathbf{S_1}$. Let us show it for $\mathbf{S_i}$ $(i = 2, \ldots, 6)$. Let $f : C^L \to D^M$ be a morphism in $\mathbf{M_{S_i}}$ and let $n \ge 5$ be a number big enough to be an index for C and big enough so that $n - 1$ is an index for f. Let N be as in the proof of Proposition 6.9. Given $u : (X, \le) \to L$ in C and $v : (Y, \le) \to M$ in D such that $f(u) \approx_N v$, define $u' : (X', \le) \to L$ as in the proof of Proposition 6.9, without caring of the fact that $(X', \le)$ might not be in $\mathbf{F_{S_i}}$. We have (see the proof of Proposition 6.9) that:

(i) $u' \approx_n u$, more specifically $(u', (x, y)) \sim_{n-1} (u, x)$ for every $(x, y) \in X'$;

(ii) the restricted second projection $p : (X', \le) \to (Y, \le)$ is a surjective frame morphism;

(iii) $(f(u), x) \sim_0 (v, y)$ for every $(x, y) \in X'$ (this is by the construction of X').

Now take $u'' : (X'', \le) \to L$ such that $u'' \approx_\infty u'$ and such that $(X'', \le) \in \mathbf{F_{S_i}}$: this is possible by Lemmas 6.10, 6.11 above and by the fact that $u' \approx_n u$ (recall that $n \ge 5$). We claim that for every $z \in X''$, $(x, y) \in X'$

(iv) $$(u'', z) \sim_\infty (u', (x, y)) \;\Rightarrow\; (f(u''), z) \sim_0 (v, y)$$

(notice that $u'' \in C$, as $u'' \approx_n u$, so it makes sense to apply f to it). In fact $(u'', z) \sim_{n-1} (u, x)$ holds by (i), so we get $(f(u''), z) \sim_0 (f(u), x)$ by the fact that $n - 1$ is an index for f and finally $(f(u''), z) \sim_0 (v, y)$ by (iii). By (ii) and the fact that $u'' \approx_\infty u'$, it is immediate to see that iterated applications of (iv) produce the stronger statement

(v) $$(u'', z) \sim_\infty (u', (x, y)) \;\Rightarrow\; (f(u''), z) \sim_\infty (v, y).$$

Again by (ii) and by the fact that $u'' \approx_\infty u'$, from (v) we get that $f(u'') \approx_\infty v$: this ends the proof of the Proposition. □

4. Exercises

The exercises 1-4, are to give the flavour of the relevant part of the results of L.L.Maksimowa, cf. [Ma1], [Ma2], [Ma3], classifying the varieties of Heyting algebras and of S4-algebras which satisfy AP. In order to simplify the matters we shall consider only those varieties that satisfy the finite model property[11].

The exercise 1 might be a help for (some cases of) exercise 2. The latter is a collection of some technical statements, concerning classes of finite rooted frames $\mathbf{F}^{\mathrm{r}}_{\mathbf{S}}$ that are closed under AP*. The exercises 3 and 4 guide the reader how to get the list of the few classes of finite posets and frames $\mathbf{F}^{\mathrm{r}}_{\mathbf{S}}$ that are closed under AP*.

1 Let f_0, f_1 are the frame morphisms with a common codomain (X, R). We say that the pair of morphisms f_0, f_1 is *thin* iff for any $x \in X$ if $|f_i^{-1}(x)| > 1$ then $|f_{1-i}^{-1}(x)| \leq 1$, for $i = 0, 1$ (here $| - |$ means cardinality). Show that if $\mathbf{F_S}$ has AP* and if f_0, f_1 is a thin pair of surjections in $\mathbf{F_S}$ then the pullback of f_0, f_1 in the category of preordered sets is a pullback in $\mathbf{F_S}$, as well.

2 Let us use the following conventions for finite frames:

- C_n is the one-cluster frame with n elements;
- $\mathbf{n}$ is the n-elements linearly ordered chain;
- $(X, \leq)^+$ is the frame obtained from $(X, \leq)$ by adding it a new root;
- $(X, \leq)^-$ is the frame obtained from $(X, \leq)$ by adding it a final element (i.e. an element e such that $x \leq e$ holds for all $x \in X$);
- $\sum_i (X_i, \leq)$ is ordinary disjoint union;
- $k(X, \leq)$ is the disjoint union of k copies of $(X, \leq)$.

Moreover '$(X_1, \leq)$ and $\cdots$ and $(X_n, \leq)$ yield $(X, \leq)$' stands for: 'there exist a frame $(Y, \leq)$ and surjective open maps $h_i : (X_i, \leq) \to (Y, \leq)$ such that any frame $(Z, \leq)$ co-amalgamating the h_i's has $(X, \leq)$ as an open quotient', i.e. we have the following commuting diagram of frames

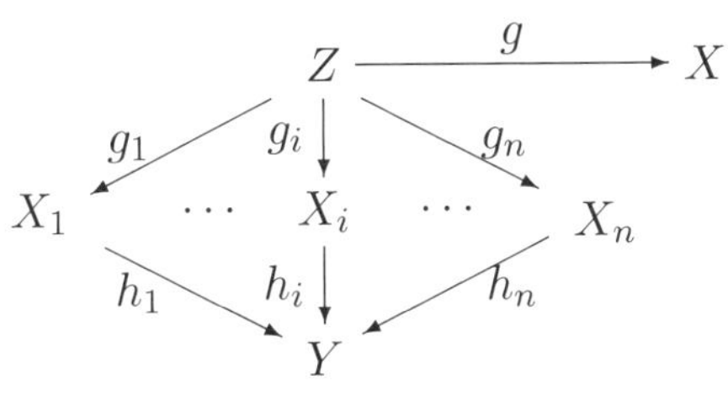

[11]The assumption of the finite model property considerably weakens the result obtained by Maksimova, since we consider a smaller collection of varieties. But as there are no varieties of either Heyting algebras or S4-algebras satisfying AP but not the finite model property both classifications remains the same as in the stronger results due to Maksimova. The arguments we are sketching in the exercises mimic to some extent the intuitions of the original proof. However they are much simpler to follow as with the additional assumption we have made: in fact, although amalgamability for finite algebras is not in general sufficient to guarantee amalgamability even in case finite model property holds, it is nevertheless implied by amalgamability in such a case (see Proposition 2.20). So in order to obtain *negative* results, we can try, by using the finite dualities, to make arguments through finite frames, where things are simple, rather than through algebras, where things tend to be more involved.

with all the morphisms being surjective. Thus, in particular, if (AP^*) holds in $\mathbf{F_S}$ and $\mathbf{F_S}$ contain all the $(X_i, \leq)$, the fact that '$(X_1, \leq)$ and $\cdots$ and $(X_n, \leq)$ yield $(X, \leq)$' implies that $\mathbf{F_S}$ contains $(X, \leq)$ as well.

Now show the following:

(i) $(X, \leq)^{++}$ and $(k\mathbf{1})^+$ yield $((i-1)\mathbf{1} + (X, \leq)^+ + (k-i+1)\mathbf{1})^+$ for all $i = 1, \ldots k$;

(ii) $((X_1, \leq)^+ + (k-1)\mathbf{1})^+$ and $\cdots$ and $((i-1)\mathbf{1} + (X_i, \leq)^+ + (k-i+1)\mathbf{1})^+$ and $\cdots$ and $((k-1)\mathbf{1} + (X_k, \leq)^+)^+$ yield $(\sum_i (X_i, \leq)^+)^+$;

(iii) $(X, \leq)^+$ and $\mathbf{3}$ yield $(X, \leq)^{++}$;

(iv) $(\mathbf{1} + (\mathbf{1} + \mathbf{1})^+)^+$ and $(\mathbf{1} + (\mathbf{1} + \mathbf{1})^+)^+$ yield $(3\mathbf{1})^+$;

(v) $(3\mathbf{1})^+$ and $(n\mathbf{1})^+$ yield $((n+1)\mathbf{1})^+$;

(vi) C_n and C_n yield C_{n+1} for $n > 2$;

(vii) $(C_n)^-$ and $(C_n)^-$ yield $(C_{n+1})^-$ for $n > 2$;

(viii) C_n and $(X, \leq)$ yield $(X_n^e, \leq)$, where $(X_n^e, \leq)$ is obtained from $(X, \leq)$ by replacing every external cluster in it by C_n;

(ix) $(C_n)^-$ and $(X, \leq)$ yield $(X_n^a, \leq)$, where $\{a\} \subseteq X$ is a singleton internal cluster and $(X_n^a, \leq)$ is obtained from $(X, \leq)$ by replacing the cluster $\{a\}$ by an n-element cluster;

(x) $\mathbf{3}_2^e$ and $\mathbf{3}_2^e$ yield $(D, \leq)$, where $(D, \leq)$ is the four elements 'diamond' poset;

(xi) $(C_2^-)^{++}$ and $(C_2^-)^{++}$ also yield $(D, \leq)$.

3 Show that there are at most 8 varieties $\mathbf{S}$ of Heyting algebras enjoying the finite model property[12] and the amalgamation property. *Argue as follows*: from Proposition 2.20, we know that in our hypotheses a necessary condition is that the amalgamation property for finite algebras must hold. We take the class $\mathbf{F}_\mathbf{S}^\mathfrak{r}$ of finite rooted posets which are dual to finite subdirectly irreducible $\mathbf{S}$-algebras; we show that there are only 8 possibilities for $\mathbf{F}_\mathbf{S}^\mathfrak{r}$. Indeed, if $\mathbf{F}_\mathbf{S}^\mathfrak{r}$ is not empty, it contains $\mathbf{1}$; if there are other elements in $\mathbf{F}_\mathbf{S}^\mathfrak{r}$, then $\mathbf{2}$ must be there. If it contains even more rooted posets (in addition to $\mathbf{1}, \mathbf{2}$), then either $\mathbf{3}$ is there or ($(\mathbf{1} + \mathbf{1})^+$ is there and $\mathbf{3}$ is not). In the latter case, either $\mathbf{F}_\mathbf{S}^\mathfrak{r} = \{\mathbf{1}, \mathbf{2}, (\mathbf{1} + \mathbf{1})^+\}$, or $\mathbf{F}_\mathbf{S}^\mathfrak{r} = \{n\mathbf{1}^+\}_{n \geq 0}$, by exercise 2(v) above. In the former case $\mathbf{n} \in \mathbf{F}_\mathbf{S}^\mathfrak{r}$ (by 2(iii)) for all n. If $\mathbf{F}_\mathbf{S}^\mathfrak{r}$ is not just $\{\mathbf{n}\}_n$, then it is easily shown that either $(\mathbf{1} + \mathbf{1})^+$ is there or ($(\mathbf{1} + \mathbf{1})^+$ is not there but the four element 'diamond' poset is there). In the former case, by 2(iii), we realize that $(\mathbf{1} + \mathbf{1})^{++}$ belongs to $\mathbf{F}_\mathbf{S}^\mathfrak{r}$, as well as $(\mathbf{1} + (\mathbf{1} + \mathbf{1})^+)^+$ (by 2(i)) and

[12]This restriction can be removed, see [Ma1].

as well as $(n\mathbf{1})^+$ (for all n, by 2(iv), 2(v)); finally by repeated applications of 2(iii), 2(i) and 2(ii), it turns out that $\mathbf{F}^{\mathtt{r}}_{\mathbf{S}}$ contains all finite trees, hence all finite rooted posets (each rooted poset is an open quotient of a tree). In the latter case, a similar argument ('attach a final point everywhere')[13] shows that $\mathbf{F}^{\mathtt{r}}_{\mathbf{S}}$ is just the class of all finite rooted posets with a final element (generating the variety axiomatized by De Morgan law).

4 Prove Theorem 6.6 assuming the finite model property. *Argue as follows*: first observe that if $\mathbf{F_S}$ enjoys (AP^*), then (AP^*) must holds in $\mathbf{F}^{\mathtt{i}}_{\mathbf{S}}$ too, where $\mathbf{F}^{\mathtt{i}}_{\mathbf{S}}$ is the class of all finite posets of $\mathbf{F_S}$ (i.e. of finite frames of $\mathbf{F_S}$ in which $\leq$ is antisymmetric). Hence, according to the previous exercise, there are only 8 possibilities for $\mathbf{F}^{\mathtt{i}}_{\mathbf{S}}$. Then apply 2(vi)-2(ix); finally, 2(x) and 2(xi) (combined with 2(viii) and 2(ix)) give the further restriction in case $\mathbf{F}^{\mathtt{i}}_{\mathbf{S}}$ is the class of all locally linear finite posets.

5 Prove that the 31 varieties listed in (*) of Theorem 6.6 all admit a model completion. (*Hint*: as height of finite frames for these varieties is bounded, condition $(\forall\exists)$ trivializes by Corollary 5.10, so just patiently check condition (AP^*)...).

6 A commutative square of monomorphisms of $\mathbf{S}$-algebras

$$\begin{array}{ccc} A_0 & \xrightarrow{f_1} & A_1 \\ {\scriptstyle f_2}\downarrow & & \downarrow{\scriptstyle g_1} \\ A_2, & \xrightarrow[g_2]{} & A \end{array}$$

is said to have the interpolation property iff for every $a_1 \in A_1, a_2 \in A_2$

$$g_1(a_1) \leq g_2(a_2) \;\Rightarrow\; \exists a_0 \in A_0\ (a_1 \leq f_1(a_0)\ \&\ f_2(a_0) \leq a_2).$$

Show that, if all morphisms in a square of monomorphisms like the above one have adjoints, then the interpolation property is equivalent to the 'Beck-Chevalley-like' condition:

$$\exists_{g_2} g_1(b) = f_2 \exists_{f_1}(b)$$

for all $b \in A_1$.[14] Conclude that the dual of a commutative square of finite frames and open surjective maps

[13] Of course, this means proving again some relevant statements from Exercise 2 under the following modification: 'replace each frame $(Y, \leq)$ appearing in the statement by $(Y, \leq)^-$'.

[14] Notice that in case all involved algebras are finite, this is not just the Beck-Chevalley condition in $\mathbf{F_S}$ because b can be any subframe of a the dual frame of A_1 and not only a generated subframe (recall that only generated subframes correspond to monomorphisms in $\mathbf{F_S}$).

$$\begin{array}{ccc} (Y, \leq) & \xrightarrow{p_1} & (X_1, \leq) \\ {\scriptstyle p_2}\downarrow & & \downarrow{\scriptstyle k_1} \\ (X_2, \leq) & \xrightarrow[k_2]{} & (X, \leq) \end{array}$$

has the interpolation property iff the canonical set-theoretical map into the ordinary pullback $Y \longrightarrow X_1 \times_X X_2$ is a surjection.

7 Prove that the following conditions are equivalent for a system $\mathbf{S}$:

(i) $\mathbf{S}$ has the Craig interpolation property, that is any square obtained by applying the free algebra functor to a pushout square of monomorphisms among finite sets has the interpolation property;

(ii) $Alg(\mathbf{S})_{fp}$ is superamalgamable, meaning that any pair of monomorphisms $f_1 : A_0 \longrightarrow A_1$, $f_2 : A_0 \longrightarrow A_2$ in that category can be filled into a commutative square having the interpolation property;

(iii) $Alg(\mathbf{S})$ itself is superamalgamable.

8 Show [Ma2] that the variety generated by the frame algebra of $(C_2^-)^e_2$ lacks the interpolation property (although it is amalgamable). (*Hint:* Use exercises 6 and 7(ii)). This exercise shows that the existence of a model completion is independent in modal logic on interpretability of second-order propositional quantifiers into ordinary propositional calculus (see Section 4.6), because the latter actually implies Craig's interpolation theorem (in a quite strong form).

Chapter 7

ALGEBRAICALLY CLOSED MODELS

1. FLEA's

In the whole Chapter we shall fix an equational theory T which is a FLEA (finitary, logical, equational, amalgamable) i.e. which is such that

(i) $Alg(T)$ has equationally definable operations (EDO) (see Section 2.5);

(ii) the amalgamation property (AP) holds for $Alg(T)_{fin}$;[1]

(iii) T has the finite model property;

(iv) the signature of T contains only finitely many function symbols.

The main examples of FLEA's are equational theories arising from propositional logics. For example, distributive lattices, Boolean algebras, Heyting algebras, interior algebras and diagonalizable algebras are FLEAs, and should be considered as the main reference examples in this Chapter. The reason for assuming (iv) is that it guarantees finite algebras to be finitely presented (a fact we shall use without explicit mention in the following).

Recall that $\mathbf{T}$ denotes the dual of the category $Alg(T)_{fp}$ of finitely presented T-algebras; we let also $\mathbf{T}^{\mathrm{f}}$ be the dual of the category $Alg(T)_{fin}$ of finite T-algebras. We first show that the jointly epi families of morphisms on both categories, denoted by $\mathbf{J}^{\infty}$ and $\mathbf{J}^{\mathrm{f}}$, respectively, form Grothendieck topologies. Moreover they provide two different presentations of the same topos. We can so reduce the problem of characterizing the geometric theory $T^{\dagger}$ classified by

[1] Note that $Alg(T)$ need not satisfy the amalgamation property, even if we assume this property for $Alg(T)_{fin}$ and even if we shall be able to prove it in the next section for $Alg(T)_{fp}$ (see [Ma4] for a counterexample).

$Sh(\mathbf{T}^{\mathrm{f}}, \mathbf{J}^{\mathrm{f}})$ to the problem of characterizing the theory classified by a site based on a customary category of finitely presented algebras. Such a characterization will be achieved in Section 7.3; we then prove in Section 7.4 that set-theoretic models of $T^{\dagger}$ correspond to algebraically closed models with a couple of additional properties (primality and local finiteness). We then examine more closely the case of Heyting and of interior algebras (in Section 7.5 and in the Exersises). The Chapter ends with some further information on existentially closed models.

2. Grothendieck Topologies on algebras

Before examining jointly epis coverings topologies on $\mathbf{T}$ and $\mathbf{T}^{\mathrm{f}}$, we need some preliminary technical information.

LEMMA 7.1 (i) *A family $\{g_i : C_i \to C\}_{i \in I}$ in a category with finite limits is jointly epic iff for every regular subobject $S \hookrightarrow C$,*

$$(\forall\, i \in I \quad \top \leq g_i^*(S)) \quad \Rightarrow \quad \top \leq S.$$

(ii) *Suppose we are given in $\mathbf{T}$ a jointly epic family $\{f_i : A_i \to A\}_i$ and a regular mono $B \hookrightarrow A$. Then the pulled back family $\{\bar{f}_i : A_i \times_A B \to B\}_i$ is jointly epic as well.*

(iii) *Suppose we are given in $\mathbf{T}$ a finite jointly epic family $\{A_i \hookrightarrow A\}_i$ entirely composed of regular monos and an arrow $f : B \to A$. Then the pulled back family is jointly epic as well.*

Proof. (i) is easy.

Ad (ii). We use (i) and the fact that T satisfies (EDPC), which means, by Proposition 2.16, that $\mathbf{T}$ has stable implications among regular subobjects.

By taking composition with $B \hookrightarrow A$, [2] a regular subobject of $S \hookrightarrow B$ can be seen as a regular subobject of A; as such it satisfies the inequality $S \leq B$. Suppose that $\top \leq \bar{f}_i^*(S)$; if we see S as a regular subobject of A, this implies $f_i^*(B) \leq f_i^*(S)$. By existence and stability of implications, we get $\top \leq f_i^*(B \to S)$. As the f_i are jointly epic, we get that B is equal to S (as a regular subobject of A), as required.

Ad (iii). By (i) and by the existence of joins among regular subobjects (see Proposition 2.16), the family $\{A_i \hookrightarrow A\}_i$ is jointly epic iff $\top \leq \bigvee_i A_i$. Thus (iii) immediately follows from the stability of joins. □

LEMMA 7.2 (i) *For a pushout square in $Alg(T)_{fp}$,*

[2] Notice that regular monic do compose in $\mathbf{T}$, because $\mathbf{T}$ is the opposite of the category of finitely presented T-algebras (recall from Proposition 2.13 that regular epis among fp algebras are exactly onto morphisms.).

$$\begin{array}{ccc} A & \xrightarrow{m} & B \\ {\scriptstyle g}\downarrow & & \downarrow{\scriptstyle g'} \\ A' & \xrightarrow[m']{} & B' \end{array}$$

we have that if A, A' are finite and m monic, then m' is monic too.

(ii) *Condition (IT) (see Section 2.6) holds in $Alg(T)_{fin}$.*

Proof. Ad (i). Since any morphism among finite algebras has a regular epi/mono factorization, we can use Lemma 7.1(ii) to reduce to the case in which g is mono, as well. Then, we can use the finite model property in order to build $h : B \to C$ such that C is finite and $h \circ m$ is still monic. The claim now follows from the amalgamation property for finite algebras, the universal property of pushouts and the fact that the first component of a mono is mono.

Ad (ii). Consider a pushout diagram like in (i): since T has the finite model property, B' can be embedded into a product of finite algebras $\{D_i\}_i$. As A' is finite, the product can be taken to be finite as well, still keeping the composite morphism $A' \xrightarrow{m'} B' \longrightarrow \Pi_i D_i$ to be mono. This proves condition (IT) in $Alg(T)_{fin}$. □

PROPOSITION 7.3 *The families of jointly epic coverings $\mathbf{J}^\infty$ and $\mathbf{J}^{\mathrm{f}}$ on categories $\mathbf{T}$ and $\mathbf{T}^{\mathrm{f}}$, respectively, form Grothendieck topologies. Moreover, the toposes $Sh(\mathbf{T}, \mathbf{J}^\infty)$ and $Sh(\mathbf{T}^{\mathrm{f}}, \mathbf{J}^{\mathrm{f}})$ are equivalent.*

Proof. Using Lemmas 7.1, 7.2 we prove that $\mathbf{J}^\infty$ is a Grothendieck topology. Let

$$\{A \xrightarrow{a_i} A_i\}_{i \in I}$$

be a jointly mono family in $Alg(T)_{fp}$ (i.e. a jointly epi family in $\mathbf{T}$) and let $A \xrightarrow{f} B$ be a morphism in $Alg(T)_{fp}$. We need to show that the family

$$\{B \xrightarrow{b_i} B_i\}_{i \in I} \tag{7.1}$$

obtained by taking pushout along f, is jointly mono in $Alg(T)_{fp}$.

First, we show that we can reduce to the special case in which B is a finite algebra. In fact, since T has the finite model property, there is a jointly monic family $\{b'_j : B \to B'_j\}_{j \in J}$, where the B'_j are finite. By stability for finite algebras (our special case), we obtain (for every $j \in J$) jointly monic families

$$\{B'_j \xrightarrow{c_{ij}} C_{ij}\}_{i \in I},$$

hence the composition $\{c_{ij} \circ b'_j : B \to C_{ij}\}_{i\in I, j\in J}$ is jointly monic too and so is the family (7.1), which is first component by the universal property of pushouts.

Now, we assume that B is finite. Then, we can form a regular epi-mono factorization in $Alg_{fp}(T)$

$$A \longrightarrow Q \longrightarrow B$$

Note that Q is finite as well. For $i \in I$ we can construct pushouts in $Alg(T)_{fp}$

$$\begin{array}{ccc} A & \longrightarrow & A_i \\ \downarrow & & \downarrow \\ Q & \longrightarrow & Q_i \\ \downarrow & & \downarrow \\ B & \longrightarrow & B_i \end{array}$$

The family

$$\{Q \longrightarrow Q_i\}_{i\in I} \tag{7.2}$$

is jointly mono by Lemma 7.1(ii) above.

Now, we show that (7.1) is jointly mono. Note that there is a finite set $I_0 \subseteq I$ such that

$$\{Q \longrightarrow Q_i\}_{i\in I_0} \tag{7.3}$$

is jointly mono. We can form pushouts for $i \in I_0$

$$\begin{array}{ccccc} Q & \longrightarrow & R_i & \longrightarrow & Q_i \\ \downarrow & & \downarrow & & \downarrow \\ B & \longrightarrow & S_i & \longrightarrow & B_i \end{array}$$

with the top row being the regular epi-mono factorization of $Q \to Q_i$. Now the family $\{B \to S_i\}$ is jointly mono by Lemma 7.1(iii) and for every i the morphism $S_i \to B_i$ is mono by Lemma 7.2(i). So the family (7.1) is jointly mono as well.

This shows that $\mathbf{J}^\infty$ is a Grothendieck topology. As we just proved that jointly monic families are stable under pushouts, it is not difficult to see, using the finite model property of T, that $\mathbf{J}^{\mathrm{f}}$ is a Grothendieck topology, as well.

Finally, since the embedding $\mathbf{T}^{\mathrm{f}} \to \mathbf{T}$ is full and faithful, a family of morphisms $\{A \xrightarrow{a_i} A_i\}_{i\in I}$ in $\mathbf{T}^{\mathrm{f}}$ is a cover in $\mathbf{J}^{\mathrm{f}}$ iff it is a cover in $\mathbf{J}^\infty$, and, by the finite model property, every object in $\mathbf{T}^\infty$ has a cover consisting of objects from $\mathbf{T}^\mathrm{f}$ the Comparison Lemma, cf. [SGA4] p. 288 (see also Appendix 226) applies. Therefore the toposes $Sh(\mathbf{T}, \mathbf{J}^\infty)$ and $Sh(\mathbf{T}^\mathrm{f}, \mathbf{J}^\mathrm{f})$ are equivalent. □

We remark that if the strong amalgamation property holds for finite algebras, then the topology $\mathbf{J}^\mathrm{f}$ is canonical; even in this case, however, (witness the case of Heyting algebras) $\mathbf{J}^\infty$ may not be canonical (see the Exercises).

3. $Sh(\mathbf{T}^\mathrm{f}, \mathbf{J}^\mathrm{f})$ as a classifying topos

By Proposition 7.3, for any FLEA T, $Sh(\mathbf{T}^\mathrm{f}, \mathbf{J}^\mathrm{f})$ is a Grothendieck topos. We shall describe below a geometric theory $T^\dagger$ classified by this topos. In the next section we shall discuss in more familiar way models of $Sh(\mathbf{T}^\mathrm{f}, \mathbf{J}^\mathrm{f})$ in Set, i.e. the ordinary models of $T^\dagger$.

An e-formula $\varphi(\vec{x})$ is *finite* iff the T-algebra $\mathcal{F}(\vec{x})/\varphi(\vec{x})$ is finite.

Let Φ_n, for $n \in \omega$, be the family of e-formulas

$$\Phi_n = \; \{\varphi(\vec{x}) | \vec{x} = x_1, \ldots, x_n \text{ and } \varphi(\vec{x}) \text{ is finite } \}$$

The theory $T^\dagger$ has the following axioms:

(i) axioms of T

(ii) $\Rightarrow \bigvee \Phi_n, \quad \text{for } n \in \omega$

(iii) $\psi(\vec{x}) \Rightarrow \exists \vec{y} \varphi(\vec{x}, \vec{y})$,
for every monomorphism $\mathcal{F}(\vec{x})/\psi(\vec{x}) \longrightarrow \mathcal{F}(\vec{x}, \vec{y})/\varphi(\vec{x}, \vec{y})$ between finite T-algebras that sends $\vec{x}$ to $\vec{x}$

(iv) $J_0 \Rightarrow x_1 = x_2$

(v) $J(x_1, x_2, x_3, x_4) \Rightarrow (x_1 = x_2 \vee x_3 = x_4)$

where $J(x_1, x_2, x_3, x_4)$ and J_0 are the e-formulas introduced in section 2.5. Notice that the second family of axioms require $L_{\omega_1, \omega}$ (coherent) language, being a countable family of countable disjunctions.

Let $E_1 = \bigwedge_{i=1}^n t_i = t_i'$, $E_2 = \bigwedge_{j=1}^m u_j = u_j'$ be e-formulas. By $\bar{J}(E_1, E_2)$ we denote the following e-formula

$$\bigwedge_{i=1}^{n} \bigwedge_{j=1}^{m} J(t_i, t_i', u_j, u_j')$$

This formula expresses joins of regular subobjects in $\mathbf{T}$ (see the way joins are introduced in the proof of Lemma 2.15), hence we obtain

LEMMA 7.4 *For any e-formulas E_1, E_2, E we have*

$$\vdash_T \bar{J}(E_1, E_2) \Rightarrow E \quad \textit{iff} \quad (\vdash_T E_1 \Rightarrow E \textit{ and } \vdash_T E_2 \Rightarrow E) \quad \square$$

We have

THEOREM 7.5 *$Sh(\mathbf{T}^{\mathbf{f}}, \mathbf{J}^{\mathbf{f}})$ classifies the theory $T^{\dagger}$.*

Proof. By Proposition 7.3, $Sh(\mathbf{T}^{\mathbf{f}}, \mathbf{J}^{\mathbf{f}})$ is equivalent to $Sh(\mathbf{T}, \mathbf{J}^{\infty})$. So we need to show that for any cocomplete topos $\mathcal{E}$ the left exact functors from $(\mathbf{T}, \mathbf{J}^{\infty})$ to $\mathcal{E}$ that are $\mathbf{J}^{\infty}$-continuous (i.e. send $\mathbf{J}^{\infty}$-covers to epimorphic families in $\mathcal{E}$), correspond to the models of the theory specified in the Theorem. Clearly, left exact functors from $\mathbf{T}$ to $\mathcal{E}$ correspond to T-models in $\mathcal{E}$. We shall show that a left exact functor $H : \mathbf{T} \to \mathcal{E}$ is $\mathbf{J}^{\infty}$-continuous iff it preserves the following four types of covers corresponding to the axioms of the theory $T^{\dagger}$:

$$\{\mathcal{F}(\vec{x}) \to \mathcal{F}(\vec{x})/\varphi(\vec{x}) \mid \varphi(\vec{x}) \in \Phi_n \} \tag{7.4}$$

where $n \in \omega$ and $\vec{x} = x_1, \ldots, x_n$;

$$\{m : \mathcal{F}(\vec{x})/\varphi(\vec{x}) \longrightarrow \mathcal{F}(\vec{x}, \vec{y})/\psi(\vec{x}, \vec{y})\} \tag{7.5}$$

where φ and ψ are finite e-formulas and m is a mono;

$$\text{empty cover on } \mathcal{F}(\emptyset)/J_0 \tag{7.6}$$

$$\{\mathcal{F}(\vec{x}, \vec{y})/J(\vec{x}, \vec{y}) \longrightarrow \mathcal{F}(\vec{x}, \vec{y})/x_i = y_i\}_{i=\{1,2\}} \tag{7.7}$$

where $\vec{x} = x_1, x_2$ and $\vec{y} = y_1, y_2$.

Clearly all the above covers are $\mathbf{J}^{\infty}$-covers. Let us fix a left exact functor $H : \mathbf{T} \to \mathcal{E}$ that preserves the covers (7.4), (7.5), (7.6), (7.7). By an *H-cover* we shall mean a cover in $\mathbf{T}$ which is sent by H to a jointly epimorphic family in $\mathbf{E}$. Thus, we need to show that any $\mathbf{J}^{\infty}$-cover is an H-cover.

Let

$$\{\mathcal{F}(\vec{x})/\psi(\vec{x}) \longrightarrow \mathcal{F}(\vec{x}, \vec{y_i})/\psi_i(\vec{x}, \vec{y_i})\}_{i\in I} \tag{7.8}$$

be an arbitrary $\mathbf{J}^{\infty}$-cover, where $\vec{x} = x_1, \ldots, x_n$ for some $n \in \omega$.

Since (7.4) is an H-cover, it is enough to show that for $\varphi \in \Phi_n$, the cover

$$\{\mathcal{F}(\vec{x})/\psi(\vec{x}), \varphi(\vec{x}) \longrightarrow \mathcal{F}(\vec{x}, \vec{y_i})/\psi_i(\vec{x}, \vec{y_i}), \varphi(\vec{x})\}_{i\in I} \tag{7.9}$$

is an H-cover. Since $\varphi(\vec{x})$ is finite, the cover (7.9) contains a finite subfamily which is a $\mathbf{J}^{\infty}$-cover. Thus, since T has the finite model property, it is enough to show that the covers of form (7.8) with ψ, ψ_i and I all finite are H-covers. As each morphism between finite T-algebras has an epi/regular mono factorization in $\mathbf{T}$ and (7.5) in an H-cover, to show that (7.8) is an H-cover is enough to show that the $\mathbf{J}^{\infty}$-cover (entirely formed by onto morphisms)

$$\{\mathcal{F}(\vec{x})/\psi(\vec{x}) \longrightarrow \mathcal{F}(\vec{x})/\psi_i(\vec{x})\}_{i=1,\ldots,n} \tag{7.10}$$

with ψ, ψ_i, I all finite, is a H-cover (notice also that with respect to (7.8), the additional variables $\vec{y_i}$ are dropped).

We shall show by induction on $n \in \omega$ that the $\mathbf{J}^\infty$-covers (7.10) are H-covers.

If $n = 0$ then (7.10) is the empty cover on $\mathcal{F}(\vec{x})/\psi(\vec{x})$. As every $\mathbf{J}^\infty$-cover is jointly mono, $\mathcal{F}(\vec{x})/\psi(\vec{x})$ must be embeddable in the terminal one-element algebra, so it must be itself the terminal one-element algebra.[3] Hence $\mathcal{F}(\vec{x})/\psi(\vec{x})$ is isomorphic to $\mathcal{F}(\emptyset)/J_0$. As (7.6) is an H-cover so is (7.10) in this case.

If $n = 1$ then (7.10) contains just one isomorphism and hence it is preserved by any functor.

If $n > 1$, then to show that (7.10) is an H-cover, it is enough to show that the following two covers

$$\{\mathcal{F}(\vec{x})/\bar{J}(\psi_1, \psi_2) \longrightarrow \mathcal{F}(\vec{x})/\psi_i\}_{i=\{1,2\}} \tag{7.11}$$

and

$$\{\mathcal{F}(\vec{x})/\psi(\vec{x}) \longrightarrow \mathcal{F}(\vec{x})/\psi'_i(\vec{x})\}_{i=2,\dots,n} \tag{7.12}$$

where

$$\psi'_i(\vec{x}) \quad = \quad \begin{cases} \bar{J}(\psi_1, \psi_2) & \text{if } n = 2 \\ \psi_i(\vec{x}) & \text{if } n > 2 \end{cases}$$

are H-covers. The cover (7.12) is jointly surjective as a consequence of Lemma 7.4, thus, as it contains less than n elements, it is an H-cover by inductive hypothesis.

The fact that (7.11) is an H-cover is the content of the following Claim.

Claim. For any e-formulas $E(\vec{x})$ and $F(\vec{x})$ the cover

$$\{\mathcal{F}(\vec{x})/\bar{J}(E,F) \longrightarrow \mathcal{F}(\vec{x})/E, \;\; \mathcal{F}(\vec{x})/\bar{J}(E,F) \longrightarrow \mathcal{F}(\vec{x})/F\} \tag{7.13}$$

is an H-cover. Identifying e-formulas with the corresponding regular subobjects, this means that in the lattice of subobjects of $H(\mathcal{F}(\vec{x}))$, we must have $H(E) \vee H(F) = H(J(E,F))$.

Proof of the Claim. Suppose that $E = \bigwedge_{i=1}^m e_i$ and $F = \bigwedge_{j=1}^n f_j$, where e_i and f_j are equations between terms in the variables among $\vec{x}$. If e.g. e_i is $t_i = t'_i$ and f_j is $u_j = u'_j$, we write $J(e_i, f_j)$ for $J(t_i, t'_i, u_j, u'_j)$. Since H-covers are stable under pushouts and (7.7) is an H-cover, we have $H(e_i) \vee H(f_j) = H(J(e_i, f_j)) = H(e_i \vee f_j)$ (recall that J expresses joins of subobjects in $\mathbf{T}$). As H preserves meets of subobjects and as distributivity law holds both in $\mathbf{T}$ and in $\mathcal{E}$, we have

$$H(E) \vee H(F) = H(\bigwedge_i e_i) \vee H(\bigwedge_j f_j) = \bigwedge_{i,j}(H(e_i) \vee H(f_j)) =$$

[3] Due to the existence of an e-formula without free variables J_0, the language of T must contain at least one constant.

$$= \bigwedge_{i,j} H(e_i \vee f_j) = H(\bigwedge_{i,j}(e_i \vee f_j)) = H(E \vee F)$$

as claimed. □

4. Models in *Set*

In view of Theorem 7.5, we know that the models of $Sh(\mathbf{T^f}, \mathbf{J^f})$ in any Grothendieck topos $\mathcal{E}$, i.e. the geometric morphisms from $\mathcal{E}$ to $Sh(\mathbf{T^f}, \mathbf{J^f})$, are exactly the internal models of the geometric infinitary theory $T^\dagger$ in $\mathcal{E}$. In the case $\mathcal{E} = Set$, we obtain that models of $Sh(\mathbf{T^f}, \mathbf{J^f})$ in Set are (correspond to) ordinary models of $T^\dagger$. In this section we develop this correspondence and as a by-product we shall describe the models of $T^\dagger$ in Set in more explicit terms.

Since the topos $Sh(\mathbf{T^f}, \mathbf{J^f})$ is equivalent to the topos $Sh(\mathbf{T}, \mathbf{J}^\infty)$, the models of $Sh(\mathbf{T^f}, \mathbf{J^f})$ correspond to those finite limit preserving functors from $\mathbf{T}$ to Set, that sends $\mathbf{J}^\infty$-covers to epimorphic families in Set.

First we recall the correspondence between T-algebras and finite limit preserving functors from $\mathbf{T}$ to Set. We first of all have an equivalence of categories

$$\overline{(-)} : Alg(T) \longrightarrow Lex(\mathbf{T}, Set)$$

$$A \mapsto \overline{A}$$

where $Lex(\mathbf{T}, Set)$ denotes the category of finite limits preserving functors from $\mathbf{T}$ to Set and natural transformations between them, and the functor

$$\overline{A} : \mathbf{T} \longrightarrow Set$$

is a hom-functor, i.e. it sends a morphism

$$\mathcal{F}(\vec{x})/E(\vec{x}) \longrightarrow \mathcal{F}(\vec{x}, \vec{y})/F(\vec{x}, \vec{y})$$

in $\mathbf{T}$ to a function

$$Hom(\mathcal{F}(\vec{x}, \vec{y})/F(\vec{x}, \vec{y}), A) \longrightarrow Hom(\mathcal{F}(\vec{x})/E(\vec{x}), A). \tag{7.14}$$

As a homomorphism from $\mathcal{F}(\vec{x})/E(\vec{x})$ to A corresponds to a tuple of elements $\vec{a}$ from A such that $A \models E(\vec{a})$, the function (7.14) can be identified with the function

$$\{\vec{a}, \vec{b} \in A : A \models F(\vec{a}, \vec{b})\} \longrightarrow \{\vec{a} \in A : A \models E(\vec{a})\}$$

such that

$$\vec{a}, \vec{b} \mapsto \vec{a}.$$

The functor $\overline{A}$ preserves the cover

$$\{\mathcal{F}(\vec{x})/E(\vec{x}) \longrightarrow \mathcal{F}(\vec{x}, \vec{y^i})/F_i(\vec{x}, \vec{y^i})\}_{i \in I}$$

iff the family of functions

$$\{\{\vec{a}, \vec{b^i} \in A : A \models F_i(\vec{a}, \vec{b^i})\} \longrightarrow \{\vec{a} \in A : A \models E(\vec{a})\}\}_{i \in I}$$

is jointly surjective, i.e. iff for all $\vec{a} \in A$ we have

$$A \models E(\vec{a}) \;\Rightarrow\; \text{there exist } i \in I \text{ and } \vec{b^i} \in A \text{ s.t. } A \models F_i(\vec{a}, \vec{b^i}),$$

which means that A must validate the infinitary geometric axiom

$$E(\vec{x}) \;\Rightarrow\; \bigvee_{i \in I} \exists \vec{y}\, F_i(\vec{x}, \vec{y}).$$

$\overline{A}$ preserves all $\mathbf{J}^\infty$-covers iff it preserves any set of covers that generates the topology $\mathbf{J}^\infty$. In the proof of Theorem 7.5 we have specified a set of generating covers that corresponds to the axioms of the theory $T^\dagger$. Thus in order to describe the models of $T^\dagger$, we need to identify those T-algebras A for which the functor $\overline{A}$ preserves the covers (7.4), (7.5), (7.6), (7.7).

If $\overline{A}$ preserves the cover (7.4), then for any $n \in \omega$ and any n-tuple $\vec{a}$ in A there is a finite formula $\varphi(\vec{x})$ with n variables such that $A \models \varphi(\vec{a})$. This means that every finite tuple of elements of A generates a finite subalgebra of A. A T-algebra with this property is called *locally finite*.

A model M of an equational theory T is *algebraically closed* iff any finite set of equations with parameters from M having a solution in an extension of M has also a solution in M itself. In Chapter 3, we have shown that this condition is elementarily expressible if $\mathbf{T}$ is r-regular, i.e. iff taking epis and regular monos we get a stable factorization system in $\mathbf{T}$. However, for locally finite algebras, as we shall see below, this condition is equivalent to the following elementary condition with no further assumptions on the theory T. We say that M has the *finite lifting property* iff for every monomorphism $h : B \to C$ among finite T-algebras and for every homomorphism $f : B \to M$, there is a homomorphisms $g : C \to M$, such that the diagram

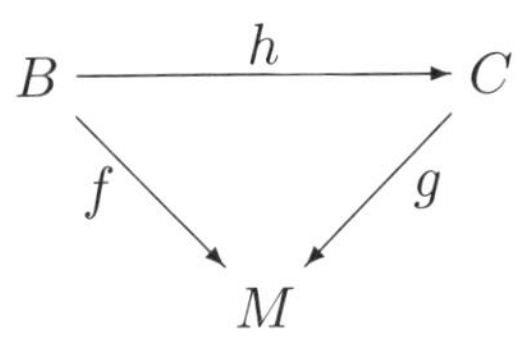

commutes (we say in this case that g is a lifting of f). Thus A has the finite lifting property iff $\overline{A}$ preserves the $\mathbf{J}^\infty$-cover (7.5).

We have

PROPOSITION 7.6 *If the T-algebra A is locally finite, then A is algebraically closed iff it has the finite lifting property.*

Proof. We fix for the proof of this Propossition a locally finite algebra A. Suppose that A has the finite lifting property and consider a monomorphism $m : A \to B$ and an e-formula $\varphi(\vec{x}, \vec{y})$ such that $B \models \varphi(m(\vec{a}), \vec{b})$. As A is locally finite, the subalgebra generated by $\vec{a}$ is of the form $\mathcal{F}(\vec{x})/\psi(\vec{x})$, where ψ is a finite formula. As our aim consists in looking for a solution to $\varphi(\vec{a}, \vec{y})$ in A, we can freely assume that $\vdash_T \varphi(\vec{x}, \vec{y}) \Rightarrow \psi(\vec{x})$ (otherwise we can take just $\varphi \& \psi$ as a new φ). We thus have a commutative square

$$\begin{array}{ccc} \mathcal{F}(\vec{x})/\psi(\vec{x}) & \xrightarrow{\ i\ } & A \\ {\scriptstyle h}\downarrow & & \downarrow{\scriptstyle m} \\ \mathcal{F}(\vec{x}, \vec{y})/\varphi(\vec{x}, \vec{y}) & \xrightarrow[\ g\ ]{} & B \end{array}$$

where i, h, g are the morphisms identified by the correspondences $i(\vec{x}) = \vec{a}$, $h(\vec{x}) = \vec{x}$ and $g(\vec{x}) = m(\vec{a}), g(\vec{y}) = \vec{b}$, respectively. Notice that h is mono as the first component of a monomorphism. By the finite model property, we can find a finite algebra C and a morphism $f : \mathcal{F}(\vec{x}, \vec{y})/\varphi(\vec{x}, \vec{y}) \to C$ such that $f \circ h$ is a monomorphism and then we can apply the definition of the finite lifting property in order to find a solution in A to $\varphi(\vec{a}, \vec{y})$.

Suppose conversely that A is algebraically closed, that $m : B \to C$ is a monomorphism among finite algebras and that $f : B \to A$ is any homomorphism. Take the pushout

$$\begin{array}{ccc} B & \xrightarrow{\ f\ } & A \\ {\scriptstyle m}\downarrow & & \downarrow \\ C & \longrightarrow & P \end{array}$$

and then give to $B \xrightarrow{m} C$ a presentation of the kind $\mathcal{F}(\vec{x})/\psi(\vec{x}) \longrightarrow \mathcal{F}(\vec{x}, \vec{y})/\varphi(\vec{x}, \vec{y})$; now it is sufficient to apply the next Lemma in order to find a lifting for f. □

LEMMA 7.7 *The pushout in $Alg(T)$ of a mono $m : B \to C$ along any morphism $f : B \to A$ is mono again, in case A is locally finite and B, C are finite.*

Proof. First notice that the Lemma holds in case A is finite as a consequence of property (IT_f) (see Lemma 7.2(ii)). If A is locally finite, it is a filtered colimit of finite algebras. Hence the claim follows from the above case and from the fact that a filtered colimit of monos is mono in any category of algebras. □

The claim in the proof of Theorem 7.5 shows that if $\overline{A}$ preserves the cover (7.7), then it preserves the more general covers (7.13), as well, i.e. that for any pair of e-formulas $E_1(\vec{x})$ and $E_2(\vec{x})$, the pair of inclusions

$$\{\{\vec{a} \in A : A \models E_i(\vec{a})\} \longrightarrow \{\vec{a} \in A : A \models \bar{J}(E_1(\vec{a}), E_2(\vec{a})\}\}_{i \in \{1,2\}}$$

is jointly surjective. The condition

$$A \models \bar{J}(E_1(\vec{a}), E_2(\vec{a}))$$

means that the congruence on A generated by $\bar{J}(E_1(\vec{a}), E_2(\vec{a}))$ is the identity and similarly the condition $A \models E_i(\vec{a})$ means that the congruence on A generated by $E_i(\vec{a})$ is the identity. On the other hand, we know that the congruence generated by $\bar{J}(E_1(\vec{a}), E_2(\vec{a}))$ is the intersection of the congruences generated by $E_1(\vec{a})$ and $E_2(\vec{a})$. Thus preservation of (7.13) means that whenever the intersection of two finitely generated congruences is the identity congruence, then one of the two congruences must be the identity congruence as well.

$\overline{A}$ preserves (7.6) iff the set $\overline{A}(\mathcal{F}(\emptyset)/J_0)$ is empty, i.e. iff there are no homomorphisms from $\mathcal{F}(\emptyset)/J_0$ to A. As J_0 generates the maximum congruence on any T-algebra (namely the intersection of the empty set of congruences, which is the total trivial congruence), this means that the maximum congruence on A is not the identity congruence, i.e. the identity congruence on A is not the intersection of the empty set of congruences on A. In other more explicit words, A must not be degenerate.

We say that a T-algebra A is *prime* iff whenever the intersection of finitely many finitely generated congruences is the identity congruence, then one of such congruences is the identity congruence. So the above considerations shows that A is prime iff $\overline{A}$ preserves covers (7.6) and (7.7).

In this way, from Theorem 7.5 and the above observations, we get the following characterization of ordinary models of $T^\dagger$.

COROLLARY 7.8 *The models of $Sh(\mathbf{T}^\mathrm{f}, \mathbf{J}^\mathrm{f})$ in Set, i.e. the category $GTop(Set, Sh(\mathbf{T}^\mathrm{f}, \mathbf{J}^\mathrm{f}))$ of geometric morphisms from Set to $Sh(\mathbf{T}^\mathrm{f}, \mathbf{J}^\mathrm{f})$ and natural transformations, is equivalent to the category of locally finite, algebraically closed and prime T-algebras.*

Before we give some examples illustrating this corollary, we prove a characterization of countable models of $T^\dagger$ which provides a method for constructing such models. We recall that an algebra is *subdirectly irreducible* iff it has a maximum non trivial congruence.

PROPOSITION 7.9 *A countable T-algebra A is locally finite, prime and algebraically closed iff it is the colimit of a chain*

$$A_0 \longrightarrow A_1 \longrightarrow A_2 \longrightarrow \ldots \longrightarrow A_n \longrightarrow \ldots \qquad (7.15)$$

of finite subdirectly irreducible T-algebras and monomorphisms with the following property:

- *for any finite subdirectly irreducible T-algebra B and any monomorphism $h : A_n \to B$ there are $m \geq n$ and a morphism $g : B \to A_m$ such that $g \circ h : A_n \to A_m$ is the embedding as in the chain (7.15).*

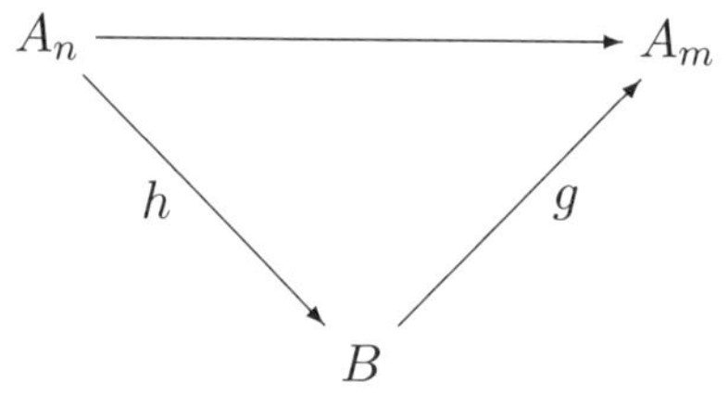

Proof. A is locally finite iff it is a filtered colimit of monos among finite algebras; by countability, we can restrict to a colimit over a chain like (7.15) and by primality the finite algebras A_n must be prime too (notice that to be prime is the same as to be subdirectly irreducible as far as finite algebras are concerned). We need only to show that the finite lifting property is the same as the property mentioned in the claim. One side is trivial: simply take the inclusion of A_n into the colimit A as the morphism to be lifted. For the converse, take finite algebras C_1, C_2, a monomorphism $h : C_1 \to C_2$ and a morphism into A, which we can suppose to factor through a certain $f : C_1 \to A_n$. By (IT_f) (see Lemma 7.2 (ii)) we can produce a commutative square

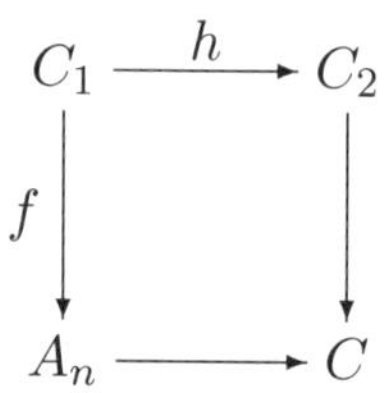

with C finite and the bottom arrow monic. Consider now a maximal congruence $\approx$ on C such that the composition

$$A_n \longrightarrow C \longrightarrow C/\approx$$

is still monic. Now notice that $C/\approx$ is subdirectly irreducible because A_n is such: in fact, any two proper congruences of $C/\approx$ restrict to proper congruences over A_n, so their intersection, containing the minimum proper congruence of A_n, cannot be the identity congruence. Having established that $C/\approx$ is subdirectly irreducible, the property mentioned in the claim of the Proposition applies, thus producing the lifting we are interested in. □

Example 1. Consider **S5**-algebras, which we recall are $S4$-algebras satisfying the further equation $\Diamond a \leq \Box\Diamond a$. Prime algebras are those algebras satisfying the condition

$$1 = \Box a_1 \vee \cdots \vee \Box a_n \quad \Rightarrow \quad \exists i\,(1 = a_i),$$

for $n \in \omega$. Hence they are non-degenerate Boolean algebras in which $\Diamond a$ is taken to be 0 if $a = 0$ and 1 if $a \neq 0$. However, morphisms among them must be *injective* Boolean morphisms, because $\Diamond$ has to be preserved. Thus, by applying Proposition 7.9, [4] we easily realize that prime, algebraically closed and locally finite **S5**-algebras are nothing but *atomless Boolean algebras*. The opposite category of finitely presented **S5** algebras is the category of finite $S5$-frames, i.e. it is the category having as objects finite sets endowed with an equivalence relation and as arrows the functions mapping each equivalence class onto an equivalence class. So sheaves over this category endowed with the canonical topology classify atomless Boolean algebras.

Example 2. Consider Dummett algebras, which are Heyting algebras satisfying the condition $1 = (a \to b) \vee (b \to a)$. Prime Dummett algebras are simply chains with a minimum and a maximum element. Morphisms $f : C_1 \to C_2$ among such chains are order-preserving maps which also preserve implication, i.e. they must satisfy the requirement

$$f(a) = f(b) \quad \Rightarrow \quad (a = b \text{ or } f(a) = 1).$$

This means that such maps are injective, apart from the fact that they can collapse an upper segment to the top element. By applying Proposition 7.9, we realize that *there are only two* countably infinite, prime, algebraically closed and locally finite Dummett algebras (up to isomorphism), namely the rational unit interval and the rational unit interval endowed with an extra element larger than the number 1. The two models differ for the existence of a penultimate element (which is an elementary but not a coherent condition). In the next section we shall see that for Heyting algebras the situation is surprisingly the same.

5. The case of Heyting algebras

We shall be interested in countable, prime, algebraically closed and locally finite Heyting algebras. Any such algebra is the colimit of a chain diagram of finite algebras and monomorphisms

$$A_0 \xrightarrow{m_0} A_1 \xrightarrow{m_1} A_2 \xrightarrow{m_2} \cdots A_i \xrightarrow{m_i} \cdots \tag{7.16}$$

We recall that the category of finite Heyting algebras is dual to the category of finite posets and open maps, see Theorem 2.1. Hence we may equivalently

[4] Proposition 7.9 holds for the uncountable case too, by using arbitrary colimits instead of only chain colimits.

replace chain (7.17) by a chain of finite posets and open surjective maps going in the opposite direction

$$(C) \qquad \cdots\ P_{i+1} \xrightarrow{f_i} P_i\ \cdots\ P_2 \xrightarrow{f_1} P_1 \xrightarrow{f_0} P_0$$

In fact, if $A_i \simeq \mathcal{D}(P_i)$ and if $m_i \simeq \mathcal{D}(f_i)$, then the colimit of (7.17) is just the colimit of[5]

$$\mathcal{D}(P_0) \xrightarrow{\mathcal{D}(f_0)} \mathcal{D}(P_1) \xrightarrow{\mathcal{D}(f_1)} \mathcal{D}(P_2) \xrightarrow{\mathcal{D}(f_2)} \cdots \mathcal{D}(P_i) \xrightarrow{\mathcal{D}(f_i)} \cdots$$

We are consequently in the condition of dualizing Proposition 7.9. From now on all posets are assumed to be *finite*. A poset is said to be *non-degenerate* iff it has more than one element; we recall that a poset is rooted iff it has a minimum element. Notice that $\mathcal{D}(P)$ is subdirectly irreducible (as a finite Heyting algebra) iff P is rooted.

PROPOSITION 7.10 *Countably infinite, prime, locally finite and algebraically closed Heyting algebras are exactly colimits of duals of chains (C) of non-degenerate rooted posets and open surjective morphisms satisfying the following condition:*

$(*)$ *for every rooted poset Q, for every i, for every open surjective morphism $q : Q \to P_i$, there are $j \geq i$ and an open morphism $r : P_j \to Q$ such that $q \circ r = f_{ij}$ (where $f_{ij} = f_i \circ \cdots \circ f_{j-1}$).*

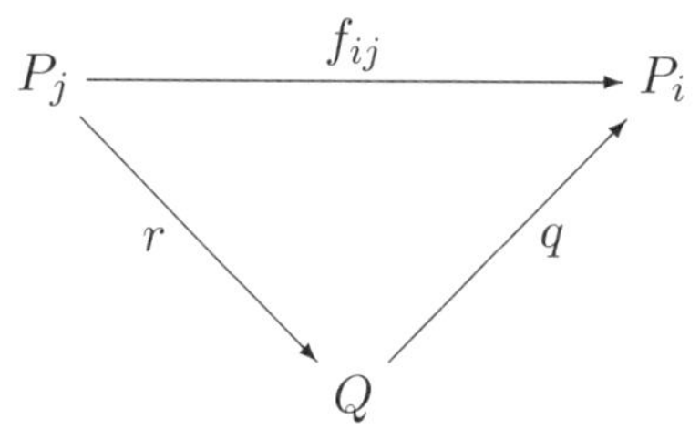

Proof. Immediate by Proposition 7.9 (the P_i are taken to be non-degenerate in order to avoid the colimit to be the two-element Heyting algebra). □

We shall first make some comments about the above Proposition.

Remarks

1. r needs not be surjective, however (by the definition of openness, which implies that an open morphism between rooted posets is surjective just in

[5] Notice however that duality breaks down at the infinite level: for instance, the colimit of the $\mathcal{D}(P_i)$ needs not to be isomorphic to the Heyting algebra of downward closed subsets of the (inverse) limit of the P_i.

case the root is in the image) it must be surjective in case q is what we call an *expansion morphism* (i.e. a surjective open morphism among rooted posets in which the root of the domain is the only element mapped onto the root of the codomain).

2. In case the property $(*)$ is satisfied with the additional requirement that r is surjective, we say that the chain (C) is *strong*.

3. In view of the fact that we are interested in the limit of the dual diagram of (C), we consider any infinite subchain of (C) isomorphic to (C) itself. Moreover, we consider (C) and (C') isomorphic in case there exists for every i, morphisms $h_i : P'_i \to P_i$ and $k_i : P_{i+1} \to P'_i$ such that $h_i \circ k_i = f_i$ and $k_i \circ h_{i+1} = f'_i$ (this is the dual of the customary 'double chain argument', cf. [CK]).

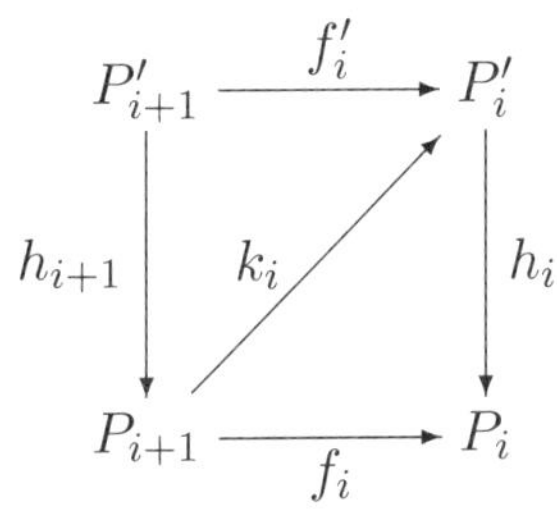

4. We say that (C) is of *the first kind* whenever all but finitely many of the f_i's are expansion morphisms, we say that it is of *the second kind* otherwise. Notice that chains of the second kind, up to isomorphisms as in remark 3, can be assumed not to contain expansion morphisms at all.

We shall prove that all chains of the same kind are isomorphic.

PROPOSITION 7.11 *All the chains of the first kind are isomorphic.*

Proof. We can freely suppose that in (C) and (C'), both of the first kind, all the f_i are expansion morphisms (we can do that by erasing an initial segment). We can also suppose that P'_0 is the two-element poset; as P_1 has more than one element, there is an expansion morphism $k_0 : P_1 \to P'_0$. Build $h_1 : P'_1 \to P_1$ such that $k_0 \circ h_1 = f'_0$ by using $(*)$ (we remove during the construction some edges of the chain, so we can freely suppose that P'_1 is the domain of h_1). By remark 1, h_1 is surjective; moreover it is an expansion morphism (because it is the first component of the expansion morphism f'_0). Build now $k_1 : P_2 \to P'_1$ in the same way and go on. Finally, apply remark 3. $\square$

Before proving that also chains of the second kind are all isomorphic, we need a little preliminary work. For a rooted poset P, P^+ is the rooted poset obtained from P by adding it a new root and $c_P : P^+ \to P$ is the morphism (called contraction morphism) which collapses the old and the new root.

LEMMA 7.12 *Any open surjective morphism $f : Q \to P$ among rooted posets which is not an expansion morphism factors through an expansion morphism $\bar{f}$ followed by the contraction morphism c_P.*

Proof. Define $\bar{f} : Q \to P^+$ by $\bar{f}(p) = f(p)$ if p is not the root, $\bar{f}(p)$ equal to the root of P^+ otherwise. □

LEMMA 7.13 *The pullback (taken in the category of posets) of an expansion morphism along any open surjective morphism among rooted posets is an expansion morphism as well.* □

PROPOSITION 7.14 *All the chains of the second kind are isomorphic.*

Proof. Notice that two strong chains are isomorphic (by a simplified version of the proof of Proposition 7.11). So it is sufficient to show that any chain of the second kind is strong. In order to prove this, consider a second kind chain (C) and a surjective morphism $q : Q \to P_i$, with rooted domain. If q is an expansion morphism, see remark 1. Otherwise, let us apply Lemma 7.12 to both q and f_i (we can suppose that f_i is not an expansion morphism, because the chain is of the second kind, see remark 4). Thus we get morphisms $\bar{q}$ and $\bar{f}_i$ such that $q = c_{P_i} \circ \bar{q}$ and $f_i = c_{P_i} \circ \bar{f}_i$. Now apply Lemma 7.13 to the pullback of $\bar{q}$ and $\bar{f}_i$: we obtain that $p_2 : Q \times_{P_i^+} P_{i+1} \to P_{i+1}$ is an expansion morphism, hence by remark 1, there is for some j a surjective morphism $h : P_j \to Q \times_{P_i^+} P_{i+1}$ such that $p_2 \circ h = f_{i+1,j}$.

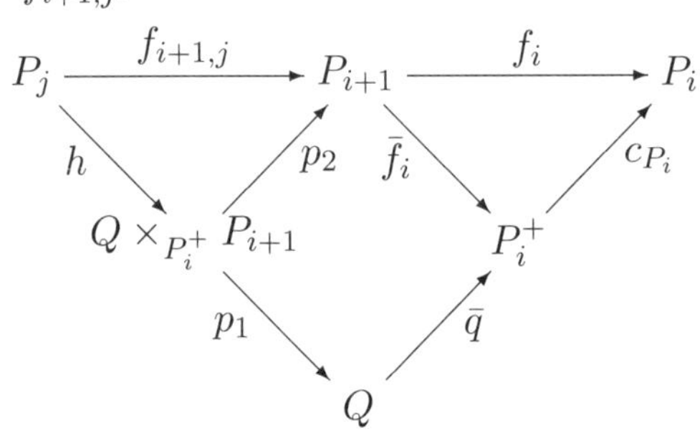

Now taking $p_1 \circ h$ we get a surjective morphism with the desired property. □

Thus there are at most two Heyting algebras satisfying our requirements. Notice that chains of the first kind and of the second kind cannot be isomorphic because they produce non isomorphic Heyting algebras in the colimit: such

Heyting algebras differ because only one of them has a penultimate element (i.e. it is subdirectly irreducible).

THEOREM 7.15 *There are exactly two countably infinite, prime, locally finite and algebraically closed Heyting algebras (up to isomorphism).*

Proof. By Propositions 7.11 and 7.14, it is sufficient to prove the existence of a chain of the first kind and of a chain of the second kind. For a given poset Q, let $\{q_i^Q : R_i^Q \to Q\}_{i\geq 1}$ be an enumeration of the open surjective maps with target Q and for a given rooted poset P, let $\{r_i^P : R_i^P \to P\}_{i\geq 1}$ be an enumeration of the open surjective maps with target P and having a rooted poset as a domain.

We first produce a strong chain as follows: we begin with the two element poset as P_0 and at stage $i > 0$ we build P_i as an iterated pullback[6] as follows. First we take $r_i^{P_0} : R_i^{P_0} \longrightarrow P_0$ and consider its pullback (Q_0, p_0', p_0) with $f_{0,i-1} : P_{i-1} \longrightarrow P_0$; we then take the pullback (Q_1, p_1', p_1) of $r_{i-1}^{P_1} : R_{i-1}^{P_1} \longrightarrow P_1$ with $f_{1,i-1} \circ p_0 : Q_0 \longrightarrow P_{i-1} \longrightarrow P_1$, etc. P_i is finally Q_{i-1} and the map $P_i \to P_{i-1}$ is $p_0 \circ \cdots \circ p_{i-1}$. Condition $(*)$ is clearly satisfied, with r surjective (the map $q : Q \to P_i$ will appear as $r_k^{P_i}$ for some k and will be considered at stage $i + k$, so that the required r will be a suitable composition of projections at that stage).

In order to produce a first kind chain, let us first produce a chain

$$(C) \qquad \cdots\ P_{i+1} \xrightarrow{f_i} P_i\ \cdots$$

as above but by using the enumeration $\{q_i^Q\}_i$ of non necessarily rooted posets. Such chain will be made of non necessarily rooted posets and will satisfy condition $(*)$ with r surjective and Q non necessarily rooted. Consider now the endo-functor $(-)^+$ of the category of posets and order-preserving functions: this functor adds a new root to each object and expands morphisms by mapping the new root into the new root (notice that it gives open maps if applied to surjective open maps). Apply the functor $(-)^+$ to the chain (C). This modified chain is of the first kind: in fact, given a surjective open map $q : Q \to P_i^+$, we can find r as required by $(*)$ by arguing as follows. First consider a minimal $q \in Q$ mapped onto the root of P_i^+; let $\bar{Q} \hookrightarrow Q$ be the open inclusion of the downward closed subset $\bar{Q} = \{q' \in Q \mid q' < q\}$ into Q and let $\bar{q}$ be q restricted to $\bar{Q}$ in the domain and to P_i in the codomain. By the construction of (C), there are j and $\bar{r} : P_j \to \bar{Q}$ open and surjective such that $\bar{q} \circ \bar{r} = f_{ij}$. The required r is obtained by composing $\bar{r}^+$ with the inclusion of $\bar{Q}^+$ into Q. □

[6] Here we obviously mean the pullback in the category of posets and order-preserving maps. Recall that in the pullback (P, p', p) of (g', g) the projection maps p', p are open surjective in case g', g are such. Moreover P is rooted if the domains of g', g are rooted.

Notice that Theorem 7.15 is a quite peculiar result which may fail in other rather similar contexts (e.g. in the case of $S4$-algebras, see the Exercises).

6. Existentially closed algebras

As we know from Chapter 3, for a theory T which is a FLEA, the existence of a model completion is equivalent to the condition that $\mathbf{T}$ is an r-Heyting category. Thus, this condition describes when the class of existentially closed models of a FLEA is elementary. However, it is hard to give a general structure theorem even for countable existentially closed algebras of such theories since there might be too many of them. For example in an exercise the reader will be asked to show that there are $2^{\aleph_0}$ countable existentially closed Heyting algebras.

On the other hand, the model completion of the theory of Heyting algebras T_H^*, and many similar theories, has a prime model which can be fairly explicitly constructed as a colimit of a chain of finite Heyting algebras. This model, since T_H^* is almost complete, provides a way of verifying the axioms of T_H^*. We sketch the proof of such a result for theories which are slightly more general than FLEA's.

PROPOSITION 7.16 *Let T be an equational theory in a finite language, with the finite model property and the amalgamation property for finite algebras. Consider any finite algebra A_0 which is a quotient of the initial algebra. There is a unique model U of T such that*

1. *U is countable;*
2. *U is locally finite;*
3. *U is an existentially closed model of $T \cup Diag(A_0)$.*

Moreover, if T has a model completion T^*, then this model is the prime model of $T^* \cup Diag(A_0)$.

Proof. We claim that to show the existence of U it is enough to show the existence of a chain

$$A_0 \longrightarrow A_1 \longrightarrow A_2 \longrightarrow \ldots \longrightarrow A_n \longrightarrow \ldots \tag{7.17}$$

of finite T-algebras and monomorphisms with the following property P:

for any finite T-algebra B and any monomorphism $h : A_n \to B$ there is an $m \geq n$ and a monomorphism $g : B \to A_m$ such that $g \circ h : A_n \to A_m$ is the embedding as in the chain above.

The colimit of a chain with the property P is the model U. Clearly it satisfies first two conditions. In order to show that it satisfies the last condition, one can repeat (with minor modifications) the proof of the first part of Proposition 7.6.

In order to construct a chain (7.17) with the property P, first notice that by the amalgamation property for finite algebras, we can amalgamate any finite number of algebras and keep the amalgam finite as well. So it is possible to repeat the enumeration strategy we used in the proof of Theorem 7.15. Uniqueness of the model so built follows from a double chain argument. The last statement of the Proposition follows from the fact that the existence of a model completion implies the amalgamation property. □

7. Exercises

1 Show that the jointly epic coverings topology $\mathbf{J}^\infty$ on the opposite category of finitely presented Heyting algebras HA_{fp}^{op} is not subcanonical. (*Argue as follows*: suppose it is and consider the full and faithful Yoneda embedding; we know from Chapter 4 that the free Heyting algebra on one generator F is a subobject classifier in HA_{fp}^{op} and that the Yoneda embedding preserves it (because so does the functor Φ of Section 4.3 and because the equivalence between $Sh(HA_{fp}^{op}, \mathbf{J}^\infty)$ and $Sh(HA_{fin}^{op}, \mathbf{J}^{\mathbf{f}})$ is induced by the inclusion of categories $HA_{fin} \subseteq HA_{fp}$). Now for any fp Heyting algebra A we have the following chain of isomorphisms

$$Sub_{HA_{fp}^{op}}(A) \simeq Hom_{HA_{fp}^{op}}(A, F) \simeq Hom(Y(A), Y(F)) =$$

$$= Hom(Y(A), \Omega) \simeq Sub(Y(A))$$

which cannot be, for instance because there are fp Heyting algebras which are not complete lattices, see e.g. [Be]).

2 Show that a prime Dummett algebra is algebraically closed iff it satisfies the following condition

$$xRy \Rightarrow \exists z\,(xRz \,\&\, zRy)$$

where xRy is defined as $(x \le y) \,\&\, (y \to x \le x)$.

3 Given an $S4$ algebra $(B, \Box)$, its *spectrum* is the set $s(B)$ of its ultrafilters preordered by the relation

$$U \le V \quad \text{iff} \quad \forall x \in B\,(\Box x \in U \Rightarrow x \in V)$$

Prove that $(B, \Box)$ embeds into the powerset $S4$-algebra associated to the frame $(s(B), \le)$.

4 [GZ4] Prove that there are infinitely many non isomorphic locally finite, countably infinite, prime and algebraically closed $S4$-algebras. (*Hint*: Keep in mind that two $S4$-algebras cannot be isomorphic in case they have different spectra and that the spectrum of the colimit of the dual of a chain like

(C) in Section 7.5 is the inverse limit of the P_i. So just produce infinitely many *suitable* chains of finite preordered sets having non isomorphic inverse limit).

5 Given a finite rooted poset P, let $k_P : P \longrightarrow \mathbf{2}$ be the constant evaluation with value 1 and let $\chi_P : P \longrightarrow \mathbf{2}$ be the evaluation defined by $\chi_P(p) = 0$ iff $p = \rho$ (here ρ is the root of P, we shall use hereafter results and notations from Chapter 4). For every irreducible evaluation $v : P \longrightarrow \mathbf{2}$, let $\psi_v \in \mathcal{S}(h_{\mathbf{2}} \times h_{\mathbf{2}})$ be

$$\pi_0^{-1}(\downarrow_\infty v) \to \pi_1^{-1}(\downarrow_\infty (k_P)) \cup \bigcup_{p \neq \rho} \pi_0^{-1}(\downarrow_\infty v_p)$$

where $\pi_0, \pi_1 : h_{\mathbf{2}} \times h_{\mathbf{2}} \longrightarrow h_{\mathbf{2}}$ are the two projections.

(a) Show that if $w : Q \longrightarrow \mathbf{2}$ is an irreducible evaluation, then $\langle w, \chi_Q \rangle \notin \psi_v$ iff $w = v$;

(b) deduce from the previous point and from the existence of infinitely many distinct irreducible $\mathbf{2}$-evaluations that the free Heyting algebra on two generators has $2^{\aleph_0}$ filters;

(c) conclude that there are $2^{\aleph_0}$ countable existentially closed Heyting algebras.

6 Fill the missing details of the proof of the Proposition 7.16.

7 Let $\varphi(\vec{p})$ be a formula of IpC and let $\mathcal{F}(\vec{p})/\varphi(\vec{p})$ denote the free Heyting algebra over $\vec{p}$ divided by the equation $\varphi(\vec{p}) = \top$.

(a) Show that the morphism

$$f : \mathcal{F}(p,q)/(p \vee q) \longrightarrow \mathcal{F}(p) \times \mathcal{F}(q)$$

such that $f(p) =< p, \top >$ and $f(q) =< \top, q >$ is a monomorphism between finitely presented Heyting algebras.

(b) Let

$$\exists_f, \forall_f : \mathcal{F}(p) \times \mathcal{F}(q) \longrightarrow \mathcal{F}(p,q)/(p \vee q)$$

be the monotone functions given for $\varphi(p) \in \mathcal{F}(p)$ and $\psi(q) \in \mathcal{F}(q)$ by

$$\exists_f(\varphi(p), \psi(q)) = (\varphi(p) \wedge q) \vee (p \wedge \psi(q))$$

$$\forall_f(\varphi(p), \psi(q)) = (p \to \psi(q)) \wedge (q \to \varphi(p))$$

Prove that $\exists_f$ and $\forall_f$ are left and right adjoints to f, i.e. that $\exists_f \dashv f \dashv \forall_f$.

(c) Let A be a Heyting algebra. Show that the functor $\overline{A}$ (notation as in Section 7.4) preserves the cover $\{f\}$ iff for any $a, b \in A$ if $a \vee b = \top$ then there are $a' \leq a$ and $b' \leq b$ such that $a' \wedge b' = \perp$ and $a' \vee b' = \top$.

8 A functor between r-regular categories is *r-regular* iff it preserves finite limits and epimorphisms. A category is *r-coherent* iff it is r-regular and has stable finite joins of regular subobjects. A functor between r-coherent categories is *r-coherent* iff it is r-regular and preserves finite joins of regular subobjects. Let T by an equational theory with EDO that admits a model-completion.

(a) Show that the category $rReg(\mathbf{T}, Set)$, consisting of r-regular functors from $\mathbf{T}$ to Set and natural transformations between them, is equivalent to the category of algebraically closed T-algebras. Conclude, that the models of the topos $Sh(\mathbf{T}, \mathbf{J}^{one})$ in Set are algebraically closed T-algebras, where $\mathbf{J}^{one}$ is the Grothendieck topology on $\mathbf{T}$ generated by epimorphisms in $\mathbf{T}$.

(b) Show that the category $rCoh(\mathbf{T}, Set)$, consisting of r-coherent functors from $\mathbf{T}$ to Set and natural transformations between them, is equivalent to the category of algebraically closed, prime T-algebras. Conclude that the models of the topos $Sh(\mathbf{T}, \mathbf{J}^{fin})$ in Set are algebraically closed, prime T-algebras, where $\mathbf{J}^{fin}$ is the Grothendieck topology on $\mathbf{T}$ generated by finite jointly epimorphic families of morphisms in $\mathbf{T}$.

Chapter 8

OPEN PROBLEMS

In this short Chapter we state some relevant open problems that are related to the topics developed in the book.

Axiomatization In Chapter 3 we presented a theorem (Theorem 3.8), characterizing some equational theories, that admits a model completion. The proof of this theorem provides an explicit axiomatization of the model completion. But as usual in case of such general constructions the axiomatization given there is, it seems, far from being optimal. Thus it would be very interesting to see how it can be simplified in some special cases such as the theory of Heyting algebras, Magari algebras, Grzegorczyk algebras, etc. This problem is related to the problem of axiomatization of second order propositional logics corresponding to these theories. By an axiomatization,we mean a proof system which is sound and complete with respect to the semantics in sheaves, as explained in Section 4.6 for intuitionistic logic.

Models A meaningful axiomatization would also be helpful in understanding existentially closed models of these theories. In particular, it could help to find natural explicit examples of existentially closed models (e.g. in the case of Heyting algebras), contributing more information than that considered in Chapter 7.

Extension of classification results In Chapters 4 and 6 we gave a full description of the subvarieties of the varieties of Heyting algebras and of **S4**-algebras whose theory admits a model completion. There might be similar results concerning other logics. For example, is it possible to provide a similar classification result for the varieties of **K4**- or **G**-algebras? Or, if not, is it possible to show that there is no such classification? An as yet unattacked problem deserving special attention would be the investigation of cases in which only a model companion (not necessarily a model completion) exists.

Interpretation of second order quantifiers The original statement of the above mentioned Pitts' Theorem states that there exists an interpretation of second order intuitionistic propositional logic IpC^2 into intuitionistic propositional logic IpC which preserves propositional connectives and provability. This formulation is equivalent to the fact that the dual of the category of finitely presented Heyting algebras is an r-Heyting category. However if the finitely generated congruences in finitely generated free algebras do not correspond to all formulas (e.g. as in case of modal algebras) then the two statements are different (see exercise 8 from Chapter 6 for a counterexample). It would be interesting to have a broader classification of modal systems in which such interpretations are possible.

Exactness properties and definability This book shows how exactness properties of the category of finitely presented algebras are related to model-theoretic properties of the corresponding theories and to the definability problems of the corresponding propositional logics.

As we saw in Chapter 3, if $\mathbf{T}$ (the dual of the category of finitely presented algebras of an equational theory T) is an r-Heyting category then the theory admits model completion. Is it possible to derive some other interesting model-theoretic properties of the theory T, and/or definability properties of the related logic, from some other categorical properties of $\mathbf{T}$, such as $\mathbf{T}$ being regular, exact, coherent, topos, etc.?

For examples, if T is regular and all the regular epimorphisms in $\mathbf{T}$ are effective descent morphisms, i.e. $\mathbf{T}$ is *regular*$^+$, than the corresponding logic enjoys a certain definability property concerning 'separation of two sets of independent variables'. The problem whether the opposite category of finitely presented Heyting algebras is regular$^+$ or not clearly deserves a solution.

Pitts' problem The original motivation of A.M.Pitts for his search for a counterexample to what eventually became his theorem (Theorem 1.1) was to show that some Heyting algebras cannot be the algebra of global sections of an elementary topos. Thus the problem, whether every Heyting algebra is the algebra of global sections of an elementary topos, which A.M.Pitts tried to solve negatively, is still open. This problem might be more interesting than it seems, if the solution is positive, because it may possibly provide a new method of constructing elementary toposes.

Dualities and sheaf representations Finally, there are also some problems concerning the dualities that we developed in this book and the sheaf representations related to them. Clearly, our techniques for getting such duality for finitely presented T-algebras, require T to have the finite model property. Are there any other reasonable conditions on T that are sufficient to obtain such dualities and sheaf representations?

As we already said in the introduction, the dual object in these dualities have mixed geometric (sheaf) and combinatorial (games) structure. However the relation between these structures is not yet well understood. For example we do not know whether the sheaf structure determines the game structure uniquely (or at least how to recover it from the sheaf structure). In more categorical terms, we can ask what properties **S** should satisfy so that the functor

$$\mathbf{\Psi_S} : \mathbf{M_S} \longrightarrow Sh(\mathbf{F_S}, \mathbf{J_S}),$$

see pages 87, 135, is full on isomorphisms. Is there another more convenient way to determine the essential image of the functor $\mathbf{\Psi_S}$?

Chapter 9

APPENDIX

In this appendix, we collect the definitions, facts and constructions from category theory and categorical logic that we use in the book. For the proofs and more on Grothendieck toposes the reader is referred to MacLane-Moerdijk book [CWL].

1. Glossary of basic categorical notions

Adjoint functors Let $\mathbf{F} : \mathbf{C} \to \mathbf{D}$, $\mathbf{G} : \mathbf{D} \to \mathbf{C}$. be two functors. An *adjunction* is given by a triple $< \mathbf{F}, \mathbf{G}, \varphi >$ where φ is a natural isomorphism between functors

$$\mathbf{C}^{op} \times \mathbf{D} \underset{\mathbf{D}(\mathbf{F}(-), +)}{\overset{\mathbf{C}(-, \mathbf{G}(+))}{\rightrightarrows}} Set$$

If $< \mathbf{F}, \mathbf{G}, \varphi >$ is an adjunction, then we say that $\mathbf{F}$ is a *left adjoint* to $\mathbf{G}$ and $\mathbf{G}$ is a *right adjoint* to $\mathbf{F}$, and we denote it by $\mathbf{F} \dashv \mathbf{G}$.

Each adjoint determines the other uniquely up to a natural isomorphism. The adjunction $< \mathbf{F}, \mathbf{G}, \varphi >$ can be uniquely determined by some other data. For example, if we have a natural transformation

$$\eta : 1_{\mathbf{C}} \longrightarrow \mathbf{F} \circ \mathbf{G}$$

such that for any objects C in $\mathbf{C}$, D in $\mathbf{D}$ and for any $f : C \to \mathbf{G}(D)$ there is a unique $g : \mathbf{F}(C) \to D$ such that the triangle in the following diagram

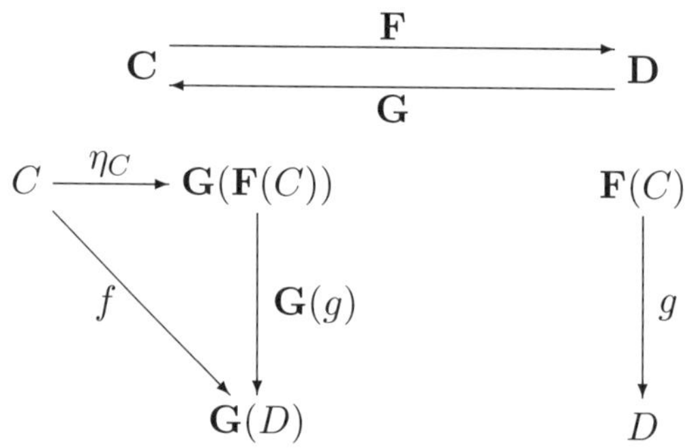

commutes, then putting $\varphi(g) = \mathbf{G}(g) \circ \eta_C$ for $g : \mathbf{F}(C) \to D$ we get an adjunction. Such natural transformation η is called *unit of adjunction*.

If both $\mathbf{C}$ and $\mathbf{D}$ are posets and hence $\mathbf{F}$ and $\mathbf{G}$ are monotone functions then $\mathbf{F} \dashv \mathbf{G}$ iff for any objects $C \in \mathbf{C}$ and $D \in \mathbf{D}$ we have

$$C \leq \mathbf{G}(\mathbf{F}(C)) \text{ and } \mathbf{F}(\mathbf{G}(D)) \leq D$$

SEE ALSO: **Category**, **Functor**, **Hom functor**, **Natural transformation**.

Beck-Chevalley condition Let $\mathbf{C}$ be a category with finite limits (or at least with pullbacks). Moreover, let $f : A \to B$ be a morphism in $\mathbf{C}$ such that the pullback functor

$$f^* : Sub(B) \longrightarrow Sub(A)$$

has a left adjoint

$$\exists_f : Sub(A) \longrightarrow Sub(B)$$

Then $\exists_f$ satisfies the *Beck-Chevalley* iff whenever the square

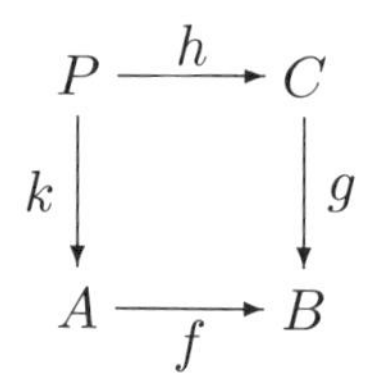

is a pullback, we have

$$g^* \circ \exists_f = \exists_h \circ k^*$$

SEE ALSO: **Pullback**, **Adjoint functors**, **Subobject**.

Binary product If A and B are objects of a category $\mathbf{C}$ then the diagram

$$A \xleftarrow{\pi} P \xrightarrow{\pi'} B$$

is a *(binary) product diagram* iff for any pair of morphisms $a : Z \to A$ and $b : Z \to B$ there is a unique morphism $c : Z \to P$ (usually denoted $< a, b >$) such that $\pi \circ c = a$ and $\pi' \circ c = b$.

We often call an object P a product of A and B when π and π' are understood. We usually draw product diagrams as follows

$$A \xleftarrow{\pi_A} A \times B \xrightarrow{\pi_B} B$$

and we call π_A and π_B the *projections*.

SEE ALSO: **Limit**.

Category A *category* **C** consists of the following data:

1 A class of objects $\mathbf{C}_0$.

2 For any pair of objects A, B in $\mathbf{C}_0$ a set $\mathbf{C}(A, B)$ of morphisms with *domain* A and *codomain* B; when $f \in \mathbf{C}(A, B)$ we write $f : A \to B$ or $A \xrightarrow{f} B$.

3 For each ordered triple $< A, B, C >$ of objects a function

$$\circ : \mathbf{C}(A, B) \times \mathbf{C}(B, C) \longrightarrow \mathbf{C}(A, C)$$

called *composition*, and written $< f, g > \mapsto g \circ f$, for $f \in \mathbf{C}(A, B)$ and $g \in \mathbf{C}(B, C)$.

4 For each object A, a morphism $1_A : A \to A$, called *identity* on A (also denoted by id_A).

These data are subject to the following conditions

1 *Associativity* For any morphisms $A \xrightarrow{f} B \xrightarrow{g} C \xrightarrow{h} D$, we have

$$h \circ (g \circ f) = (h \circ g) \circ f.$$

2 *Unit law* For any morphism $A \xrightarrow{f} B$, we have

$$1_B \circ f = f = f \circ 1_A.$$

If $< A, B >$ and $< C, D >$ are different pairs of objects we treat the sets $\mathbf{C}(A, B)$ and $\mathbf{C}(C, D)$ as if they were disjoint even if (set-theoretically) they are not.

A category $\mathbf{C}$ is *small* iff $\mathbf{C}_0$ is a set. $\mathbf{C}$ is *essentially small* iff it is equivalent to a small category.

The notion we have described above is the notion of a *locally small* category, i.e. a category in which the morphisms between any pair of objects form a set. See Chapter 1 of [CWM] for more on the notion of a category.

SEE ALSO: **Equivalence of categories**.

Coequalizer The diagram

$$C \overset{g}{\underset{h}{\rightrightarrows}} A \xrightarrow{f} B$$

in a category $\mathbf{C}$ is a *coequalizer* iff $f \circ g = f \circ h$ and for any morphism $k : A \to D$ such that $k \circ g = k \circ h$ there is a unique morphism $l : B \to D$ such that $l \circ f = k$.

If a diagram, as above, is a coequalizer then f is epi.

SEE ALSO: **Colimit**.

Colimit The notion of colimit is dual to the notion of limit.

SEE ALSO: **Limit**, **Dual notion**, **coequalizer**, **Pushout**.

Concrete category A *concrete category* is a category C equipped with a faithful functor $U : C \to Set$. Usually, in a concrete category each object A comes equipped with and underlying set $U(A)$ and every morphism $f : A \to B$ is an actual function $U(f) : U(A) \to U(B)$. For example the category $Poset$ of posets and monotone maps is concrete, with $U(X, R) = X$ for any poset (X, R).

SEE ALSO: **Category**, **Faithful functor**, **Surjection**, **Injection**, **Inclusion**.

Conservative functor The functor $\mathbf{F} : \mathbf{C} \to \mathbf{D}$ is *conservative* iff it reflects isomorphisms, i.e. for any morphism f in $\mathbf{C}$, f is an isomorphism whenever $\mathbf{F}(f)$ is.

SEE ALSO: **Functor**, **Isomorphism**.

If $\mathbf{C}$ has finite limits and $\mathbf{F}$ is a conservative functor preserving them, then $\mathbf{F}$ faithful.

Contravariant functor A functor with domain $\mathbf{C}^{op}$ is sometimes called a *contravariant* functor on $\mathbf{C}$.

SEE ALSO: **Category**, **Opposite Category**, **Functor**, **Presheaf**.

Coregular factorization The notion of a *coregular factorization* is dual to the notion of a regular factorization.

SEE ALSO: **Regular factorization**, **Dual notion**.

Dual notion The *dual notion* is a concept applied to the opposite category. The dual notions have often the same names as the original ones preceded by 'co-', e.g the notion of a *colimit* is the dual notion to the notion of a *limit*. However there is many exceptions, e.g. *pushout* is dual to *pullback* and *epi* is dual to *mono*.

SEE ALSO: **Category**, **Opposite Category**.

Epimorphism A morphism $f : A \to B$ in a category $\mathbf{C}$ is an *epimorphism* (or *epi*) iff for any pair of morphisms $g, h : B \to C$ in $\mathbf{C}$, we have that $g = h$ whenever $g \circ f = h \circ f$.

It can be shown that $f : A \to B$ is epi iff the square

$$\begin{array}{ccc} A & \xrightarrow{f} & B \\ {\scriptstyle f}\downarrow & & \downarrow{\scriptstyle 1_B} \\ B & \xrightarrow[1_B]{} & B \end{array}$$

is a pushout.

SEE ALSO: **Monomorphism**.

Equalizer The diagram

$$A \xrightarrow{f} B \underset{h}{\overset{g}{\rightrightarrows}} C$$

in a category $\mathbf{C}$ is an *equalizer* iff $g \circ f = h \circ f$ and for any morphism $k : D \to B$ such that $g \circ k = h \circ k$ there is a unique morphism $l : D \to A$ such that $f \circ l = k$.

If a diagram, as above is an equalizer, then f is mono.

SEE ALSO: **Coequalizer**, **Limit**.

Equivalence of categories The functor $\mathbf{F} : \mathbf{C} \to \mathbf{D}$ is an *equivalence of categories* iff it is full, faithful and essentially surjective.

SEE ALSO: **Category**, **Functor**, **Conservative functor**, **Full functor**, **Essentially surjective functor**.

Equivalence relation A parallel pair of morphisms

$$R \underset{b}{\overset{a}{\rightrightarrows}} X$$

in a category **C** with finite limits is an *equivalence relation* iff

1 $< a, b >: R \longrightarrow X \times X$ is mono.

2 *Reflexivity.* The diagonal morphism $\delta =< 1_X, 1_X >: X \to X \times X$ factors through $< a, b >$, i.e. there is a morphism $f : X \to R$ such that $< a, b > \circ f = \delta$.

3 *Symmetry.* There is a morphism $s : R \to R$ such that $s \circ a = b$ and $s \circ b = a$.

4 *Transitivity.* If the square

$$\begin{array}{ccc} P & \xrightarrow{\pi_{12}} & R \\ {\scriptstyle \pi_{01}}\downarrow & & \downarrow{\scriptstyle a} \\ R & \xrightarrow[b]{} & X \end{array}$$

is a pullback, then $< a \circ \pi_{01}, b \circ \pi_{12} >: P \longrightarrow X \times X$ factors through $< a, b >$.

Any kernel pair is an equivalence relation.

SEE ALSO: **Binary product**, **Pullback**, **Kernel pair**.

Essentially small category SEE: **Category**.

Essentially surjective functor The functor $\mathbf{F} : \mathbf{C} \to \mathbf{D}$ is *essentially surjective* iff for any object D in **D** there is an object A of **C** such that the objects $\mathbf{F}(A)$ and D are isomorphic.

SEE ALSO: **Functor**.

Exact category A category **C** is *exact* iff **C** is a regular category and every equivalence relation in **C** is a kernel pair.

SEE ALSO: **Regular category**, **Equivalence relation**, **Kernel pair**.

Faithful functor The functor $\mathbf{F} : \mathbf{C} \to \mathbf{D}$ is *faithful* iff for any objects A, B of **C** the function

$$\mathbf{F}_{A,B} : \mathbf{C}(A, B) \longrightarrow \mathbf{D}_0(\mathbf{F}_0(A), \mathbf{F}_0(B))$$

is one-to-one.

SEE ALSO: **Functor**.

Full functor The functor $\mathbf{F} : \mathbf{C} \to \mathbf{D}$ is *full* iff for any objects A, B of $\mathbf{C}$ the function

$$\mathbf{F}_{A,B} : \mathbf{C}(A, B) \longrightarrow \mathbf{D}_0(\mathbf{F}_0(A), \mathbf{F}_0(B))$$

is surjective.

SEE ALSO: **Functor**.

Functor A *functor* $\mathbf{F}$ from a category $\mathbf{C}$ to a category $\mathbf{D}$, denoted by $\mathbf{F} : \mathbf{C} \to \mathbf{D}$, is an *object function* $\mathbf{F}_0 : \mathbf{C}_0 \longrightarrow \mathbf{D}_0$ together with a collection of functions

$$\mathbf{F}_{A,B} : \mathbf{C}(A, B) \longrightarrow \mathbf{D}(\mathbf{F}_0(A), \mathbf{F}_0(B))$$

such that for each object A in $\mathbf{C}$, $\mathbf{F}_{A,A}(1_A) = 1_{\mathbf{F}_0(A)}$ and for any objects A, B, C in $\mathbf{C}$ the diagram

$$\begin{array}{ccc}
\mathbf{C}(A,B) \times \mathbf{C}(B,C) & \xrightarrow{\quad\circ\quad} & \mathbf{C}(A,C) \\
{\scriptstyle \mathbf{F}_{A,B} \times \mathbf{F}_{B,C}} \big\downarrow & & \big\downarrow {\scriptstyle \mathbf{F}_{A,C}} \\
\mathbf{D}(\mathbf{F}(A),\mathbf{F}(B)) \times \mathbf{D}(\mathbf{F}(B),\mathbf{F}(C)) & \xrightarrow[\circ]{} & \mathbf{C}(\mathbf{F}(A),\mathbf{F}(B))
\end{array}$$

commutes. Usually, we drop the subscripts in $\mathbf{F}_0$ and $\mathbf{F}_{A,B}$.

SEE ALSO: **Category**.

Functor category If $\mathbf{C}$ and $\mathbf{D}$ are categories and $\mathbf{C}$ is essentially small then we can consider a *functor category* $Cat(\mathbf{C}, \mathbf{D})$, also denoted by $\mathbf{D}^{\mathbf{C}}$, whose objects are functors from $\mathbf{C}$ to $\mathbf{D}$ and whose morphisms are natural transformations.

SEE ALSO: **Category**, **Functor**, **Natural transformation**.

Heyting category A category $\mathbf{C}$ is a *Heyting category* iff

1 $\mathbf{C}$ has finite limits.

2 For any object A in $\mathbf{C}$, the poset $Sub(A)$ is a distributive lattice.

3 The pullback functors have both left and right adjoints satisfying the Beck-Chevalley condition.

It can be shown that the Beck-Chevalley condition holds for left adjoints $\exists_f$ iff it holds for right adjoints $\forall_f$.

SEE ALSO: **Category**, **Adjoint functors**, **Limit**, **Subobject**, **Beck-Chevalley condition**, **Regular category**.

Hom functor If $\mathbf{C}$ is a category and A, B are object of $\mathbf{C}$ then the association $(A, B) \mapsto \mathbf{C}(A, B)$ is the object part of the *hom-functor*

$$\mathbf{C}(-,+) : \mathbf{C}^{op} \times \mathbf{C} \longrightarrow Set$$

which assigns to a pair of morphism $f : A' \to A$, $g : B \to B'$, i.e. a morphism from (A, B) to (A', B') in $\mathbf{C}^{op} \times \mathbf{C}$, a function

$$\mathbf{C}(f, g) : \mathbf{C}(A, B) \longrightarrow \mathbf{C}(A', B')$$

$$h \mapsto g \circ h \circ f$$

SEE ALSO: **Category**, **Functor**.

Inclusion We say that the morphism $i : A \to B$ in a concrete category C equipped with a faithful functor $U : C \to Set$, is an *inclusion* iff the function $U(i) : U(A) \to U(B)$ is an inclusion. By abuse of language, we say that a natural transformation $\tau : X \to Y$ between set-valued functors X and Y, is an inclusion iff all the components of τ are inclusions.

SEE ALSO: **Concrete category**, **Faithful functor**, **Surjection**, **Injection**.

Initial object An object I in a category $\mathbf{C}$ is an *initial object* iff for any object A in $\mathbf{C}$ there is exactly one morphism from I to A.

I is an initial object in $\mathbf{C}$ iff it is a colimit of a functor from the empty category.

SEE ALSO: **Colimit**.

Injection We say that the morphism $i : A \to B$ in a concrete category C equipped with a faithful functor $U : C \to Set$, is an *injection* iff the function $U(i) : U(A) \to U(B)$ is an injection. By abuse of language, we say that a natural transformation $\tau : X \to Y$ between set-valued functors X and Y, is an injection iff all the components of τ are injections.

SEE ALSO: **Concrete category**, **Faithful functor**, **Surjection**, **Inclusion**.

Injective object The notion of an *injective object* is dual to the notion of a projective object.

SEE ALSO: **Projective object**, **Dual notion**.

Isomorphism A morphism $f : A \to B$ in a category $\mathbf{C}$ is an *isomorphism* (or *iso*) iff there is a morphism $g : B \to A$ in $\mathbf{C}$, such $g \circ f = 1_A$ and $f \circ g = 1_B$.

SEE ALSO: **Category**.

Jointly surjective family The family of morphisms $\{f_i : X_i \to X\}_{i \in I}$ in a concrete category C equipped with a faithful functor $U : C \to Set$, is a *jointly surjective* iff the sum of images of functions $U(f_i)$, for $i \in I$, is equal to $U(X)$.

SEE ALSO: **Surjection**, **Jointly mono family**, **Jointly surjective family**, **Concrete category**.

Jointly epi family The family of morphisms $\{f_i : X_i \to X\}_{i \in I}$ in a category C is a *jointly epi* iff for any pair of morphism $g, h : X \to Y$ in C, $g = h$ whenever $g \circ f_i = H \circ f_i$, for all $i \in I$.

SEE ALSO: **Epimorphism**, **Jointly mono family**, **Jointly surjective family**.

Jointly mono family The notion of a *jointly mono family* of morphism is dual to the notion of jointly epi family of morphism.

SEE ALSO: **Monomorphism**, **Jointly epi family**, **Jointly surjective family**, **Dual notion**.

Kernel pair The parallel pair of morphisms

$$C \underset{h}{\overset{g}{\rightrightarrows}} A$$

is a *kernel pair* iff there is a morphism f such that the diagram

$$\begin{array}{ccc} C & \xrightarrow{g} & A \\ {\scriptstyle h}\downarrow & & \downarrow{\scriptstyle f} \\ A & \xrightarrow[f]{} & B \end{array}$$

is a pullback. In that case the diagram

$$C \underset{h}{\overset{g}{\rightrightarrows}} A \xrightarrow{f} B$$

is called a *kernel diagram*.

Any kernel pair is an equivalence relation.

SEE ALSO: **Pullback**, **Equivalence relation**.

Limit A *limit of a functor* $\mathbf{F} : \mathbf{J} \to \mathbf{C}$ is a constant functor $L_A : \mathbf{J} \to \mathbf{C}$ (sending all objects of $\mathbf{J}$ to A and all morphisms of $\mathbf{J}$ to 1_A) together with a natural transformation $\sigma : L_A \to \mathbf{F}$ such that for any other constant functor $L_B : \mathbf{J} \to \mathbf{C}$ and transformation $\tau : L_B \to \mathbf{F}$ there is a unique natural transformation $\varphi : L_B \to L_A$ (a morphism in $\mathbf{C}$) such that $\tau = \sigma \circ \varphi$. The pair (σ, L_A) is called a *limiting cone* and for each $j \in \mathbf{J}$, the j-th component σ_j of the natural transformation σ is called a *projection*.

A limit is *finite* if the category $\mathbf{J}$ is finite (i.e. has finitely many objects and morphisms). The category $\mathbf{C}$ has *(finite) limits* iff every functor from a small (finite) category $\mathbf{J}$ to $\mathbf{C}$ has a limit.

Existence of all finite limits is equivalent to existence of terminal object and either pullbacks or (binary products and equalizers).

SEE ALSO: **Functor**, **Binary product**, **Pullback**, **Equalizer**, **Terminal object**.

Locally small category SEE: **Category**.

Monomorphism A morphism $f : A \to B$ in a category $\mathbf{C}$ is a *monomorphism* (or *mono*) iff for any pair of morphisms $g, h : C \to A$ in $\mathbf{C}$, we have that $g = h$ whenever $f \circ g = f \circ h$.

It can be shown that $f : A \to B$ is mono iff the square

$$\begin{array}{ccc} A & \xrightarrow{1_A} & A \\ {\scriptstyle 1_A}\downarrow & & \downarrow{\scriptstyle f} \\ A & \xrightarrow[f]{} & B \end{array}$$

is a pullback.

SEE ALSO: **Epimorphism**, **Subobject**.

Natural transformation If $\mathbf{F}, \mathbf{G} : \mathbf{C} \longrightarrow \mathbf{D}$ are functors, a *natural transformation* σ from $\mathbf{F}$ to $\mathbf{G}$, denoted by $\sigma : \mathbf{F} \to \mathbf{G}$, is a family of morphisms in $\mathbf{D}$

$$\{\sigma_A : \mathbf{F}(A) \to \mathbf{G}(A)\}_{A \in \mathbf{C}}$$

such that for any morphism $f : A \to B$ in $\mathbf{C}$ the square

$$\begin{array}{ccc} \mathbf{F}(A & \xrightarrow{\sigma_A} & \mathbf{G}(A) \\ {\scriptstyle \mathbf{F}(f)}\downarrow & & \downarrow{\scriptstyle \mathbf{G}(f)} \\ \mathbf{F}(B) & \xrightarrow[\sigma_B]{} & \mathbf{G}(B) \end{array}$$

commutes. Natural transformations can be composed pointwise, i.e. if $\mathbf{H} : \mathbf{C} \longrightarrow \mathbf{D}$ is another functor and if $\tau : \mathbf{G} \rightarrow \mathbf{H}$ is another natural transformation then $\tau \circ \sigma$ is given by

$$(\tau \circ \sigma)_A = \tau_A \circ \sigma_A$$

for $A \in \mathbf{C}_0$. The family $\{1_{\mathbf{F}(A)}\}_{A \in \mathbf{C}_0}$ is the *identity transformation*, denoted by $1_{\mathbf{F}}$. A natural transformation $\sigma : \mathbf{F} \rightarrow \mathbf{G}$ is *natural isomorphism* if there is another transformation $\tau : \mathbf{G} \rightarrow \mathbf{F}$ such that $\tau \circ \sigma = 1_{\mathbf{F}}$ and $\sigma \circ \tau = 1_{\mathbf{G}}$.

SEE ALSO: **Category**, **Functor**.

Natural isomorphism SEE: **Natural transformation**.

Opposite category If **C** is a category, then the *opposite category* $\mathbf{C}^{op}$ has the same objects as **C** and for any objects A, B in $\mathbf{C}^{op}$

$$\mathbf{C}^{op}(A, B) = \mathbf{C}(B, A);$$

composition $g \circ f$ in $\mathbf{C}^{op}$ is $f \circ g$ in **C**, and $\mathbf{C}^{op}$ has the same identities as **C**.

SEE ALSO: **Category**.

Poset A *poset* is a small category such that there is at most one morphism between any two objects and any two isomorphic objects are equal.

SEE ALSO: **Category**.

Presheaf If C is a small category, then a contravariant functor from C to Set is called a *presheaf* on **C**. The category $Set^{C^{op}}$ is called the *category of presheaves* on C or a *presheaf category*.

SEE ALSO: **Opposite category**, **Contravariant functor**, **Functor category**.

Preservation of limits A functor *preserves (finite) limit* iff it sends all (finite) limiting cones to limiting cones.

SEE ALSO: **Category**, **Functor**, **Limit**.

Projective object An object X in a category C is *(regular) projective* iff the representable functor

$$\mathbf{C}(X, +) : \mathbf{C} \longrightarrow Set$$

sends (regular) epis to surjections.

SEE ALSO **Epimorphism**, **Regular epimorphism**, **Representable functor**, **Injective object**.

Projection SEE: **Limit**, **Binary product**.

Pullback The commutative square

$$\begin{array}{ccc} P & \xrightarrow{h} & Y \\ {\scriptstyle k}\downarrow & & \downarrow{\scriptstyle f} \\ Z & \xrightarrow[g]{} & X \end{array}$$

in a category $\mathbf{C}$ is a *pullback* iff for any pair of morphisms $a : T \to Y$, $b : T \to Z$ such that $f \circ a = g \circ b$ there is a unique morphism $c : T \to X$ such that $h \circ c = a$ and $k \circ c = b$. We also say that k is a pullback of f along g.

SEE ALSO: **Category**, **Limit**.

Pullback functor Both monomorphisms and regular monomorphisms are pullback stable, i.e. in any pullback square

$$\begin{array}{ccc} P & \xrightarrow{g} & C \\ {\scriptstyle n}\downarrow & & \downarrow{\scriptstyle m} \\ A & \xrightarrow[f]{} & B \end{array}$$

in any category $\mathbf{C}$, if m is a (regular) mono, so is the morphism n, the pullback of m along f. Thus for any morphism $f : A \to B$ in a category $\mathbf{C}$ with finite limits we can define a *pullback functor*

$$f^* : Sub(B) \longrightarrow Sub(A)$$

sending a mono $m : C \to B$ to its pullback along f. By what we said above, it restricts to a *pullback functor operating on regular subobjects*

$$f^* : Sub_r(B) \longrightarrow Sub_r(A)$$

Both functors are homomorphisms of meet-semilattices.

SEE ALSO: **Pullback**, **Functor**.

Pushout The notion of *pushout* is dual to the notion of pullback.

SEE ALSO: **Dual notion**, **Pullback**.

Regular category A category $\mathbf{C}$ is *regular* iff

1 $\mathbf{C}$ has finite limits.

2 Every kernel pair in **C** has a coequalizer.

3 Regular epis in **C** are pullback stable, i.e. pullback of a regular epi along any morphism is a regular epi.

It can be shown[1] that **C** is a regular category iff

1 **C** has finite limits.

2 For any morphism f in **C** the pullback functor f^* has a left adjoint $\exists_f$ satisfying the Beck-Chevalley condition.

SEE ALSO: **Category**, **Limit**, **Kernel pair**, **Coequalizer**, **Regular epimorphism** .

Regular epimorphism A morphism $f : A \to B$ in a category **C** is a *regular epimorphism* iff it is a coequalizer, i.e. if there are morphism $g, h : C \to A$ such that the diagram

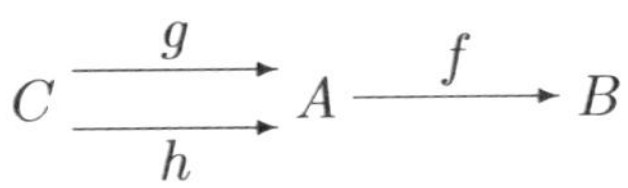

is a coequalizer.

SEE ALSO: **Coequalizer**, **Epimorphism**.

Regular factorization The factorization (e, m) of a morphism f in a category C is a *regular factorization* if e is a coequalizer of the kernel pair of f and m is the unique morphism making the triangle

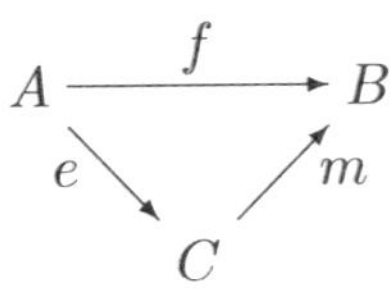

commute.

SEE ALSO: **Kernel pair**, **Coequalizer**, **Coregular factorization**.

Regular injective object The notion of a *regular injective object* is dual to the notion of a regular projective object.

SEE ALSO: **Regular projective object**, **Projective object**, **Dual notion**.

[1] See Chapters 2 and 3 of [MR1], particularly Corollary 3.3.3' on page 115.

Regular monomorphism A morphism $f : A \to B$ in a category $\mathbf{C}$ is a *regular monomorphism* iff it is an equalizer, i.e. if there are morphism $g, h : B \to C$ such that the diagram

$$A \xrightarrow{\ f\ } B \overset{g}{\underset{h}{\rightrightarrows}} C$$

is an equalizer.

SEE ALSO: **Equalizer**, **Monomorphism**, **Subobject**.

Regular projective object SEE: **Projective object**.

Regular subobject SEE: **Subobject**.

Representable functor If A is an object of the category $\mathbf{C}$, then by fixing either first or second variable in the hom-functor we can define the following two functors

$$\mathbf{C}(A, +) : \mathbf{C} \longrightarrow Set$$
$$B \mapsto \mathbf{C}(A, B)$$

and

$$\mathbf{C}(-, A) : \mathbf{C}^{op} \longrightarrow Set$$
$$B \mapsto \mathbf{C}(B, A).$$

Both functors are defined on morphisms in the obvious way, by composition. A functor $\mathbf{F}$ from $\mathbf{C}$ (or $\mathbf{C}^{op}$) to Set is *representable* if there is an object A in $\mathbf{C}$ such that $\mathbf{F}$ is naturally isomorphic to $\mathbf{C}(A, +)$ (resp. to $\mathbf{C}(-, A)$).

SEE ALSO: **Category**, **Functor**.

Small category SEE: **Category**.

Split epimorphism The notion of *split epimorphism* is dual to the notion of split monomorphism.

SEE ALSO: **Dual notion**, **Spit monomorphism**.

Split monomorphism A morphism $m : A \to B$ in a category $\mathbf{C}$ is a *split monomorphism* (or *split mono* iff there is a morphism $f : B \to A$ in a category $\mathbf{C}$, such that $f \circ m = 1_A$. Any split mono is a regular mono.

SEE ALSO: **Regular monomorphism**.

Subcategory A category $\mathbf{D}$ is a *subcategory* of a category $\mathbf{C}$ if $\mathbf{D}_0$ is a subclass of $\mathbf{C}_0$ and for any objects A, B in $\mathbf{D}$, $\mathbf{D}(A, B) \subseteq \mathbf{C}(A, B)$ and the composition functions in $\mathbf{D}$ are restriction of composition functions in $\mathbf{C}$.

D is a *full subcategory* of **C** if **D** is a subcategory of **C** and for any objects A, B in **D**, $\mathbf{D}(A,B) = \mathbf{C}(A,B)$.

If **D** is a (full) subcategory of **C** the inclusion functor $\mathbf{D} \to \mathbf{C}$ is (full and) faithful.

SEE ALSO: **Category**.

Subobject In any category **C**, we define a *subobject* of an object A to be an equivalence class of monomorphisms $m : B \to A$, where $m : B \to A$ and $n : C \to A$ are identified iff there is an isomorphism $k : B \to C$ such that $n \circ k = m$. A subobject is *regular* if it is represented by a regular monomorphism. It is a common practice to identify subobjects with their representatives or even domains of representatives if the morphisms are understood. We say that $m : B \to A$ is *smaller* then $n : C \to A$, notation $m \leq n$ or $B \leq C$, if there is a morphism $k : B \to C$ such that $n \circ k = m$. Thus in any category **C** the class $Sub(A)$ of subobjects of A and the class $Sub_r(A)$ of regular subobjects of A are partially ordered.

A category **C** is *well-powered* if for any object A in **C**, $Sub(A)$ is a set. Since all the examples of categories considered in this book are well-powered, we always assume that both $Sub(A)$ and $Sub_r(A)$ are posets.

If **C** has finite limits then both $Sub(A)$ and $Sub_r(A)$ are meet semilattices with meet defined by pullback of subobjects.

SEE ALSO: **Monomorphism**, **Regular monomorphism**.

Surjection We say that the morphism $i : A \to B$ in a concrete category C equipped with a faithful functor $U : C \to Set$, is a *surjection* iff the function $U(i) : U(A) \to U(B)$ is a surjection. By abuse of language, we say that a natural transformation $\tau : X \to Y$ between set-valued functors X and Y, is a surjection iff all the components of τ are surjections.

SEE ALSO: **Concrete category**, **Faithful functor**, **Injection**, **Inclusion**.

Terminal object On object T is a category **C** is a *terminal object* iff for any object A in **C** there is exactly one morphism from A to T.

T is a terminal object in **C** iff it is a limit of a functor from the empty category.

SEE ALSO: **Limit**.

Yoneda embedding If C is a small category then C embeds into the presheaf category on C:

$$Y : C \longrightarrow Set^{C^{op}}$$

so that $Y(X) = hom(-, X)$, i.e. $Y(X)$ is the contravariant functor on C represented by X. Y on morphisms acts in obvious way by composition.

This functor is called the *Yoneda embedding*. The Yoneda Lemma states that this embedding is full and faithful. Y preserves all the limits that exist in C.

SEE ALSO: **Functor**, **Faithful functor**, **Full functor**, **Functor category**, **Presheaf**, **Limit**.

2. Internal algebras in categories

Let L be a first order language consisting of functional symbols of finite arity, T an equational theory in L. The traditional definition of a T-algebra can be described as follows. An *L-structure* $\mathcal{A}$ is a *set* A together with a *n-ary function* $f^{\mathcal{A}} : A^n \longrightarrow A$ for each function symbol f from L. Then we can define an interpretation of a term t over L with variables among $x_1, \ldots, x_m$ as a function $t^{\mathcal{A},m} : A^m \longrightarrow A$. This is done by induction on complexity of t. $\mathcal{A}$ satisfies an equation $t_1 = t_2$ iff $t_1^{\mathcal{A},m} = t_2^{\mathcal{A},m}$ (where m is large enough so that all variables of t_1 and t_2 are among $x_1, \ldots, x_m$). Finally, $\mathcal{A}$ is a *T-algebra* iff it is an L-structure satisfying all equations of T.

Now, let $\mathbf{C}$ be a category with finite products. The notion of an *internal T-algebra* $\mathcal{A}$ in $\mathbf{C}$ is obtained by replacing in the above definition sets by objects of $\mathbf{C}$ and functions (from cartesian products) by morphisms of $\mathbf{C}$ (from products in $\mathbf{C}$). More specifically, a L-structure in $\mathbf{C}$ consists of an object A of $\mathbf{C}$ (a universe of $\mathcal{A}$) and a morphism $f^{\mathcal{A}} : A^n \longrightarrow A$ in $\mathbf{C}$ for each function symbol f of L. The interpretation $t^{\mathcal{A},m} : A^m \longrightarrow A$ of a term t with variables among $x_1, \ldots, x_m$ is given inductively:

(1) if $t = x_i$ with $i \leq m$ then $t^{\mathcal{A},m}$ is the i-th projection from the product A^m;

(2) if $t = f(t_1, \ldots, t_k)$ with f a function symbol of arity k and t_i a term for $i \leq k$, then $t^{\mathcal{A},m}$ is the following composition of morphisms in $\mathbf{C}$

$$A^m \xrightarrow{< t_1^{\mathcal{A},m}, \ldots, t_k^{\mathcal{A},m} >} A^k \xrightarrow{f^{\mathcal{A}}} A$$

An L-structure $\mathcal{A}$ in $\mathbf{C}$ satisfies an equation $t_1 = t_2$ iff $t_1^{\mathcal{A},m} = t_2^{\mathcal{A},m}$, (with m chosen as in former case). We say that $\mathcal{A}$ is an *internal T-algebra in* $\mathbf{C}$ iff it is an L-structure in $\mathbf{C}$ satisfying all equations of T. The homomorphisms of internal T-algebras can be defined in an analogous way.

The reader may easily verify now, that the usual T-algebras are nothing but internal T-algebras in the category Set.

Let $\mathcal{A}$ be an internal T-algebra in $\mathbf{C}$ with universe A, C an object of $\mathbf{C}$. Then, there is a structure of a T-algebra $\mathcal{A}_C$ on the set $\mathbf{C}(C, A)$. For an n-ary operation f of L, and $g_1, \ldots, g_n : C \longrightarrow A$ we put $f^{\mathcal{A}_C}(g_1, \ldots, g_n)$ to be equal to

$$f^{\mathcal{A}_C}(g_1, \ldots, g_n) : C \xrightarrow{<g_1, \ldots, g_n>} A^n \xrightarrow{f^{\mathcal{A}}} A$$

The fact that the L-structure $\mathcal{A}_C$ satisfies equations of T follows from the fact that $\mathcal{A}$ is an internal T-algebra in $\mathbf{C}$. Moreover, the composition with any morphism $h : D \longrightarrow C$ in $\mathbf{C}$ induces a function $\mathbf{C}(h, A) : \mathbf{C}(C, A) \longrightarrow \mathbf{C}(D, A)$. Since, for any $g_1, \ldots, g_n$ in $\mathbf{C}(C, A)$ we have

$$\mathbf{C}(h, A)(f^{\mathcal{A}_C}(g_1, \ldots, g_n)) = f^{\mathcal{A}_D}(\mathbf{C}(h, A)(g_1), \ldots, \mathbf{C}(h, A)(g_n))$$

$\mathbf{C}(h, A)$ is in fact a homomorphism of T-algebras. Clearly, the association $h \mapsto \mathbf{C}(h, A)$ is functorial. Thus, any internal T-algebra $\mathcal{A}$ induces a (contravariant) functor

$$\mathbf{C}(-, A) : \mathbf{C}^{op} \longrightarrow Alg(T)$$

We shall describe an example of an internal T-algebra $\mathcal{G}_T$ in $Alg(T)^{op}_{fp}$ which is a *generic T-algebra* in the categories with finite limits. The universe of $\mathcal{G}_T$ is the $\mathcal{F}_T(x)$, i.e. the free T-algebras on one generator x. An n-ary function symbol f of T is interpreted as a homomorphism in $Alg(T)_{fp}$

$$\mathcal{F}_T(x) \longrightarrow \mathcal{F}_T(x_1, \ldots, x_n) \tag{9.1}$$

$$x \mapsto f(x_1, \ldots, x_n)$$

i.e. an operation from n-th power of $\mathcal{F}_T(x)$ to $\mathcal{F}_T(x)$ in $Alg(T)^{op}_{fp}$. Using the fact that any $\mathcal{F}_T(x_1, \ldots, x_n)$ is a T-algebra one can verify that $\mathcal{G}_T$ is indeed an internal T-algebra in $Alg(T)^{op}_{fp}$. This algebra is universal in the following sense. If M is a model of T in a category with finite limits $\mathcal{E}$ then we can define an association $\overline{M}$ sending morphisms of $Alg(T)^{op}_{fp}$ of form (9.1) to a morphism

$$M(f) : M^n \longrightarrow M$$

in $\mathcal{E}$. Since objects of $Alg(T)_{fp}$ are coequalizers of pairs of morphisms between free objects and homomorphisms in $Alg(T)_{fp}$ are given by tuples of terms of T (see description of $Alg(T)_{fp}$ in section 2.4), the association $\overline{M}$ extends uniquely (up to a natural equivalence) to a finite limits preserving functors, denoted by $\overline{M} : Alg(T)^{op}_{fp} \to \mathcal{E}$.

Since any homomorphism of internal T-algebras $h : M \to N$ induces a natural transformation $\overline{h} : \overline{M} \longrightarrow \overline{N}$ we get a functor

$$\overline{(-)} : Alg_{\mathcal{E}}(T) \longrightarrow Lex(Alg(T)^{op}_{fp}, \mathcal{E})$$

from the category of internal T-algebras $Alg_{\mathcal{E}}(T)$ in $\mathcal{E}$ to the category $Lex(Alg(T)^{op}_{fp}, \mathcal{E})$ of finite limits preserving functors from $Alg(T)^{op}_{fp}$ to $\mathcal{E}$ and natural transformations. This functor is an equivalence of categories. In this way any T-algebra M in $\mathcal{E}$ corresponds to a unique (up to isomorphism) finite limit preserving functor $\overline{M}$ such that $\overline{M}(\mathcal{G}_T)$ is isomorphic to M.

3. Grothendieck topologies

The notion of a site that we introduce here is a general notion of a category equipped with covering families.

Let $\mathbf{C}$ be a small category. A *sieve* S on object C in $\mathbf{C}$ is a family of morphisms in $\mathbf{C}$, all with codomain C, such that

$$f \in S \quad \text{implies} \quad f \circ g \in S$$

whenever the composition makes sense. If S is a sieve on C and $h : D \longrightarrow C$ is a morphism in $\mathbf{C}$,

$$h^*(S) = \{g \mid cod(g) = D \text{ and } h \circ g \in S\}$$

is a sieve on D.

A *Grothendieck topology* on a category $\mathbf{C}$ is a function $\mathbf{J}$ which assigns to each object C of $\mathbf{C}$ a collection $\mathbf{J}(C)$ of sieves on $\mathbf{C}$, in such a way that

(i) the maximal sieve $t_C = \{f \mid cod(f) = C\}$ is in $\mathbf{J}(C)$;

(ii) *(stability)* if $S \in \mathbf{J}(C)$ and $h : D \longrightarrow C$ is in $\mathbf{C}$ then $h^*(S) \in \mathbf{J}(D)$;

(iii) *(transitivity)* if $S \in \mathbf{J}(C)$ and R is a sieve on C such that $h^*(R) \in \mathbf{J}(D)$ for all $h : D \longrightarrow C \in S$, then $R \in \mathbf{J}(C)$.

Often, it is more convenient to work with a basis for a topology rather than with the whole topology.

A *basis* (for a Grothendieck topology) on a category $\mathbf{C}$ is a function $\mathbf{K}$ which assigns to each object C of $\mathbf{C}$ a collection $\mathbf{K}(C)$ of $\mathbf{K}$*-covers of* C consisting of families of morphisms with codomain C, such that

(i') if $f : C' \longrightarrow C$ is an isomorphism, then $\{f\} \in \mathbf{K}(C)$;

(ii') *(stability)* if $\{f_i : C_i \longrightarrow C\}_{i \in I} \in \mathbf{K}(C)$ and $h : D \longrightarrow C$ is in $\mathbf{C}$ then there is a cover $\{g_j : D_j \longrightarrow D\}_{j \in I'} \in \mathbf{K}(D)$ such that for each $j \in I'$ there is $i \in I$ and morphism α_{ji} making the diagram

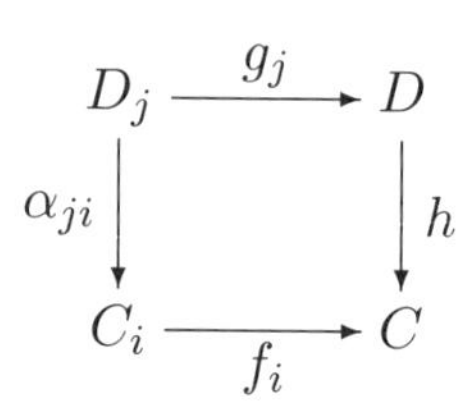

commute;

(iii') *(transitivity)* If $\{f_i : C_i \longrightarrow C\}_{i\in I} \in \mathbf{K}(C)$ and if for each $i \in I$ one has a family $\{g_{ij} : D_{ij} \longrightarrow C_i\}_{j\in I_i} \in \mathbf{K}(C_i)$ then the family of composites $\{f_i \circ g_{ij} : D_{ij} \longrightarrow C\}_{i\in I\ j\in I_i}$ belongs to $\mathbf{K}(C)$.

If **K** is a basis on **C**, then **K** *generates* a topology **J** by

$$S \in \mathbf{J}(C) \text{ iff there is } R \text{ in } \mathbf{K}(C) \text{ such that } R \subseteq S;$$

i.e. a sieve is a **J**-cover iff it contains a **K**-cover. It is not difficult to verify that **J** so defined is a topology on **C**. We say that **K** is a basis for **J**.

The pair $(\mathbf{C}, \mathbf{J})$, where **C** is a small category and **J** is a topology or a basis for a topology on **C**, is called a *site*.

4. Sheaves

Let **C** be a small category and **J** a topology on **C**. A *presheaf* on **C** is a functor $P : \mathbf{C}^{op} \longrightarrow Set$, i.e. it is a contravariant functor from **C** to Set. Presheaves on **C** together with natural transformations form a functor category denoted by $Set^{\mathbf{C}^{op}}$. Any object C in **C** give rise to a presheaf

$$\mathbf{C}(-, C) : \mathbf{C}^{op} \longrightarrow Set$$

such that for $f : D' \longrightarrow D$, $\mathbf{C}(f, C)(\alpha) = \alpha \circ f$ for $\alpha \in \mathbf{C}(D, C)$.

If P is a presheaf and S a covering sieve on C, a *compatible family* for S of elements of P is a set $\{x_f\}_{f\in S}$ such that for $f : D \longrightarrow C \in S$, $x_f \in P(D)$ and

$$P(g)(x_f) = x_{f\circ g} \quad \text{for all } g : E \longrightarrow D \text{ in } \mathbf{C}$$

An *extension* of a compatible family $\{x_f\}_{f\in S}$ is an element $x \in P(C)$ such that

$$P(f)(x) = x_f \quad \text{for all} f \in S$$

P is a (**J**-)*sheaf*, or as we sometimes say has sheaf property, iff any compatible family for any **J**-cover of any object of **C** has a unique extension.

If **K** is a basis for the topology **J**, the definition of a **J**-sheaf can be expressed purely in terms of the basis **K**, as follows. Let $R = \{f_i : C_i \longrightarrow C\}_{i\in I}$ be a **K**-cover of C, a family of elements $\{x_i\}_{i\in I}$ such that $x_i \in P(C_i)$ for all $i \in I$ is a *compatible family* on R iff for any $i, j \in I$ and any morphisms $h_i : D \longrightarrow C_i$, $h_j : D \longrightarrow C_j$, if $f_i \circ h_i = f_j \circ h_j$ then $P(h_i)(x_i) = P(h_j)(x_j)$. An *extension* for $\{x_i\}_{i\in I}$ on R is an element $x \in P(c)$ such that $x_i = P(f_i)(x)$ for $i \in I$.

One can verify that if **K** is a basis for **J** and P is a presheaf then P is a **J**-sheaf iff any compatible family on any **K**-cover has a unique extension.

If **J** is a (basis for a) topology on **C** such that all representable presheaves on **C** are sheaves, we will call **J** *subcanonical*. We shall spell this definition in details in terms of covers. If a cover $R = \{h_i : C_i \longrightarrow C\}_{i\in I}$ of an object C in $\mathbf{C}$ belongs to a basis for a subcanonical topology then the following holds. For any object D in $\mathbf{C}$ and a compatible family $\{f_i : C_i \longrightarrow D\}_{i\in I}$ of elements of the representable functor $\mathbf{C}(-, D) : \mathbf{C}^{op} \longrightarrow Set$ there is a unique morphism $h : C \longrightarrow D$ such that $h \circ h_i = f_i$ for $i \in I$. The condition that $\{f_i\}_{i\in I}$ is compatible family on $\{h_i\}_{i\in I}$ means in this case that for any pair of morphisms $g_i : C' \longrightarrow C_i$, $g_j : C' \longrightarrow C_j$ in $\mathbf{C}$ with $i, j \in I$ (i, j are *not* necessarily different) $f_i \circ g_i = f_j \circ g_j$ whenever $h_i \circ g_i = h_j \circ g_j$. The *canonical topology* is the largest subcanonical topology on $\mathbf{C}$.

The sheaves on a site $(\mathbf{C}, \mathbf{J})$ form a category, where the maps are the natural transformations, i.e. maps of presheaves. So the category of sheaves, for which we write $Sh(\mathbf{C}, \mathbf{J})$, is a full subcategory of the functor category $Set^{\mathbf{C}^{op}}$,

$$\mathbf{i} : Sh(\mathbf{C}, \mathbf{J}) \longrightarrow Set^{\mathbf{C}^{op}}$$

A *Grothendieck topos* is a category which is equivalent to the category $Sh(\mathbf{C}, \mathbf{J})$ of sheaves on some site $(\mathbf{C}, \mathbf{J})$.

Notice that it may happen that different sites give equivalent Grothendieck toposes (i.e. that, when passing to sheaves, they produce equivalent categories); we recall here a useful sufficient criterion (known as *Comparison Lemma* cf. [SGA4] p. 288)[2] to recognize such an equivalence. Suppose we are given a site $(\mathbf{C}, \mathbf{J})$ and a full subcategory $\mathbf{C}'$ of $\mathbf{C}$. We may induce a topology $\mathbf{J}'$ on $\mathbf{C}'$ by declaring that $S \in \mathbf{J}'_D$ iff the sieve generated by S in $\mathbf{C}$ belongs to $\mathbf{J}_D$. We now have that *$Sh(\mathbf{C}, \mathbf{J})$ is equivalent to $Sh(\mathbf{C}', \mathbf{J}')$ in case every object D in $\mathbf{C}$ is $\mathbf{J}$-covered by* (the sieve generated by) *a family of the kind $\{C_i \longrightarrow D\}_i$ with $C_i \in \mathbf{C}'$.*

5. Associated sheaf functor

Let $(\mathbf{C}, \mathbf{J})$ be a site. Then the inclusion functor $\mathbf{i} : Sh(\mathbf{C}, \mathbf{J}) \longrightarrow Set^{\mathbf{C}^{op}}$ has a left adjoint $\mathbf{a} : Set^{\mathbf{C}^{op}} \longrightarrow Sh(\mathbf{C}, \mathbf{J})$. This is a way of saying that for any presheaf $P : \mathbf{C}^{op} \longrightarrow Set$ there is a 'closest' $\mathbf{J}$-sheaf $\mathbf{a}(P) : \mathbf{C}^{op} \longrightarrow Set$ in the following sense: there is a morphism $\eta_P : P \longrightarrow \mathbf{a}(P)$ in $Set^{\mathbf{C}^{op}}$ and for any other morphism $f : P \longrightarrow A$ in $Set^{\mathbf{C}^{op}}$ into a $\mathbf{J}$-sheaf A there is a unique morphism $\bar{f} : \mathbf{a}(P) \longrightarrow A$ making the triangle

[2] See also [KM] for a more refined statement.

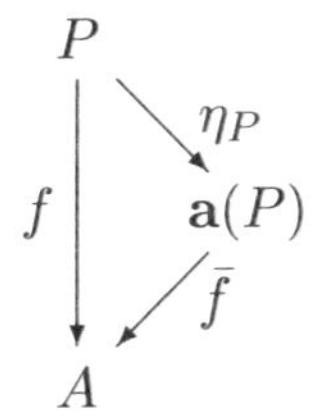

commute. The whole construction of the associated sheaf functor **a** can be found in [CWL]. We shall describe here an easy special case of the construction of **a** needed in the paper.

Let Q be a **J**-sheaf and P a subpresheaf of Q. The the sheaf $\mathbf{a}(P)$ is a subpresheaf of Q such that for C in **C** we have

$$x \in \mathbf{a}(P)(C) \text{ iff } x \in Q(C) \text{ and there is a cover } \{f_i : C_i \longrightarrow C\}_{i \in I}$$
$$\text{such that } \; Q(f_i)(x) \in P(C_i) \; \text{ for } \; i \in I$$

6. Properties of Grothendieck toposes

Grothendieck toposes are elementary toposes, but we need less. We shall show that they are Heyting categories with all monos regular. Thus they are r-Heyting categories, as well.

Fix a site $(\mathbf{C}, \mathbf{J})$. Then the finite limits in $Sh(\mathbf{C}, \mathbf{J})$ are computed as in the category of presheaves, pointwise.

Recall that in a category $\mathcal{E}$ with finite limits, a *subobject classifier* is a monomorphism $true : 1 \longrightarrow \Omega$, such that to every monomorphism $m : A \longrightarrow B$ in $\mathcal{E}$ there is a unique morphism χ_m making the square

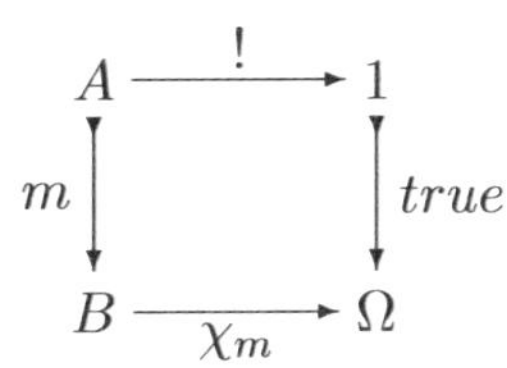

a pullback. In any category $\mathcal{E}$, is a subobject classifier exists it is unique up to an isomorphism.

We shall describe the subobject classifier in $Sh(\mathbf{C}, \mathbf{J})$. A sieve S on C is **J**-*closed* iff for any morphism $f : D \longrightarrow C$ in **C**, if $f^*(S) \in \mathbf{J}(D)$ then $f \in S$. Note that, if S is **J**-closed and $g : C' \longrightarrow C$ then $g^*(S)$ is a **J**-closed, as well. Thus we have a presheaf

$$\Omega : \mathbf{C}^{op} \longrightarrow Set$$

such that $\Omega(C) = \{S \,|\, S \text{ is a } \mathbf{J}\text{-closed sieve on } C\}$ for C in **C** and $\Omega(g)(S) = g^*(S)$ for $g : C' \longrightarrow C$, $S \in \Omega(C)$. In fact Ω is a **J**-sheaf. The natural transformation

$$true : 1 \longrightarrow \Omega$$

such that $true_C(*) = t_C$, is a subobject classifier in $Sh(\mathbf{C}, \mathbf{J})$. To see this, let $m : A \longrightarrow B$ be a mono in $Sh(\mathbf{C}, \mathbf{J})$. Then the morphism $\chi_m : B \longrightarrow \Omega$ such that

$$(\chi_m)_C(b) = \{f : D \longrightarrow C \mid \exists_{a \in A(D)} m(a) = B(f)(b)\}$$

is the required morphism making the above square a pullback.

The diagram

$$A \overset{m}{\rightarrowtail} B \underset{\chi_m}{\overset{true \circ !}{\rightrightarrows}} \Omega$$

is an equalizer. Thus any mono in $Sh(\mathbf{C}, \mathbf{J})$ is regular.

The subobject classifier carries a structure of an internal Heyting algebra in $Sh(\mathbf{C}, \mathbf{J})$. We have the following natural transformations

$$\mathbf{0}, \mathbf{1} : 1 \longrightarrow \Omega$$

$$\wedge, \vee, \rightarrow : \Omega \times \Omega \longrightarrow \Omega$$

where $\mathbf{0}$, $\mathbf{1}$ picks the minimal and the maximal sieve (respectively), and for $S, R \in \Omega(C)$ we have

$$S \wedge R = S \cap R$$

$$S \vee R = \{f : D \longrightarrow C \mid f^*(S \cup R) \in \mathbf{J}(D)\}$$

(i.e. $S \vee R$ is the $\mathbf{J}$-closure of the sieve $S \cup R$) and

$$S \rightarrow R = \{f : D \longrightarrow C \mid \text{for all } g, \text{ if } f \circ g \in S \text{ then } f \circ g \in R\}$$

The above transformations gives an internal Heyting algebra structure on Ω.

For any A in $Sh(\mathbf{C}, \mathbf{J})$ we have an isomorphism

$$Sub(A) \cong Hom(A, \Omega)$$

Thus, in view of the above and first subsection of this appendix, there is a Heyting algebra structure on $Sub(A)$. In fact, $Sub(A)$ is a complete Heyting algebra. The infinite meets are just meets computed pointwise and infinite joins are the associated sheaves to the pointwise joins. Moreover, if $f : Y \longrightarrow X$ is a morphism in $Sh(\mathbf{C}, \mathbf{J})$ then the pullback functor

$$f^* : Sub(X) \longrightarrow Sub(Y)$$

preserves all joins and meets. Therefore, it has both the left $\exists_f$ and the right $\forall_f$ adjoints

$$Sub(Y) \underset{\forall_f}{\overset{\exists_f}{\rightleftarrows}} Sub(X) \quad (f^* : Sub(X) \longrightarrow Sub(Y))$$

$\exists_f \dashv f^{-1} \dashv \forall_f$. We shall describe $\exists_f$ and $\forall_f$ explicitly. For A in $Sub(Y)$, $\exists_f(A)$ is a subobject of X such that for C in $\mathbf{C}$ and $x \in X(C)$ $x \in \exists_f(A)(C)$ iff the sieve

$$\{h : D \longrightarrow C | \text{ there is } a \in A(D) \text{ such that } f_D(a) = X(h)(x)\}$$

$\mathbf{J}$-covers C, and $x \in \forall_f(A)(C)$ iff for each $h : D \to C$ in $\mathbf{C}$ and $y \in Y(D)$,

$$\text{if } f_D(y) = X(h)(x) \text{ then } y \in A(D)$$

We shall verify that Beck-Chevalley condition holds in $Sh(\mathbf{C}, \mathbf{J})$. Let

$$\begin{array}{ccc} Y \times_X Z & \xrightarrow{\pi_2} & Z \\ {\scriptstyle \pi_1}\downarrow & & \downarrow{\scriptstyle g} \\ Y & \xrightarrow[f]{} & X \end{array}$$

be a pullback in $Sh(\mathbf{C}, \mathbf{J})$. We need to show that $g^* \circ \exists_f = \exists_{\pi_2} \circ \pi_1^*$. Let $A \in Sub(Y)$, C be an object of $\mathbf{C}$ and $x \in Z(C)$. Then

$$\begin{array}{c}
x \in g^* \circ \exists_f(A)(C) \\ \hline
g_C(x) \in \exists_f(A)(C) \\ \hline
\{h : D \to C | \exists\, a \in A(D),\ \ f_D(a) = X(h)(g_C(x))\} \text{ covers C} \\ \hline
\{h : D \to C | \exists\, a \in A(D),\ \ f_D(a) = g_D(Z(h)(x))\} \text{ covers C} \\ \hline
\{h : D \to C | \exists\, z \in Y \times_X Z(D),\ \ (\pi_1)_D(z) \in A(D) \text{ and } \\ (\pi_2)_D(z) = Z(h)(x)\} \text{ covers C} \\ \hline
\{h : D \to C | \exists\, z \in \pi_1^*(A)(D),\ \ (\pi_1)_D(z) \in A(D) \text{ and } \\ (\pi_2)_D(z) = Z(h)(x)\} \text{ covers C} \\ \hline
x \in \exists_{\pi_2} \circ \pi_1^*(A)(C)
\end{array}$$

as required. This shows that any Grothendieck topos is a Heyting as well as r-Heyting category.

7. Classifying toposes

In this section by a topos we mean a Grothendieck topos. Toposes are 'generalized universes of variable sets'. They have limits, colimits, exponential objects and an object of truth values of internal logic called subobject classifier. However, they can be also treated as generalized (topological) spaces. The latter approach suggest a very useful notion of a morphisms of toposes, i.e.

the notion of a geometric morphism. If $\mathcal{E}$ and $\mathcal{H}$ are toposes then a *geometric morphism* $f : \mathcal{E} \to \mathcal{H}$ is a pair of adjoint functors

$$\mathcal{E} \underset{f_*}{\overset{f^*}{\rightleftarrows}} \mathcal{H}$$

$f^* \dashv f_*$ such that f^* preserves finite limits. We call f^* the *algebraic part* and f_* the *geometric part* of the geometric morphism f. f^* preserves the *geometric logic* i.e. the formulas, called *geometric formulas*, which are built up from atomic formulas of a first order language with equality by means of the existential quantifier and the following connectives: $\top$, $\bot$, &, $\bigvee$ only ($\bigvee$ is infinitary disjunction of formulas having altogether finitely many variables). Each topos is a topos generated by a universal model of a *geometric theory* T, i.e. one having axioms of form $\forall_{\vec{x}}(\varphi(\vec{x}) \Rightarrow \psi(\vec{x}))$, where both $\varphi(\vec{x})$ and $\psi(\vec{x})$ are geometric formulas. In other words any topos is a classifying topos of a geometric theory. By this we mean that for any topos $\mathcal{E}$ there is a geometric theory T and a *generic model* $\mathcal{G}_T$ of T in $\mathcal{E}$ such that for any other topos $\mathcal{H}$ the geometric morphisms $\mathcal{H} \to \mathcal{E}$ correspond to models of T in $\mathcal{H}$, so that for any model M of T in $\mathcal{H}$ there is a unique (up to isomorphism) geometric morphism $f : \mathcal{H} \to \mathcal{E}$ such that $f^*(\mathcal{G}_T) \cong M$. This correspondence is natural in $\mathcal{H}$. The converse is also true. For any geometric theory T there is a topos $\mathcal{E}(T)$ classifying T.

Dealing with particular toposes it is always interesting to identify the theory they classify.

If a theory T is equational then $\mathcal{E}(T)$ is particularly simple, namely it is the topos of presheaves on the dual of the category of finitely presented T-algebras $Alg(T)^{op}_{fp}$, i.e. it is $Set^{Alg(T)_{fp}}$. The generic model $\mathcal{G}_T$ in $Set^{Alg(T)_{fp}}$ is just the image of the generic model $\mathcal{G}_T$ in $Alg(T)^{op}_{fp}$ described in section 9.2 under the Yoneda embedding $Y : Alg(T)^{op}_{fp} \longrightarrow Set^{Alg(T)_{fp}}$.

If a geometric theory T' extends T by some additional geometric axioms in the same language then $\mathcal{E}(T')$ is a topos of sheaves $Sh(Alg(T)^{op}_{fp}, J)$ for some Grothendieck topology J generated by the additional axioms. For example, for an axiom

$$\forall_{\vec{x}}\,(\varphi(\vec{x}) \Rightarrow \bigvee_{i\in I} \exists_{\vec{y}}\,\psi_i(\vec{x},\vec{y}))$$

with $\varphi(\vec{x})$ and $\psi_i(\vec{x},\vec{y})$ for $i \in I$ e-formulas we add a cover of $\mathcal{F}(\vec{x})/\varphi(\vec{x})$

$$\{\mathcal{F}(\vec{x})/\varphi(\vec{x}) \to \mathcal{F}(\vec{x},\vec{y})/\psi_i(\vec{x},\vec{y}) \wedge \varphi(\vec{x})\}_{i\in I}$$

and the topology J is the least topology on $Alg(T)^{op}_{fp}$ containing all such covers.

So putting it the other way around a topos $Sh(Alg(T)^{op}_{fp}, J)$ classifies some special models of T, namely those for which the corresponding finite limit preserving functors from $Alg(T)_{fp}^{op}$ preserves J-covers. Thus the additional set axioms of a geometric theory T' containing T classified by $Sh(Alg(T)_{fp}^{op}, J)$ is any set of geometric axioms which generates the topology J. If J is a subcanonical topology then the generic model $\mathcal{G}_T$ of T in $Set^{Alg(T)_{fp}}$, defined above, is the generic model for T' in $Sh(Alg(T)_{fp}^{op}, J)$, as well.

References

[*] Following common practice, the entries [SGA4], [CWM], [CWL] in this bibliograpy are quoted by the shorthands of the title and listed by the name of the first author (M.Artin, S.MacLane).

[SGA4] M. Artin, A. Grothendieck, J.L. Verdier, *Séminaire de Géometrie Algébrique 4*, Springer LNM 269 (1972).

[BH] R. Balbes, A. Horn *Injective and projective Heyting algebras*, Trans. Amer. Mat. Soc., 148, pp.549-559 (1970).

[Be] F. Bellissima, *Finitely generated free Heyting algebras*, Journal of Symbolic Logic, 51, pp. 152-165 (1986).

[BP] W.J. Blok, D. Pigozzi, *A finite basis theorem for quasivarieties*, Algebra Universalis, 22, pp. 1-13 (1986).

[Bo] G. Boolos, *The unprovability of consistency*, Cambridge University Press (1979).

[Bu] R.A. Bull, *All normal extensions of $S4.3$ have the finite model property*, Zeitschr. für Math. Log. und Grundl. der Math., 12, pp. 341-344 (1966).

[CJKP] A. Carboni, G. Janelidze, G.M. Kelly, R. Paré, *On localization and stabilization for factorization systems*, preprint, University of Sidney (1995).

[CK] C.C. Chang, H.J. Keisler, *Model theory*, North Holland, 3rd edition (1990).

[CZ] A. Chagrov, M. Zakharyaschev, *Modal Logic*, Oxford University Press, Oxford (1997).

[Dr] A. Dragalin, *Mathematical Intuitionism, introduction to proof theory*, Translations of Mathematical Monographs, vol.67, American Mathematical Society (1988).

[vD] D. van Dalen, *Intuitionistic Logic*, in (Gabbay-Günther eds.) 'Handbook of Philosophical Logic', vol. III, ch. 4, pp. 225-339, Reidel (1986).

[Fi1] K. Fine, *Logics containing K4, Part I*, Journ. of Symb. Logic, 34, pp. 31-42 (1974).

[Fi2] K. Fine, *Logics containing K4, Part II*, Journ. of Symb. Logic, 50, pp. 619-651 (1985).

[FK] P.J. Freyd, G.M. Kelly, *Categories of continuous functors I*, Journ. of Pure and Appl. Algebra, 2, pp. 169-191 (1972).

[GU] P. Gabriel, F. Ulmer, *Lokal prasentierbare Kategorien*, Springer LNM 221 (1971).

[GG] Z. Gleit, W. Goldfarb, *Characters and fixed points in provability logic*, Notre Dame Journal of Formal Logic, 31, pp. 26-36 (1990).

[Gh1] S. Ghilardi, *Free Heyting algebras as bi-Heyting algebras*, Math. Rep. Acad. Sci. of Canada, XIV, 6, pp. 240-244 (1992).

[Gh2] S. Ghilardi, *Irreducible models and and definable embeddings*, in 'Logic Colloquium 92 (Proceedings)', CSLI Publications, pp. 95-213 (1995).

[Gh3] S. Ghilardi, *An algebraic theory of normal forms*, Annals of Pure and Applied Logic, 71, pp. 189-245 (1995).

[Gh4] S. Ghilardi, *Unification in intuitionistic logic*, J. of Symbolic Logic, 64, pp.859-880 (1999).

[Gh5] S. Ghilardi, *Best solving modal equations*, Annals of Pure and Applied Logic, 102, pp.183-198 (2000).

[GZ1] S. Ghilardi, M. Zawadowski, *A sheaf representation and duality for finitely presented Heyting algebras*, Journ. of Symb. Logic, vol 60, pp. 911-939 (1995).

[GZ2] S. Ghilardi, M. Zawadowski, *Undefinability of propositional quantifiers in the modal system* $S4$, Studia Logica, vol 55, pp. 259-271 (1995).

[GZ3] S. Ghilardi, M. Zawadowski, *Model completions and r-Heyting categories*, Annals of Pure and Applied Logic, 88, pp. 27-46 (1997).

[GZ4] S. Ghilardi, M. Zawadowski, *From Bisimulation Quantifiers to Classifying Toposes*, in F. Wolter, H. Wansing, M. de Rijke and M. Zakharyaschev (eds.) "Avances in Modal Logic, vol. III", CSLI Publications, to appear.

[Ha] D. Harel, *Dynamic Logic*, in Gabbay D. M., Günther F. (eds.) 'Handbook of Philosophical Logic', Reidel, pp. 497-604 (1984).

[JaTh] G. Janelidze, W. Tholen, *Facets of Descent, I*, Applied Categorical Structures, 2, pp. 245-281, (1994).

[Jo] P.T. Johnstone, *Stone Spaces*, Cambridge University Press, (1982).

[KM] A. Kock, I, Moerdijk *Representation of étendues*, Cahiers Top. Geom. Diff. vol.32, n.2, pp. 145-164 (1991).

[Kö] P. Köhler, *Brouwerian semilattices*, Transactions of AMS, vol 268, no 1, pp. 103-126 (1981).

[KP] P. Köhler, D. Pigozzi, *Varieties with equationally definable principal congruences*, Algebra Universalis, 11, pp. 213-219 (1980).

[Kr] M. Kracht, *Tools and techniques in Modal Logic*, North Holland (1999).

[K] G. Kreisel, *Explicit definability in intuitionistic logic*, Journal of Symbolic Logic, 25, pp. 389-390 (1960).

[La] F.W. Lawvere, *Intrinsic co-Heyting boundaries and the Leibniz rule in certain toposes*, in *Category Theory, Proceedings Como 1990*, Springer LNM 1488, pp. 279-281 (1991).

[Li] P. Lipparini, *Existentially complete closure algebras*, Boll. Un. Mat. Ital., serie VI, vol. I-D, n.1, pp. 13-19 (1982).

[MI] A. MacIntyre, *Model completeness*, in Barwise J. (ed.) 'Handbook of Mathematical Logic', North Holland, pp. 139-180 (1977).

[CWM] S. MacLane, *Categories for the working mathematician*, Springer, Berlin (1971).

[CWL] S. MacLane , I.Moerdijk, *Sheaves in geometry and logic*, Springer (1992).

[MR1] M. Makkai , G.E. Reyes, *First order categorical logic*, Springer LNM 611 (1977).

[MR2] M. Makkai, G.E. Reyes, *Completeness results for intuitionistic and modal logic in a categorical setting*, Annals of Pure and Appl. Logic, vol.72, n.1, pp. 25-101 (1995).

[MM1] M. Makkai, *Duality and Definability in First order Logic*, Memoirs of AMS, 503 (1993).

[MaPi] M.Makkai, A.M. Pitts, *Some results on locally finitely presentable categories*, Transactions of AMS, 299, pp. 473-495 (1987).

[Ma1] L.L. Maksimova, *The Craig theorem in superintuitionistic logics and amalgamable varieties of pseudo-Boolean algebras*, Algebra i Logika, vol.16, n.6, pp. 643-681 (1977).

[Ma2] L.L. Maksimova, *Interpolation theorems in modal logics and amalgamable varieties of topological Boolean algebras*, Algebra i Logika, vol. 18, n.5, pp. 556-586 (1979).

[Ma3] L.L. Maksimova, *Interpolation theorems in modal logics - Sufficient conditions*, Algebra i Logika, vol.19, n.2, pp. 194-213 (1980).

[Ma4] L.L. Maksimova, *Absence of the interpolation property in the consistent normal extensions of the Dummett logic*, Algebra i Logika, vol. 21, n.6, pp. 690-694 (1982).

[Ma5] L.L. Maksimova, *An analog of Beth's theorem in normal extensions of the modal logic $K4$*, Siberian Journ. of Math., vol.33, n.6, pp. 118-130 (1992).

[Ma6] L. L. Maksimova, *Intuitionistic Logic and Implicit Definability* , to appear in Annals of Pure and Applied Logic.

[Mar] S. Mardaev, *Least Fixed Points in Modal Logic*, Proceedings of the International Conference on Mathematical Logic to the memory of A.I. Mal'tsev, august 1999, to appear.

[Pi1] A.M. Pitts, *Conceptual completeness for first order intuitionistic logic: an application of categorical logic*, Annals of Pure and Applied Logic, 41, pp. 33-81 (1989).

[Pi2] A.M. Pitts, *On an interpretation of second order quantification in first order intuitionistic propositional logic*, Journ. of Symb. Logic, 57, 1, pp. 33-52 (1992).

[Pr] H.A. Priestley, *Representation of Distributive Lattices by means of Ordered Stone Spaces*, Bul. London Math. Soc. 2 pp. 186-190 (1970).

[Ro] A. Robinson, *Complete Theories*, North-Holland, Amsterdam, (1956).

[Ry] V.V. Rybakov, *Admissibility of inference rules with parameters in intuitionistic logic and intuitionistic Kripke models* - Dokl. Akad. Nauk. SSSR 312, 1, pp. 42-45 (1990).

[Sa] G. E. Sacks, *Saturated Model Theory*, Reading, Massachussets: W.A. Benjamin (1972).

[Sh] V.Yu. Shavrukov, *Subalgebras of diagonalizable algebras of theories containing arithmetic*, Dissertationes Mathematicae, CCCXXIII, Polska Akademia Nauk, Instytut Matematyczny, Warsaw (1993).

[SoTh] M. Sobral, W. Tholen, *Effective descent morphisms and effective equivalence relations*, Canadian Mathematical Society Conference Proceedings (AMS, Providence), pp. 421-431 (1992).

[Ur] A. Urquhart, *Free Heyting algebras*, Algebra Universalis, 3, pp. 94-97 (1973).

[Vi] A. Visser, *Bisimulations, Model Descriptions and Propositional Quantifiers*, Logic Group Preprint Series, n. 161, Department of Philosophy, Utrecht University, Utrecht (1996).

[Wh] W.H. Wheeler, *Model companions and definability in existentially complete structures*, Israel Journ. of Math., vol. 25, pp. 305-330 (1976).

[Za] M.W. Zawadowski, *Descent and duality*, Annals of Pure and Applied Logic, 71 pp. 131-188 (1995).

Glossary of Notation

(AP) 9, 41
$(AP)_{st}$ 41
$(AP)^*$ 41
$(AP)^*_{st}$ 41
$Alg(\mathbf{S})$ 116
$Alg(\mathbf{S})_{fin}$ 116
$Alg(\mathbf{S})_{fp}$ 116
$A/E(\vec{a})$ 33
(BPD) 47
(BP) 4, 34
(CEP) 41
$(CEP)^*$ 41
$\mathcal{D}(P)$ 8, 17
$\mathcal{D}(f)$ 20
(EDO) 7, 37
$(EDPC)$ 7, 37
$I(x_1, x_2, x_3, x_4)$ 37
$(EDPM)$ 7, 37
$J(x_1, x_2, x_3, x_4)$ 37
J_0 37
(EQ) 25
$\mathbf{F}$ 116
$(FLEA)$ 12, 181
$\mathbf{F_S}$ 10, 117
$\mathbf{G}$ 117
$\mathbf{Grz}$ 117
HA^{op}_{fp} 3
(IT) 41
$(IT)^*$ 41
IpC 1, 96
IpC^2 1, 96
IpC_2 96
$\mathbf{J}$ 224
$\mathbf{J_0}$ 8, 78
$\mathbf{J_S}$ 10, 129
$Sh(\mathbf{C}, \mathbf{J})$ 226
$\mathbf{a}$ 226
$Set^{\mathbf{C}^{op}}$ 226
$\mathbf{J^f}$ 181
$\mathbf{J}^{\infty}$ 12, 181
$\mathbf{K4}$ 9, 18, 116
$\mathbf{M_H}$ 8, 82
$\mathbf{M_S}$ 10, 133
$\mathbf{M_T}$ 5
Ω 79, 137, 227
χ_S 79, 227
$true$ 79, 227
$\mathbf{P_0}$ 8, 78
$\mathbf{\Phi_H}$ 8, 87
$\mathbf{\Phi_S}$ 10, 159
$\mathbf{\Phi_T}$ 5
$\mathbf{\Psi_H}$ 87
$\mathbf{\Psi_S}$ 11, 135
$\mathbf{\Psi_T}$ 5
$[x]_R$ 123
$R^{(\infty)}$ 101
R^+ 123
$\asymp_R$ 123
$\mathbf{S}$ 9, 116
$\mathbf{S4}$ 9, 18, 117
$\mathbf{S4.2}$ 117
$\mathbf{S4.3}$ 117
$\mathbf{S5}$ 117
(SBP) 46
$(SBPD)$ 47
$\mathbf{T_H}$ 83
$\mathbf{T_S}$ 11, 137
(X, R) 9, 19, 116
$\delta^e(X, \leq)$ 169
$\delta^i(X, \leq)$ 169
X^e 169
$e(X, \leq)$ 169
X^i 169
$wt(X, \leq)$ 169
Y^X 116
$\mathbf{F} \dashv \mathbf{G}$ 207
$Alg(T)$ 32
$Alg(T)_{fin}$ 32

Subject Index

TRENDS IN LOGIC

1. G. Schurz: *The Is-Ought Problem.* An Investigation in Philosophical Logic. 1997 ISBN 0-7923-4410-3
2. E. Ejerhed and S. Lindström (eds.): *Logic, Action and Cognition.* Essays in Philosophical Logic. 1997 ISBN 0-7923-4560-6
3. H. Wansing: *Displaying Modal Logic.* 1998 ISBN 0-7923-5205-X
4. P. Hájek: *Metamathematics of Fuzzy Logic.* 1998 ISBN 0-7923-5238-6
5. H.J. Ohlbach and U. Reyle (eds.): *Logic, Language and Reasoning.* Essays in Honour of Dov Gabbay. 1999 ISBN 0-7923-5687-X
6. K. Došen: *Cut Elimination in Categories.* 2000 ISBN 0-7923-5720-5
7. R.L.O. Cignoli, I.M.L. D'Ottaviano and D. Mundici: *Algebraic Foundations of many-valued Reasoning.* 2000 ISBN 0-7923-6009-5
8. E.P. Klement, R. Mesiar and E. Pap: *Triangular Norms.* 2000 ISBN 0-7923-6416-3
9. V.F. Hendricks: *The Convergence of Scientific Knowledge.* A View From the Limit. 2001 ISBN 0-7923-6929-7
10. J. Czelakowski: *Protoalgebraic Logics.* 2001 ISBN 0-7923-6940-8
11. G. Gerla: *Fuzzy Logic.* Mathematical Tools for Approximate Reasoning. 2001 ISBN 0-7923-6941-6
12. M. Fitting: *Types, Tableaus, and Gödel's God.* 2002 ISBN 1-4020-0604-7
13. F. Paoli: *Substructural Logics: A Primer.* 2002 ISBN 1-4020-0605-5
14. S. Ghilardi and M. Zawadowki: *Sheaves, Games, and Model Completions.* A Categorical Approach to Nonclassical Propositional Logics. 2002 ISBN 1-4020-0660-8

KLUWER ACADEMIC PUBLISHERS – DORDRECHT / BOSTON / LONDON